E. Amitai Halevi

Orbital Symmetry and Reaction Mechanism

The OCAMS View

With 113 Figures

Springer-Verlag
Berlin Heidelberg New York
London Paris Tokyo
Hong Kong Barcelona
Budapest

E. Amitai Halevi
Department of Chemistry
Technion – Israel Institute of Technology
Technion City
Haifa 32000, Israel

ISBN 3-540-50164-9 Springer-Verlag Berlin Heidelberg New York
ISBN 0-387-50164-9 Springer-Verlag New York Berlin Heidelberg

Library of Congress Cataloging-in-Publication Data
Halevi, E. Amitai (Emil Amitai), 1922–.
Orbital symmetry and reaction mechanism: the OCAMS view/E. Amitai Halevi. p. cm.
Includes bibliographical references and index.
ISBN 3-540-50164-9 (Berlin: alk. paper). – ISBN 0-387-50164-9 (alk. paper)
1. Molecular orbitals. 2. Chemical reactions, Conditions and laws of. I. Title.
QD461.H169 1992 547.1'39–dc20 91-27747 CIP

Typesetting: Springer TEX in-house system
51/3140–543210 Printed on acid-free paper

In Memoriam:

Christopher Ingold and David Ginsburg

"Choose a teacher and make a friend."

Ethics of the Fathers 1,6.

"עשה לך רב וקנה לך חבר."

- פרקי אבות א',ו'

Foreword

The concepts of orbital symmetry and correlation diagrams, appreciated by theoreticians and physical chemists interested in very small molecules from their very inception in the first half of this century, were scarcely known to the rank and file organic chemists at that time, and certainly had no discernible impact on their mechanistic thinking. This changed almost overnight in the sixties, and did so with a vengeance, transforming profoundly the way in which organic chemists think about their reactions.

Nowadays, these concepts are a standard part of the organic mechanistic vocabulary, they are taught from the beginning to organic graduate students, and are at least mentioned even in undergraduate classes. Perhaps inevitably, but still no less unfortunately, their use by organic chemists sometimes suffers from a lack of rigor and deep understanding.

Professor Halevi has set out to correct the situation by making available a rigorously written text. He relies on a particularly well thought-through formulation, referred to as "Orbital Correspondence Analysis in Maximum Symmetry" (OCAMS). The playful reference to the medieval scholar, Occam, whose fabled razor all students are told about, is well chosen: this is a text of high intellectual and scholarly standards.

The reader is assumed to be a complete beginner, is taken very gently by the hand, and is led with infinite patience through the minefield of the group theoretical description of the symmetry aspects of organic reactions (at the very end, even some inorganic ones). Ultimately, even the most formidable looking cases turn out to be actually not all that difficult to analyze.

No stone is left unturned along this journey. Numerous footnotes point out pitfalls, exceptions, and limitations of the statements made. Mastery of the material by a novice will require patience, but the rewards will be great. The author distinguishes carefully between firm conclusions and mere conjectures based on symmetry arguments. Although some of the theoretically inclined expert readers may consider a conjecture unlikely here and there, all stand to learn from a careful examination of the arguments, and will enjoy the rigor with which the subject is treated. All of us who are interested in orbital symmetry will feel indebted to Professor Halevi for having written such a very thoughtful book.

Boulder, Colorado, USA
February 29, 1992

Josef Michl

Preface

Physical Organic Chemistry can be defined as the application of the concepts and techniques of physical chemistry to the study of organic compounds and their reactions. The incorporation of modern instrumental and computational techniques into the discipline has been impressively rapid, but the absorption of new theoretical concepts has been slower. Thus, the criteria of orbital symmetry conservation, that had such a profound influence on mechanistic thinking immediately after its introduction into organic chemistry, are commonly applied today in much the same manner as they were a quarter of a century ago.

My object in writing this book is to present a coherent set of operational rules for the analysis of reaction mechanism in terms of symmetry, and to explore their scope and reliability. It is written from the viewpoint of *Orbital Correspondence Analysis in Maximum Symmetry – OCAMS* for short, hence its subtitle, but not with the intention of touting a homemade prodct! The procedural details of *OCAMS* are of secondary importance; its advantage lies in the provision of a coherent overview of the relation between symmetry and mechanism: it allows us to see the forest without losing sight of the trees. For reasons of self-consistency, the book remains within the framework of molecular orbital theory; reformulation of the *OCAMS* approach in valence-bond terms has not been attempted.

This book is neither a textbook nor a monograph, but perhaps a little of both. Part I is a critical – necessarily sketchy and admittedly personal – survey of earlier developments. Part II provides the necessary theoretical background as informally as possible. The familiar thermal reactions, to which qualitative symmetry- and topology-based arguments have been repeatedly applied, are analyzed in Part III and the results compared with those of the earlier methods, as well as with available experimental and computational evidence. The generality of the approach is illustrated in Part IV, where it is extended to contiguous areas of chemistry.

Whenever theoretical ideas are developed informally, there is danger of oversimplification – perhaps even of misrepresentation. If these pitfalls have been avoided successfully, thanks are due in great measure to Prof. J. Katriel for a critical reading of the entire manuscript. Chapter 2, the most problematic in this respect, was kindly "refereed" by Profs. R. Pauncz and F. Weinhold as well. I am no less grateful to Prof. M.B. Rubin, whose expertise was enlisted repeatedly when my own command of organic chemistry proved inadequade.

Needless to say, the responsibility for any remaining infelicities is solely my own.

I am indebted to Mr. G. Berg for all of the hand-drawn Figures, and to Prof. J. Goldberg as well as to the the staff of the Technion Computer Center – Mr. B. Pashkoff in particular – for advice and assistance with TEX. Thanks are due to Dr. R. Stumpe and to the staff of Springer-Verlag – especially Mrs. U. Beiglböck, Mr. K. Koch and Mr. F. Holzwarth, for their friendly cooperation.

The hospitality extended by the Max Planck Institute for Medical Research in Heidelberg and by the Department of Chemistry of the University of California at Irvine while the book was being put into final form is greatly appreciated. I am grateful to the latter, as well as to the Institute for Theoretical Chemistry of the University of Heidelberg, for making their computational facilities freely available to me. Generous support from the Fund for Promotion of Research at the Technion is gratefully acknowledged.

Finally, I would like to commend my wife, Ada, for her patient forbearance during the excessively long time that it took me to write this book.

Haifa, January 1992 *E. Amitai Halevi*

Table of Contents

Part I. Preliminary Survey

Part II. Symmetry and Energy

Part III. The Classical Thermal Reactions

Part IV. Spin and Photochemistry

List of Figures

List of Tables

Part I

Preliminary Survey

Chapter 1

The Woodward-Hoffmann Rules in Perspective

1.1 Prolegomenon

In order to appreciate properly the impact of Woodward and Hoffmann's *The Conservation of Orbital Symmetry* [1] on contemporary chemical thinking, let us look back to the early nineteen-sixties and recall the state of the art of mechanism elucidation at that time.

Physical Organic Chemistry had burgeoned, developing along lines that had been laid down several decades earlier by its founding fathers [2, 3, 4] and their contemporaries. It was primarily concerned with the detection and determination of reactive intermediates, or – when their direct observation proved impossible – with the attempt to deduce their involvement or non-involvement in the mechanism of a reaction by a detailed study of its reaction kinetics. Even when the latter were so obstinately simple as to suggest that the reaction might be occurring in a single step, much could be learned about the nature of its transition state from the temperature dependence of the rate, from its sensitivity to homogeneous – notably acid and base – catalysis [3] and to isotopic substitution [5, 6], and from an analysis of solvent and substituent effects by means of one or more of the many extant Linear Free Energy Relations [7].

It had become apparent, however, that many thermal reactions, including well known molecular rearrangements, decompositions and cycloadditions, were not amenable to study by these methods. The mechanistic information which they were able to provide about such thermo-reorganization reactions was so meager that "half in jest, half in desperation [they were designated] *No Mechanism* reactions" [8].

No Mechanism reactions were also extremely accommodating as regards "electron-pushing", a favorite pastime of many organic chemists at that time. In reactions that could be presumed to have some polar character, the directionality of charge transmission could be deduced in a more or less straightforward manner. [9] In contrast, the flow of electrons in the course of the Diels-Alder reaction, for example, could be variously depicted as:

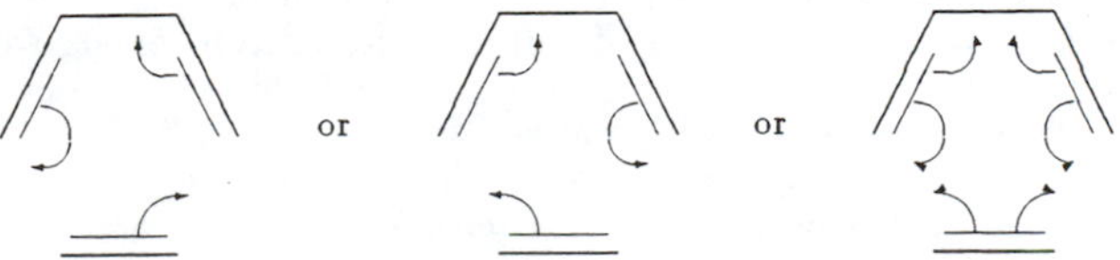

Small wonder, then, that the promulgation of the *Woodward-Hoffmann Rules*, which were eminently suitable for dealing with just these elusive thermorearrangements and were even capable of casting some light on their still more obscure photochemical counterparts, was greeted with tremendous enthusiasm. They were soon exposed to the chemical community in a number of secondary publications [10, 11, 12, 13] and rapidly became an integral part of the modern organic chemist's stock in trade.

The categorization of chemical reactions as *allowed* or *forbidden* by the symmetry of the orbitals involved, dates back to the earliest days of Quantum Chemistry. In 1928, Wigner and Witmer [14] formulated rules for the formation of diatomic molecules, based on the group theoretical characterization of the orbitals of its constituent atoms. The newly conceived molecular orbital (MO) description of diatomic molecules [15, 16] was systematized by Mulliken [17], who related their MOs to the atomic orbitals (AOs) of the separated atoms on one side of a correlation diagram and to those of a united atom on the other. Walsh [18] later extended the correlation diagram approach to polyatomic molecules, taking specific note of how the symmetry properties of the MOs of an initially linear molecule varies with molecular geometry, and showing how this variation can be used to explain the conformation preferentially adopted by the molecule.

At about the same time, Shuler [19] analyzed the mechanism of fragmentation of a simple polyatomic molecule in terms of symmetry conservation, and Griffing [20] followed with a similar study of the four- and three-center mechanisms of hydrogen atom exchange. Unlike Walsh's papers, the latter two publications had only slight immediate impact, probably as a result of the rather awesome formalism in which they were cast. The same can be said of Bader's pioneering paper of 1962, [21] in which he demonstrated formally how the energetically favored mode of fragmentation of a non-linear polyatomic molecule can be deduced from the symmetry properties of its molecular orbitals and of its vibrational coordinates.

This and parallel lines of development led to the publication, within a very short period, of a number of methods for the systematic characterization of organic reactions, principally those of the *no mechanism* type, as being either *allowed* or *forbidden*; we shall refer specifically to several of them in this chapter. The factors that gave the *Woodward-Hoffmann Rules* unchallenged pride of place among them would seem to be: their wide applicability, convincingly illustrated in the monograph [1], their beautiful simplicity[1], and the immense authority with which they were presented.

It was also becoming evident that Woodward and Hoffmann's criteria of mechanism are not cut out of whole cloth, but comprise at least three distinct features, based on different – though intimately related – sets of assumptions. [23, 24] This fact would have been of purely academic interest, were it not for

[1] "Sebastian: 'Oh yes, I believe that. It's a lovely idea.'
Charles: 'But you can't *believe* things because they're a lovely idea.'
Sebastian: 'But I *do*. That's how I believe.'" – Evelyn Waugh, *Brideshead Revisited.*

the growing number of exceptions that had been observed and the consequent loss of confidence among organic chemists, not only in the *Rules* as such but in orbital symmetry arguments in general.[2] It might also be noted that the severe criticism levelled by George and Ross [22] at the theoretical basis of the *Rules* does not apply to their various aspects with equal force. The several facets of the Woodward-Hoffmann approach will be illustrated in the following sections, using as our primary examples three familiar types of electrocyclic reaction: the Diels-Alder reaction, i.e. $[_{\pi}4 + {_{\pi}2}]$-cycloaddition, the related $[_{\pi}2 +_{\pi} 2]$-cycloaddition, and polyene cyclization.

1.2 The Suprafacial-Antarafacial Dichotomy

The *Woodward-Hoffmann Rules* for cycloaddition as well as those for electrocyclic and sigmatropic rearrangements [1, pp. 45, 117], all involve the same basic distinction, which is primarily related to the topology of the reacting system rather than to its symmetry: When a reactant molecule undergoes concerted bond-making or bond-breaking processes at two of its atoms, it is said to be reacting *suprafacially* (s) if the two bonding changes occur on the same face of the molecule and *antarafacially* (a) if they occur at opposite faces.[3]

Evidently, before this distinction can be made, the term *face* has to be understood. The plane common to the two CH_2 groups of an ethylene molecule obviously separates its upper face from its lower, but the two faces of a ketene molecule, though sharply distinguished at its methylene end, are less well defined near the carbonyl group. It is more dubious still to speak of the *faces* of a cylindrically symmetrical acetylene molecule, merely because its π system can be conceptually divided into an arbitrary pair of orthogonal π bonds, each with its own nodal plane. The most satisfactorily two-faced molecules of all are those like benzene or cyclobutadiene, in which a ring of σ-bonded atoms defines a common nodal plane for all of the p orbitals that make up π system. These two molecules, the *aromatic* benzene and the *antiaromatic* cyclobutadiene provide a good starting point for a discussion of the suprafacial-antarafacial dichotomy.

1.2.1 Ground-State Reactions

When an ethylene and a butadiene molecule are in the coplanar arrangement shown on the left side of (a) in Fig. 1.1, the similarity of the combined π system to that of benzene is inescapable. If the p orbitals are oriented so that all of their positive lobes are on one side of the nodal plane and all of their negative lobes

---

[2] Fleming's comment [25, p. 228] :"Exceptions – there are many" is a gently mocking paraphrase of Woodward and Hoffmann's pronouncement [1, p. 173]: "Violations: There are none!"

[3] The *Rules* for polyene cyclization [1, p. 45] are formulated in terms of *conrotation* and *disrotation* of the two interacting methylene groups at the ends of the reactant molecule. In the present limited context, these two modes of internal rotation – which are characterized most straightforwardly in terms of their symmetry properties – can be regarded as equivalent to the respective topological categories: *antarafacial* and *suprafacial*.

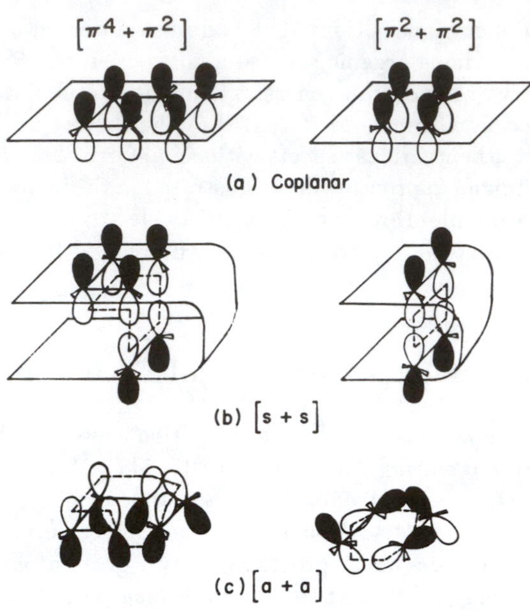

Figure 1.1a–c. Various orientations for $[_\pi 4 +_\pi 2]$- and $[_\pi 2 +_\pi 2]$-cycloaddition

are on the other, we will have produced what looks like the cyclically bonding lowest molecular orbital of benzene.

The reactants cannot retain their coplanarity during cycloaddition, but – as M.G. Evans pointed out over half a century ago – the transition state of the Diels-Alder reaction nevertheless "reduces to a six-electron problem" [26] in which "the mobile electrons ... simulate their behaviour in a benzene molecule" [27]. Following this line of reasoning, Dewar [28] extended to non-planar systems, like those illustrated in (b) and (c) of Fig. 1.1, the scope of the familiar *Hückel Rule* [29], according to which planar monocyclic molecules with $4N + 2$ mobile electrons have aromatic stability. It follows that $[_\pi 4 +_\pi 2]$-cycloaddition, which has an *aromatic* transition state, is *allowed*, whereas $[_\pi 2 +_\pi 2]$-cycloaddition is *forbidden* because – having $4N$ mobile electrons ($N = 1$) – its transition state is *antiaromatic*. It is clear from Fig. 1.1 that both reactions are $[s + s]$ or $[a + a]$, but not $[s + a]$.

The nodal plane, common to the atomic p orbitals in the prototype π systems in (a), has been distorted in (b) to a non-planar nodal surface which, however, still bisects all of them. The two bonds linking the terminal atoms of the reactants are now of σ rather than π type, but the phases of the p orbitals can still be chosen so as to produce a continuous bonding cycle, as traced by the dashed lines in (b). The reacting pair of molecules thus constitute a *Hückel system*, so the $[_\pi 4_s +_\pi 2_s]$-cycloadditon pathway is judged to be *allowed* and the $[_\pi 2_s +_\pi 2_s]$ pathway to be *forbidden*.

Alternatively, if the terminal atoms are twisted as in (c) to bring the upper lobes of one pair of interacting atoms into better overlap, and also the lower lobes of the other pair, the *Hückel Rule* again applies, characterizing the $[_{\pi}4_a +_{\pi} 2_a]$ and $[_{\pi}2_a +_{\pi} 2_a]$ modes as respectively *allowed* and *forbidden*. Here too, a line tracing a continuous cycle around the ring can be drawn in any of several ways. This line now has to cross the nodal surface (not drawn in the figure) at least twice – in general, an even number of times – but each of the crossings can be made through one of the atoms.

The criterion just enunciated, which defines a *Hückel system* as one for which a cyclically bonding molecular orbital can be constructed, evidently requires justification. Since a single MO can be occupied by at most two electrons, all π systems with more than two electrons must have additional occupied MOs that are not continuously bonding around the ring.

The special stability of rings containing $4N + 2$ mobile electrons goes back to the *Hückel Molecular Orbital (HMO)* theory of cyclic π systems [29, 30]. The continuously bonding lowest molecular orbital, ϕ_1 of Fig. 1.2, is non-degenerate in any Hückel system and can be occupied by two paired electrons, that are regarded as having filled a *closed shell*. Except for the highest antibonding MO, which is also non-degenerate whenever there is an even number (n) of p AOs in the ring – but in any case remains unoccupied – all of the MOs above the lowest come in pairs. As a result, a closed shell configuration, *i.e.* one in which all MOs with the same energy are fully occupied, must contain $4N+2$ electrons, N being the number of filled pairs of doubly degenerate MOs.

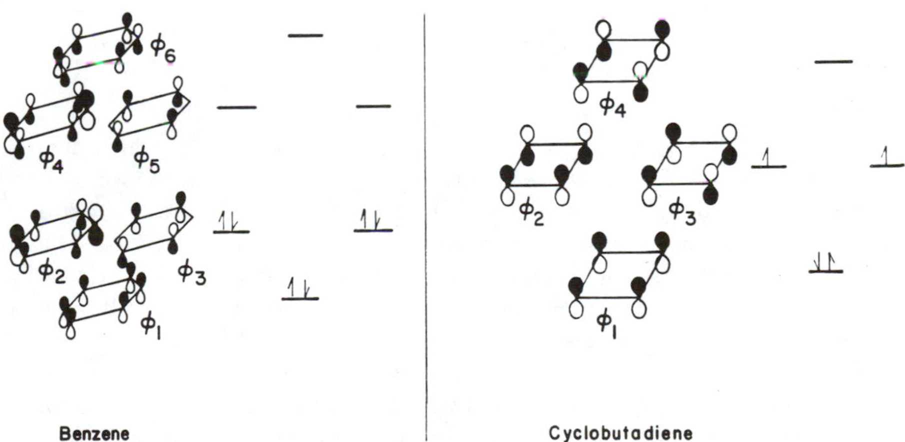

Figure 1.2. The Hückel molecular orbitals of benzene and cyclobutadiene

Conversely, molecules with $4N$ electrons are presumed to produce less stable *open shell* systems, in which – according to *Hund's Rule* [30] – two electrons with parallel spins occupy singly a degenerate pair of molecular orbitals, as in the HMO representation of C_4H_4 at the right of Fig. 1.2. Adding two electrons to cyclobutadiene would produce a closed shell, and thus convert the antiaromatic ($4N$ electron) molecule to its aromatic ($4N + 2$ electron) dianion.

In order to render a $4N$ electron molecule aromatic and a $4N + 2$ electron molecule antiaromatic, the non-degenerate lowest molecular orbital has to be eliminated. If the cyclic array of bonding atoms is converted to a non-Hückel system, *i.e.* one in which there is no way of arranging the relative phases of the atomic orbitals so as to form an unbroken bonding cycle, the lowest molecular orbital, and all above it, become doubly degenerate. Fig. 1.3 shows two ways of accomplishing this: In the first, suggested by Craig [31], one of the p AOs in the ring is replaced by a d orbital. No matter how this AO is oriented, at least one phase-discontinuity appears and foils all attempts to construct a cyclically bonding MO. A line drawn to connect orbitals of like phase cannot complete the cycle without crossing the nodal plane at least once at some point between two atoms. A similar result was shown by Heilbronner [32] to ensue when a ring of p AOs is placed on a *Möbius surface*[4], the "inside" and "outside" of which are indistinguishable.

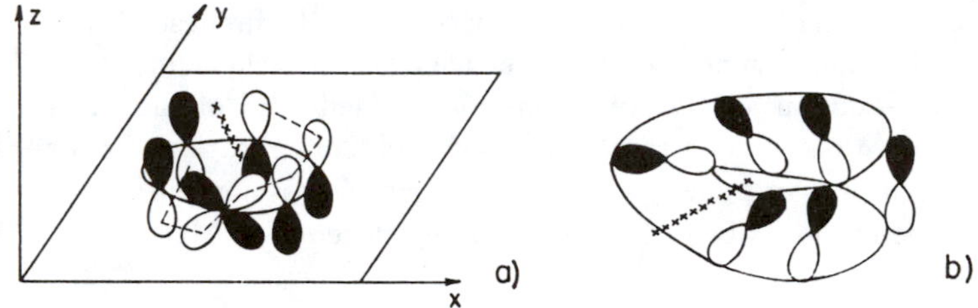

Figure 1.3a, b. Examples of non-Hückel systems. (**a**) Inclusion of a d orbital in a ring of p orbitals; (**b**) p orbitals on a Möbius strip

The idea that transition states can be of Möbius type, in which the relative stability of $4N$- and $4N$+2-electron systems is reversed, was developed and systematized by Zimmerman [33], who derived the Woodward-Hoffmann Rules for the various thermo-rearrangements in terms of the Hückel or Möbius nature of their transition states. As shown in (a) and (b) of Fig. 1.4, a cycloaddition in which one of the reaction partners reacts suprafacially and the other antarafacially mimics a Möbius surface, so $[_\pi 2_s +_\pi 2_a]$-cycloaddition is *allowed* whereas reaction along the $[_\pi 4_a +_\pi 2_s]$ pathway is *forbidden*, as is the $[_\pi 4_s +_\pi 2_a]$-cycloaddition as well. Another familiar example is illustrated in (c): The antarafacial (conrotatory) cyclization of a polyene has a Möbius type transition state, so it is an *allowed* pathway for the ground-state isomerization of a $4N$ π-electron molecule like butadiene to cyclobutene but *forbidden* for the homologous $4N + 2$ isomerization of hexatriene to cyclohexadiene.

[4] A Möbius surface is easily constructed by pasting the ends of a rectangular strip of paper to one another top-to-bottom, rather than top-to-top as one would to form an ordinary ring.

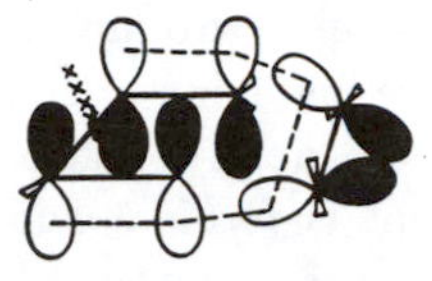

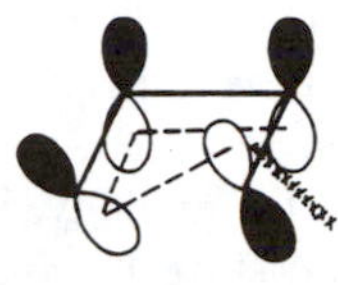

a) $[\pi 4_a + \pi 2_s]$ Cycloaddition b) $[\pi 2_s + \pi 2_a]$ Cycloaddition c) Antarafacial (conrotatory) cyclization of butadiene.

Figure 1.4. Reactions with Möbius transition states

Having shown that the *Rules*, as applied to ground-state reactions, can be derived directly from considerations of molecular topology, a few cautionary comments about their theoretical basis are in order:

1. The Hückel $4N + 2$ rule was derived for monocyclic π systems from elementary molecular orbital theory, whereas simple valence bond theory finds rings with $4N$ π electrons to be no less stable. [34, p. 33] It has long been known that real molecules lie somewhere between the two extremes. [35]

2. When the accuracy of HMO theory is improved by introducing interelectronic repulsion, the degeneracy of the highest occupied orbitals is broken in both aromatic and antiaromatic molecules. As a result, even cyclobutadiene turns out to have a plane-rectangular singlet ground-state. [36]

3. The MOs of bi- or polycyclic aromatics like naphthalene or phenanthrene do not exhibit the orbital degeneracies characteristic of benzene, even at the primitive HMO level; to cite Craig [34, p. 13]: "the [Hückel] rule has no more than vestigial force [in them]". Nor has its extension to non-planar transition states, plausible as it may appear, been proven to be generally valid.

1.2.2 Excited State Reactions

An important feature of the *Rules* is the assumption that an *allowed* pathway for a given ground-state reaction is *forbidden* to the corresponding excited state reaction and *vice versa*. The topological arguments presented so far have all been based on the ground-state stability of closed shell systems. Returning to Fig. 1.2, it may be observed that the promotion of an electron from ϕ_2 or ϕ_3 to ϕ_4 or ϕ_5 of benzene raises it to an unstable open shell state. One might then indeed feel justified in concluding that open shell transition states of Hückel type with $4N + 2$ electrons – and by analogy also open shell Möbius type transition states with $4N$ electrons – should be relatively unstable. On the other hand, promotion of one electron of cyclobutadiene from either ϕ_2 or ϕ_3 to ϕ_4, or from ϕ_1 to one of the two singly-occupied orbitals, does not transform the open shell ground-state to a closed shell excited state; it is, therefore, difficult to see why Hückel $4N$ or Möbius $4N + 2$ transition states of excited state reactions should be assigned any particular stability.

In order to rationalize the fact that photochemical $[_\pi 2_s + _\pi 2_s]$-cycloaddition is indeed a facile reaction, it is necessary to go beyond the Hückel approximation.

If cyclobutadiene can be assumed to have a closed shell singlet ground-state,[5] its first excited singlet (perhaps) and triplet (certainly) are low-lying open shell states, in which each of two electrons occupies one of two nearly degenerate MOs. By analogy, it could be argued that a reaction that is *forbidden* in the ground-state because it has an *antiaromatic* transition state becomes *allowed* on the excited state reaction surface, but only at the cost of abandoning the simple HMO model, which led to the initial conclusion that the ground-state reaction is *forbidden*!

1.3 Frontier Electrons and Frontier Orbitals

A dominant role is assigned by Woodward and Hoffmann [1, pp. 43–44] to the most loosely held electrons of the reactant, those which Fukui [37] had named the *frontier electrons*. Thermal reactions usually originate in a closed shell ground-state of the reactant and terminate in the ground-state, also ordinarily closed shell, of the product. In such a reaction, the frontier electrons, *i.e.* those that occupy the *Highest Occupied Molecular Orbital (HOMO)* of the reactant, are eventually delivered to a doubly occupied orbital – not necessarily the HOMO – of the product. The condition set by both Woodward and Hoffmann and by Fukui [37, 38, 39, 40] for classifying a thermal reaction as *allowed* is that the energy of the HOMO decreases as the reaction gets under way.

Similar considerations are applied to excited state reactions, though it has long been recognized that orbital symmetry arguments should be used with caution when considering photochemical reactions [23, p. 450]. [41, 42]. For the present, let us relax our guard sufficiently to adopt a few common assumptions:

i) The reaction originates in the lowest excited singlet state of the reactant and terminates in the lowest excited singlet of the product.[6]

ii) Each of these states is the open shell state produced by excitation of an electron from the HOMO of the ground-state to its *lowest unoccupied molecular orbital(LUMO)*.

iii) All of the molecular orbitals involved have been correctly identified.

Following Fukui [40], we refer to the lower of the reactant's two singly occupied orbitals as *SOMO′*, reserving the unprimed *SOMO* for the upper – formerly the ground-state's LUMO – which bears the lone frontier electron. Then, by the same criterion that is applied to ground-state reactions, an excited state reaction is considered to be *allowed* if the energy of the SOMO decreases as the reaction gets under way and *forbidden* if it increases. In a bimolecular reaction, the two reactant molecules are set up – by Woodward and Hoffmann as well as

[5] This too is an oversimplification; in its ground-state, cyclobutadiene has considerable biradical character, which serves to stabilize rather than destabilize it [36].

[6] Simple molecular orbital theory does not distinguish between singlet and triplet states with the same electronic configuration; it is taken for granted that the excitation takes place without change of spin multiplicity and that reaction takes place before intersystem crossing can occur.

by Fukui – in the appropriate geometry for generating the expected product. The judgement as to whether the energy of the frontier electron(s) increases or decreases is made on different grounds in the two methods[7]: Woodward and Hoffmann construct a *correlation diagram* between the occupied orbitals of the reactants and the product(s), taking their symmetry properties explicitly into account, and disregarding the unoccupied orbitals. The evaluation of this procedure is deferred to the following section.

Fukui's approach, which is shared by other perturbation treatments of reactivity [44, Chaps. 3–5,7], [45] focuses its attention primarily on the stabilization, as the reaction gets under way, of a bonding orbital of one reactant, usually its HOMO by its interaction with an antibonding orbital of the other, ordinarily its LUMO.

1.3.1 HOMO-LUMO Interaction

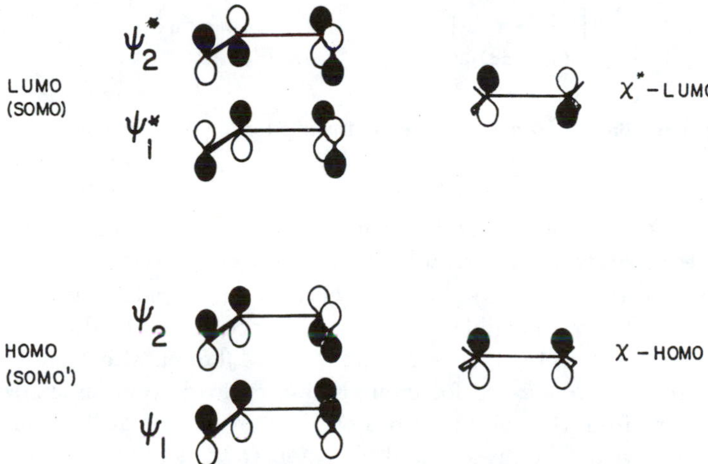

Figure 1.5. Molecular orbitals of *s-cis*-butadiene and ethylene

The HOMO and LUMO are often easy to identify, as in Fig. 1.5, where the π-orbitals of *s-cis*-butadiene are stacked in order of increasing energy, alongside those of ethylene – its reaction partner in the prototypical $[_{\pi}4+_{\pi}2]$-cycloaddition. Its HOMO, ψ_2, is less stable than χ of ethylene by virtue of the phase discontinuity between the two central atoms, so it is assumed to be the orbital bearing the *frontier electrons* when the reaction takes place on the ground-state surface. On photoexcitation, one of these two electrons is raised to ψ_1^*, the less antibonding of the two unoccupied orbitals, which becomes the SOMO. In both the ground-state and excited state reactions, the frontier orbital of butadiene, HOMO or SOMO respectively, is presumed to be stabilized by interaction with

[7] The difference is more apparent than real. [43]

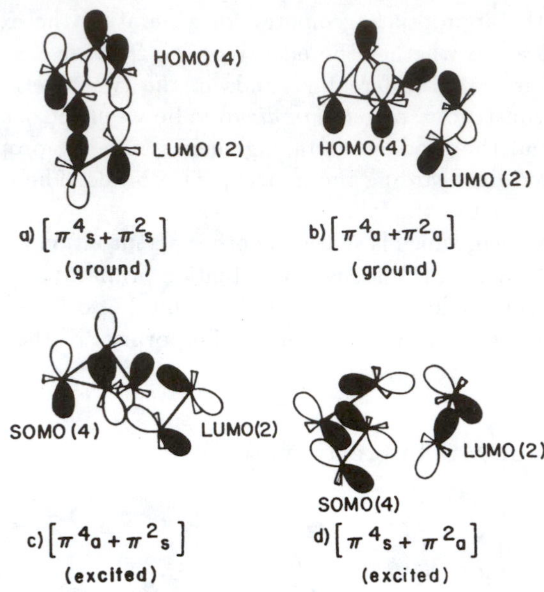

Figure 1.6a–d. *Frontier-allowed* pathways for $[{}_{\pi}4 +_{\pi} 2]$-cycloaddition

χ^*, the LUMO of ethylene, provided that the mutual orientation of the reaction partners permits favorable overlap between their two interacting orbitals.

In Fig. 1.6, the two molecules are set up in the orientations suitable for reaction along the four possible pathways for concerted cycloaddition. χ^* and ψ_2 overlap favorably in the $[{}_{\pi}4_s +_{\pi} 2_s]$ and $[{}_{\pi}4_a +_{\pi} 2_a]$ orientations, so these two pathways are *frontier allowed* for ground-state cycloaddition, as expected for Hückel systems from the considerations outlined in Section 2. The $[{}_{\pi}4_a +_{\pi} 2_s]$- and $[{}_{\pi}4_s +_{\pi} 2_a]$-cycloadditions are similarly categorized as *allowed* in the excited state, because the SOMO-LUMO overlap is favorable in both cases. An analysis of the excited state reactions in orientations a) and b), and of the ground-state cycloadditions corresponding to c) and d),[8] shows all four to be *forbidden*, in agreement with the *Rules.* [1, p. 173]

A similar analysis of $[{}_{\pi}2 +_{\pi} 2]$-cycloaddition shows the $[s + s]$ and $[a + a]$ modes to be *forbidden* in the ground-state but *allowed* in the excited state, and *vice versa* for the $[s + a]$ mode. The latter is illustrated in Fig. 1.7, in which the molecules come together in the off-orthogonal approach recommended by Woodward and Hoffmann [1, p. 69], and then turn and/or twist towards one another in order to increase the favorable HOMO-LUMO overlap.[9] Then, regardless of how the orbital phases are chosen, the reaction is always *allowed*.

[8] For example, replace the SOMO of the 4-electron component, ψ_1^* with its HOMO ψ_2, in (c) and (d) of Fig. 1.6.

[9] It might be noted that in the specificied approach the two ethylene molecules are not treated equivalently; the upper is foreordained to react antarafacially and the lower suprafacially.

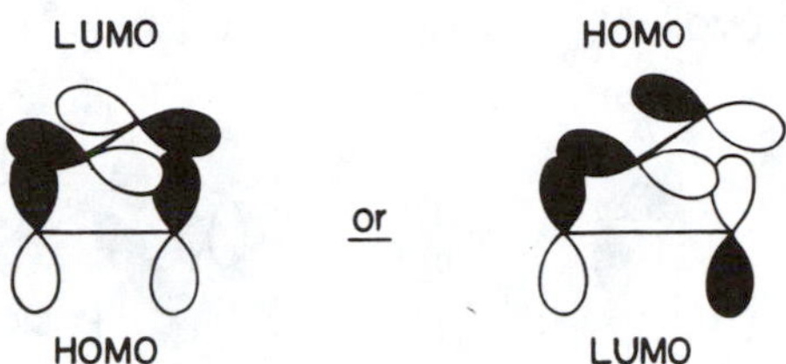

Figure 1.7. Off-orthogonal approach for $[_{\pi}2_s +_{\pi} 2_a]$-cycloaddition

In his analysis of unimolecular reactions, Fukui [40] divides the reacting molecule arbitrarily into two interacting subsystems, associating one frontier orbital with each. They are still loosely referred to as HOMO, SOMO and LUMO; though they obviously do not qualify as genuine molecular orbitals, they might perhaps be referred to as *moiety orbitals*. The method has had considerable success in dealing with thermal and photochemical fragmentations, but ambiguities remain.

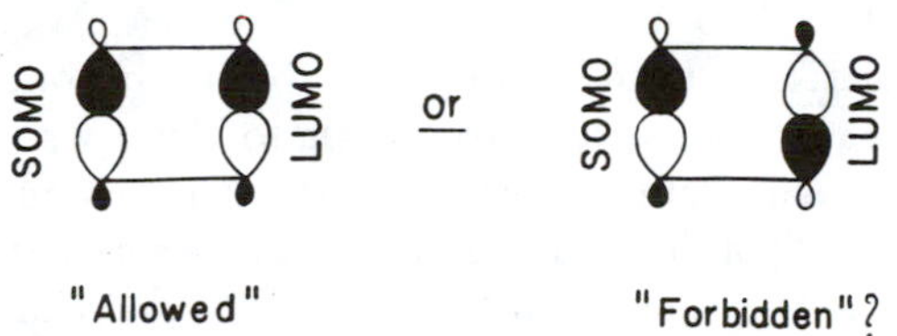

Figure 1.8. "HOMO-LUMO" interaction in $[_{\pi}2_s +_{\pi} 2_s]$-photofragmentation of cyclobutane

Consider $[_{\sigma}2_s +_{\sigma} 2_s]$-cycloreversion, which, like its converse $[_{\pi}2_s +_{\pi} 2_s]$-cycloaddition, should be *allowed* from its excited state. It can be treated within the present context by taking each of the σ bonds to be a separate subsystem. Excitation of an electron in one bond raises it to its "SOMO", which is identical to the "LUMO" of the other, unexcited bond. Then if the favorable orientation of orbital phases on the left-hand side of Fig. 1.8 is chosen, the reaction is confirmed to be *allowed*. The choice is arbitrary as long as the two bonds can be regarded as non-interacting; but can they be? The alternatives in Fig. 1.8 represent two distinct *group orbitals*, which must necessarily be assigned different energies, and the intuitive considerations which suggest that the combination on the left of the figure lies below the one on the right are by no means infallible. Nor are these two group orbitals genuine molecular orbitals, if only because they suggest that the two σ bonds included are the only ones which can be broken, whereas the other – completely equivalent – pair of σ bonds have to remain intact. In addition to these reservations, that apply to ground-state fragmentation as well, it is difficult to entertain the idea that photoexcitation of a molecule can be confined to one of four equivalent bonds.

It is perhaps less objectionable to divide the orbitals of cyclohexene into a π and a σ subsystem. Its thermal *retro*-Diels-Alder fragmentation, which the

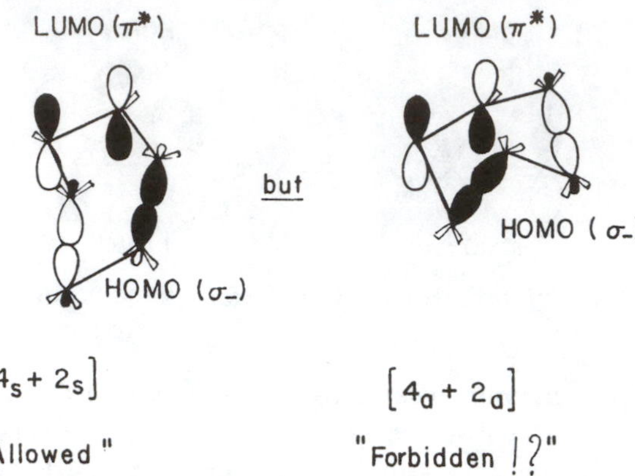

Figure 1.9. HOMO-LUMO interactions in $[4_s + 2_s]$- and $[4_a + 2_a]$-cycloreversion

Rules specify to be allowed along the $[_\sigma 4_s + {}_\sigma 2_s]$ and $[_\sigma 4_a + {}_\sigma 2_a]$ pathways, is illustrated in Fig. 1.9.

The frontier electrons are to be sought in the two σ bonds that are broken in the course of the reaction, so it is reasonable to identify σ_-, the out-of-phase combination of the σ subsystem, as the HOMO. The second frontier orbital is evidently π^*, the LUMO of the π subsystem. It can be seen that their interaction is favorable in the orientation suitable for $[_\sigma 4_s + {}_\sigma 2_s]$-cycloreversion, which is therefore *allowed*, but unfavorable in that leading to the $[_\sigma 4_a + {}_\sigma 2_a]$ pathway, which – rather embarrassingly – must be characterized as *forbidden*.

1.3.2 Superjacent and Subjacent Orbitals

The latter pathway can be rescued from *forbiddenness* if we are prepared to take not just the HOMO into account, but the orbital lying just below it as well. In this reaction mode, π^* overlaps favorably with the in-phase combination σ_+, which is no less intimately involved in the bond breaking process than σ_-, and does not lie far below it in energy. Clearly, limiting the analysis to a particular pair of *frontier orbitals* and forsaking all others can lead to error. The possible importance of the occupied MO lying just below the HOMO did not escape Fukui. [38] He recognized it explicitly in the reactions of those aromatic molecules in which the two highest occupied orbitals can be regarded as if they were a degenerate pair (e.g. ϕ_2 and ϕ_3 of benzene in Fig. 1.2) that had been split by a weak perturbation.

After it was recognized that the interaction of the LUMO with such a *subjacent* orbital, as it was named by Berson and Salem [46], can be strong enough to "allow" an otherwise *forbidden* pathway, and thus to control the stereochemistry of a reaction, the frontier electrons lost much of their uniqueness. Instances were then cited in which the *superjacent* orbital, the one lying immediately above the LUMO, seems to have a dominant influence on the course of the reaction. [47]

When we recognize that it is often necessary to consider *both* the HOMO and the LUMO of each of the two reactants, [48, 49] it becomes logically inconsistent to ignore the two subjacent and two superjacent orbitals as well. Having so broadened the working definition of *frontier orbitals* that we have to take eight of them into account when evaluating the *allowedness* of a particular reaction, doubts may assail us as to whether the approach is still practical as a qualitative diagnostic tool.

1.4 Orbital and Configuration Correlation

The criteria of "allowedness" discussed in the preceding two sections do not require the explicit consideration of *orbital symmetry*, in the sense that the symmetry elements retained along the reaction path do not enter directly into the analysis; consequently, they were not drawn in the figures. However, it is easy to ascertain from Fig. 1.1, for example, that two ethylene molecules in either the coplanar or $[s + s]$ orientation have three perpendicular mirror planes: one common to the four carbon atoms, another reflecting one molecule into the other, and a third bisecting both of them; three twofold axes of rotation (one at the intersection of each pair of mirror planes); and a center of inversion at the point where the three rotational axes intersect. After both molecules have been twisted so as to react in the $[a + a]$ mode (Fig. 1.1c), only the rotational axes remain, whereas the off-orthogonal orientation of Fig. 1.4b retains a single twofold rotational axis and no other element of symmetry.

It is also clear from Fig. 1.1 that coplanar ethylene and butadiene have one rotational axis and two mirror-planes in common. Only one of these symmetry elements, a mirror plane, remains in the $[4_s + 2_s]$ orientation, whereas only the rotational axis is retained along the $[4_a + 2_a]$ pathway. The same twofold axis is present in the $[4_a+2_s]$ orientation (Figs. 1.4a and 1.6c), whereas the topologically equivalent arrangement of Fig. 1.6d, leading to $[4_s + 2_a]$-cycloaddition, has no symmetry elements at all.[10]

The construction of an *orbital correlation diagram* requires the retention of at least one symmetry element along the reaction pathway, so none can be drawn for $[_\pi 4_s +_\pi 2_a]$-cycloaddition. Whether or not one can be constructed for $[_\pi 4_a +_\pi 2_a]$-cycloaddition depends on how we choose to interpret Woodward and Hoffmann's injunction to the effect that "the symmetry elements chosen for the analysis must bisect bonds made or broken in the process" [1, p. 31]. Looking at Fig. 1.1c or 1.6b once more, we see that a twofold rotational axis can be drawn so that it bisects both reactant molecules. In Fig. 1.9, however, the two σ bonds made in the process lie on either side of this C_2 axis and are transformed into each other by rotation about it; it bisects their symmetric and antisymmetric combinations, but not the bonds themselves. It is less easy to decide whether this axis can be regarded as bisecting the π bond of ethylene; it lies in the

[10] Strictly speaking, Woodward and Hoffmann's analysis of the latter reaction [1, p. 69] thus does not directly invoke *the conservation of orbital symmetry*.

nodal plane of the π orbital, so why – by this criterion – is it a more legitimate symmetry element than the nodal mirror plane itself?

Orbital correlation diagrams for the $[_{\pi}4_s + _{\pi}2_s]$- and $[_{\pi}4_a + _{\pi}2_s]$-cycloadditions are included in Fig. 1.10; in both cases the symmetry element retained, a mirror plane (m) and a rotational axis (C_2) respectively, unquestionably bisects the π bond of ethylene. Following Woodward and Hoffmann [1, pp. 23–24], we proceed as follows:

a) Stack the orbitals of the reactant system in order of increasing energy (Fig. 1.5).

b) Order the product MOs on the assumptions that: *i.* π orbitals are less bonding and π^* orbitals less antibonding than σ and σ^* orbitals respectively; *ii.* antisymmetric σ and σ^* orbitals are less stable than their symmetric counterparts.

c) Label each of the orbitals as being either *symmetric* (S) or *antisymmetric* (A) with respect to the appropriate symmetry operation: reflection in m or rotation about C_2.

d) Connect orbitals having the same symmetry label by *correlation lines*, taking care that lines connecting pairs of similarly labelled orbitals do not cross.

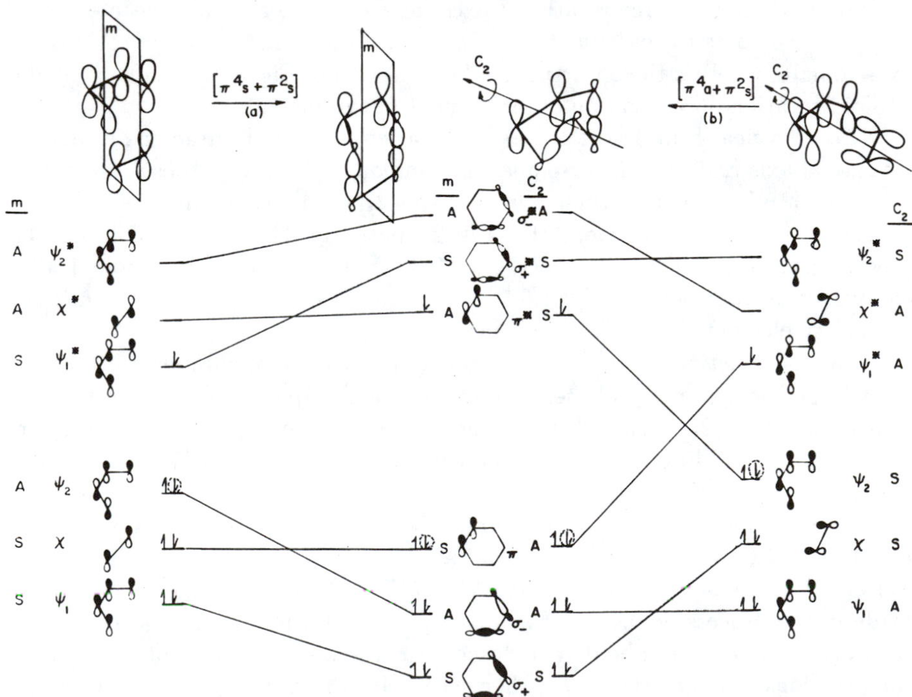

Figure 1.10a, b. Orbital correlation diagrams for [4 + 2]-cycloaddition. **(a)** $[_{\pi}4_s + _{\pi}2_s]$; **(b)** $[_{\pi}4_a + _{\pi}2_s]$

In order to allow the same diagram to be used for the analysis of both the ground- and excited state reactions, the electrons are represented by half-arrows alongside the occupied orbitals. The electron promoted from HOMO to LUMO in order to generate the excited state is enclosed in a dashed circle in the former and represented by a dashed arrow in the latter, as a reminder that it is either in one orbital or in the other – but not in both at once.[11]

1.4.1 Frontier Electron Energy

We can begin by retaining the frontier electron point of view, provided that the reactant – although it is made up of two interacting molecules – is taken to be a single entity like the product, thus putting the forward and reverse reactions on an equal footing. The HOMO on both sides of Fig. 1.10 is ψ_2 and that at the center is π. In the right-hand diagram, the energy of each of these goes up along the line connecting it with an antibonding MO, so the ground-state pathways for $[_\pi 4_a +_\pi 2_s]$-cycloadddition and $[_\sigma 4_a +_\sigma 2_s]$-cycloreversion are gratifyingly *forbidden*. Moreover, the LUMO (ψ_1^*) on the right and π^* in the center, which become the SOMOs in excited state cycloaddition and cycloreversion respectively, both go down in energy as each correlates with a bonding orbital, confirming $[4_a + 2_s]$ to be an *allowed* pathway for the excited state reaction in both directions.

The "allowedness" of $[_\pi 4_s +_\pi 2_s]$-cycloaddition in the ground-state follows directly from the correlation of ψ_2, the HOMO on the left of Fig. 1.10, with a bonding σ orbital in the center, but that of $[_\sigma 4_s +_\sigma 2_s]$-cycloreversion is less evident; here the HOMO, π, correlates with χ, the virtually isoenergetic π-orbital of ethylene, so the frontier electrons would seem to be irrelevant to the reaction. Turning perforce to the *subjacent* orbital, σ_1, and noting that its energy goes up during cycloreversion, we might be tempted to label the pathway *forbidden*, were it not for the fact that it goes through the same *aromatic* transition state as the *allowed* cycloaddition!

The corresponding excited state reactions are no less ambiguous: ψ_1^* goes up in energy so $[_\pi 4_s +_\pi 2_s]$-cycloaddition is *forbidden*, but the energy of neither the SOMO (π^*) nor the SOMO′ (π) of cyclobutene appears to be changing, so the *frontier electron* criterion is indeterminate as regards photochemical $[_\sigma 4_s +_\sigma 2_s]$-cycloreversion, just as it is for the corresponding thermal reaction.

1.4.2 Correlation of Electron Configurations

Advisedly, Woodward and Hoffmann do not rely heavily on frontier electron arguments for their conclusion that the $[4_s + 2_s]$ pathway is *allowed* in the ground-state and *forbidden* in the excited state for both the cycloaddition and its cy-

[11] The two half-arrows representing the unpaired electrons in Fig. 1.10 are drawn pointing in the opposite directions, in order to suggest the lowest excited singlet. Simple HMO theory does not distinguish between an open shell singlet and the corresponding triplet, so the correlation diagram can serve equally well for analysis of reaction on the triplet surface.

cloreversion [50], [1, pp. 23–24]. Their more rigorous criterion of "allowedness" is so nearly identical with that set forth by Longuet-Higgins and Abrahamson [51] that the two will henceforth be referred to together as the *WH-LHA Correlation Procedure.*[12]

A ground-state reaction is *allowed* whenever the *electron configuration* of the product's ground-state has the same symmetry as that of the reactant with respect to the symmetry element(s) retained along the pathway. Similarly, if the lowest excited states of reactant and product have the same symmetry properties, the photochemical reaction is *allowed*; otherwise, it is *forbidden.*

1.4.2.1 [4+2]-Cycloaddition and Cycloreversion

The correlation of two ground-states of reactant and product and the non-correlation of their excited states along the $[_\pi 4_s + _\pi 2_s]$ pathway, which preserves the mirror plane, m, can be summarized compactly as follows:

Ground-State:

$$[\psi_1(S)^2\,\chi(S)^2\,\psi_2(A)^2] \iff [\sigma_+(S)^2\,\sigma_-(A)^2\,\pi(S)^2]$$

Excited-State:

$$[\psi_1(S)^2\,\chi(S)^2\,\psi_2(A)\,\psi_1^*(S)] \not\Leftrightarrow [\sigma_+(S)^2\,\sigma_-(A)^2\,\pi(S)\,\pi^*(A)]$$

The *electron configuration*, of a particular state is a list of its occupied orbitals, a superscript 2 indicating occupancy by two electrons of opposite spin. The two ground-state configurations correlate because each has three doubly occupied orbitals, two of them symmetric and one antisymmetric with respect to m. The correlation or non-correlation of the *states* of reactant and product does not depend on the energetic ordering of their occupied orbitals, but simply on whether the number of doubly-occupied orbitals with each symmetry label is the same in both. It follows that the initial slope of the HOMO is irrelevant to this criterion of "allowedness".

Turning to the excited state reaction, we see that the singly-occupied orbitals correlate across the diagram, but correlation of the two states is prevented by the doubly-occupied orbitals; both of them are labelled S in butadiene, whereas in cyclobutene one is S and the other is A. As can be seen in Fig. 1.10, one of the former is forced to correlate with an unoccupied S orbital of the product, and the latter with an unoccupied A orbital of the reactant. The orbital correlation diagram specifies which MOs on either side correlate with one

[12] Two points at which the two treatments diverge are:

1) Longuet Higgins and Abrahamson distinguish between *state* and *configuration correlation* [52], a distinction that will be taken up in a later chapter.

2) Unlike Woodward and Hoffmann, Longuet-Higgins and Abrahamson do not specify that a symmetry element can be used only if it bisects bonds made or broken in the process.

another, but spelling out the precise orbital correlations is not essential; the "allowedness" or "forbiddenness" of the pathway is fully established by the success or failure of *configuration correlation.*

The $[_{\pi}4_a + {_{\pi}2_s}]$ pathway can be summarized similarly in terms of configuration correlation with respect to C_2.

Ground-State:

$$[\psi_1(A)^2\,\chi(S)^2\,\psi_2(S)^2] \;\not\Longleftrightarrow\; [\sigma_+(S)^2\,\sigma_-(A)^2\,\pi(A)^2]$$

Excited-State:

$$[\psi_1(A)^2\,\chi(S)^2\,\psi_2(S)\,\psi_1^*(A)] \;\Longleftrightarrow\; [\sigma_+(S)^2\,\sigma_-(A)^2\,\pi(A)\,\pi^*(S)]$$

The former is *forbidden* and the latter *allowed*, in agreement with the *Rules*. Once more, these conclusions can be reached without recourse to Fig. 1.10, which shows the actual orbital correlations.

1.4.2.2 [2+2]-Cycloaddition and Cycloreversion

The orbital correlation diagram for the formation of cyclobutane from two ethylene molecules, using two perpendicular mirror planes as diagnostic symmetry elements, [1, p. 19] is too familiar to have to be reproduced. We recall that when the reactant molecules are set up in the coplanar $[_{\pi}2_s + {_{\pi}2_s}]$ orientation (Fig. 1.1b), other symmetry elements exist as well, including two rotational axes which also "bisect bonds made or broken in the process" [1, p. 31]. The use of these symmetry elements instead of the mirror planes is also more realistic, because cyclobutane is most stable in a puckered conformation [53], in which the rotational axes are retained but the mirror planes are not.

At the extreme left of Fig. 1.11 the reacting molecules have been brought close enough for their MOs to interact and form the bonding and antibonding combinations of π orbitals, π_- and π_+, which split as shown.[13] The two π^* orbitals interact less strongly in the off-perpendicular approach than if the geometry were strictly coplanar, but they split nevertheless, provided that the approach is not perpendicular; in the latter more symmetrical geometry, π^*_- and π^*_+ become a degenerate pair of non-bonding orbitals.

Turning to the center of Fig. 1.11 and considering the reverse reaction, we set aside any momentary qualms that we might have about ignoring two of the four equivalent σ bonds of cyclobutane in the construction of σ and σ^* by assuming that the two bonds which are being specifically taken into account have been elongated slightly, the molecule having already "decided" which two bonds are going to be broken. The doubly-occupied orbitals fail to correlate across the diagram; we find – in agreement with the *Rules* as well as with the

[13] The subscripted sign indicates whether the MOs of the interacting ethylene molecules are oriented in the same or in the opposite direction: The antibonding combination is labeled π_+, because its orbital lobes point in the same direction, whereas the mutually bonding combination, in which they face one another, becomes π_-.

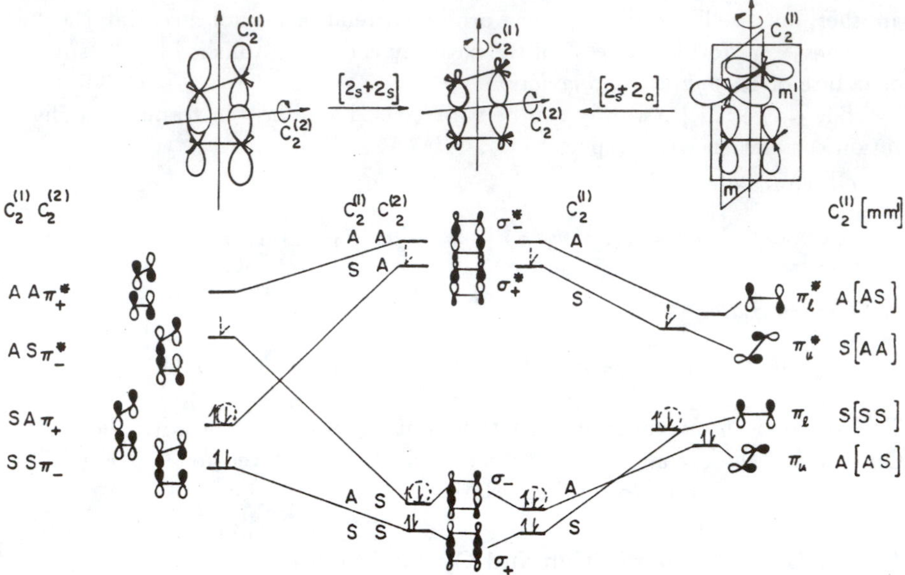

Figure 1.11. Orbital correlation diagrams for [2 + 2]-cycloaddition. *Left side*: $[_{\pi}2_s +_{\pi}2_s]$; *right side*: $[_{\pi}2_s +_{\pi}2_a]$

correlation diagram based on the mirror planes [1, p. 19], and even with the slope of the orbital bearing the frontier electron(s) – that the $[2_s + 2_s]$ pathway is *forbidden* to both cycloaddition and cycloreversion in the ground-state and *allowed* to both in the excited state.

In the correlation diagram for $[2_s + 2_a]$-cycloaddition that appears on the right side of Fig. 1.11, the reacting molecules are set up in the off-orthogonal orientation of Fig. 1.7, which leads to the *WH-LHA allowed* pathway [1, p. 69] for this mode of cycloaddition. Only one C_2 axis is retained along the pathway; though it does not bisect the σ bonds of cyclobutane and lies in the nodal plane of the π bond of the upper ethylene molecule, it does bisect that of the lower and thus qualifies as a non-trivial "diagnostic" symmetry element. In the strictly orthogonal approach there are also two mirror planes, drawn at the upper right of the figure, which are helpful in determining the initial order of orbital energies, as follows: The bonding MO of the upper molecule, π_u, is stabilized by interaction with π_l^*, the antibonding orbital of the lower, which is destabilized accordingly. In contrast, π_l and π_u^*, the HOMO and LUMO respectively, begin to interact only after the departure from the orthogonal orientation, because they differ in their symmetry properties with respect to the mirror planes.

If we can tolerate the inequity of treating two identical ethylene molecules as if they were inherently non-equivalent, completion of the correlation diagram is straightforward: $[_{\pi}2_s +_{\pi}2_a]$-cycloaddition is *allowed* in the ground state both by the negative slope of the HOMO and – more convincingly – by configuration correlation, which "allows" $[_\sigma 2_s +_\sigma 2_a]$-cycloreversion as well. Cycloaddition in the excited state appears to be *forbidden* by the non-correlation:

$$[\pi_u(A)^2\, \pi_l(S)\, \pi_u^*(S)] \not\Leftrightarrow [\sigma_+(S)^2\, \sigma_-(A)\, \sigma_+^*(S)]\,.$$

Note, however, that the configuration ascribed to the two interacting ethylenes is that of a *charge-transfer exciplex*, in which an electron has been transferred from the lower to the upper molecule. Ordinarily, excitation is presumed to occur within one of the reacting molecules before they approach closely enough to react. In that case, the question whether $[_\pi 4_s +_\pi 2_a]$-cycloaddition is *allowed* or *forbidden* in the excited state becomes more difficult to answer – or even to frame – because we have to know whether the excited molecule is destined to be the "upper" or "lower", before it has come within range of its reaction partner.

1.5 Problems and Prospects

1.5.1 Some Unanswered Questions

The application of the WH-LHA correlation procedure to the various modes of [4 + 2]- and [2 + 2]-cycloaddition has led to a reasonably consistent set of results, but questions of several kinds remain unanswered. The choice of molecular orbitals for inclusion in an orbital correlation diagram is arbitrary and their energetic ordering is not always known. For example, interaction with the two neglected σ_{CC} bonds of cyclobutane may suffice to place σ_-^* below σ_+^* in Fig. 1.11. Would such an inversion render $[_\pi 2_s +_\pi 2_s]$-cycloaddition *forbidden* in the excited state? Moreover, non-bonding orbitals, lone-pair orbitals in particular, are involved in excitation processes too often to be disregarded in excited state processes, even when they are not directly involved in the bonding changes. We will see that even σ_{CH} orbitals, which are almost invariably neglected in the analysis of reactions that do not involve the rupture or formation of CH bonds, can sometimes attain unexpected importance.

The symmetry elements on the left side of Fig. 1.11 are not the same as those used in the conventional analysis [1, pp. 23–24], but the conclusions are identical. In both cases, one element bisects the bonds broken and the other bisects the bonds formed. Was it necessary to employ both, and – if only one of them were chosen – would the result have been the same? If so, could the invariance of the result to the choice of diagnostic symmetry element(s) be cited as evidence for the inherent reliability of the method, or might it not be a consequence of the extremely simple system under study and a fortuitously happy choice of the geometry in which the reaction was analyzed? Should a correlation diagram like the above, which is based on two "legitimate" symmetry elements (four, if the approach is coplanar and the mirror planes are also used), be regarded as more reliable than the two in Fig. 1.11, each of which is based on a single symmetry element that bisects either the bonds made or those broken, but not both?

The reader can confirm easily that a correlation diagram based on the C_2 axis retained during $[4_a +2_a]$-cycloaddition, shows it to be *allowed* in the ground state and *forbidden* in the excited state.[14] Should the analysis be discarded because the C_2 axis lies in the nodal plane of the ethylene molecule and bisects no other bonds made or broken, or may we grant it legitimacy anyway, as do Silver and Karplus [54] in their *primitive symmetry classification*? Furthermore, as was pointed out by Dauben, Salem and Turro [55], the "plane containing the reaction centers" – which certainly does not bisect bonds between any two of them – is the "discriminating symmetry element" in many photochemical reactions.

Ground-state reactions are by no means exempt from many of the difficulties raised above, but the application of orbital symmetry arguments to excited state reactions, which are particularly susceptible to them, is subject to other severe limitations as well [23, Chap. 6], [42], [44, Chap. 8]. An especially serious one is the question whether a reaction originating in the first excited state of the reactant will necessarily proceed smoothly to the lowest excited state of the product [41]. May not a higher excited state of the product be more easily accessible than the first, and still low enough in energy for the photochemical reaction to occur with ease? Then too, as Woodward and Hoffmann point out, [1, p. 100] "there is no necessity to reach the excited state of the reactant"; once the *allowed* pathway to the first excited state has been entered, the reaction can proceed *via* "a radiationless transition to the ground-state of the product". The question then arises whether orbital symmetry arguments may perhaps be helpful in providing "the physical rationale of such a transition [which] is still lacking" [1, p. 100], particularly when the two states between which the transition occurs have different spin multiplicities? The "mirror image" of the latter situation occurs when a thermodynamically unstable molecule in its closed shell singlet ground-state rearranges or decomposes thermally to an excited triplet of the thermodynamically more stable product, as in the isomerization of Dewar benzene to benzene [56] or the thermolysis of tetramethyldioxetane [57].

Finally, what are the limits of reliability of orbital symmetry arguments?

1) Are they restricted to reactive systems of genuinely high symmetry, or is it legitimate to resort to "local symmetry"? If so, just how is the term to be understood?

2) Sigmatropic rearrangements typically convert the reactant into a product homomeric or enantiomeric with itself *via* a transition state that is more symmetrical than either. In such cases the electron configurations necessarily correlate, so orbital correlation diagrams are of little use, and we have to fall back on topological arguments [1, p. 114], [23, pp. 94–98]. May not ways be found of circumventing this restriction on the use of correlation diagrams for the analysis of sigmatropic reactions?

---

[14] When the ethylene molecule in Fig. 1.10b is rotated about its own molecular axis by 90°, χ becomes A and χ^* becomes S; all of the other MOs are unchanged. The configurations correlate as follows:

Ground-State: $[\psi_1(A)^2\,\chi(A)^2\,\psi_2(S)^2] \iff [\sigma_+(S)^2\,\sigma_-(A)^2\,\pi(A)^2]$

Excited-State: $[\psi_1(A)^2\,\chi(A)^2\,\psi_2(S)\,\psi_1^*(A)] \not\Leftrightarrow [\sigma_+(S)^2\,\sigma_-(A)^2\,\pi(A)\,\pi^*(S)]$

3) Some of the more highly symmetrical systems are in the inorganic domain; how efficiently can orbital symmetry deal with them? Mango and Schachtschneider's early attempt [58, 59] to extend the *Woodward-Hoffmann Rules* to transition metal catalysis of *forbidden* organic reactions foundered on experimental evidence [60] showing that several of the reactions that had been made *allowed* in this way actually proceed along stepwise pathways. However, the fact that a particular reaction follows one pathway rather than another merely shows that the first is more facile; it hardly constitutes proof that the second is *forbidden*. Still less does it justify the almost complete abandonment of attempts to deal with inorganic reactions in terms of orbital symmetry, as seems to have occurred.

4) The various approaches discussed in this chapter all stem from elementary *Hückel Molecular Orbital* theory. Why do qualitative arguments based on so approximate a set of assumptions work as well as they do? Can such naive considerations still serve a useful purpose in this day of sophisticated semi-empirical and *ab initio* multiconfigurational potential energy surface computations, and – audacious presumption! – perhaps even suggest ways of improving the efficiency and reliability of these very computations?

1.5.2 Where Do We Go From Here?

The latter chapters of this book constitute an attempt to deal with the questions raised in the preceding paragraphs. The viewpoint adopted is that of *Orbital Correspondence Analysis in Maximum Symmetry (OCAMS)* [61], which has been shown to be equivalent to the WH-LHA correlation procedure under certain clearly defined conditions [62]. It differs from the latter formally in that, instead of investigating possible pathways separately and characterizing each one as *allowed* or *forbidden*, it specifies *a priori* the nature of the reaction coordinate which is consistent with the symmetry properties of the occupied molecular orbitals of the reactant and product.[15] Its formalism is not only more compact and precise, but it can be extended conveniently in new directions and affords greater insight into the energetic and conformational factors that govern reactivity.

An elementary exposition of OCAMS [65], including a detailed analysis of $[_{\pi}4 +_{\pi} 2]$-cycloaddition, was cast in the familiar S,A notation employed in this chapter. Symmetry properties are expressed so much more concisely and unambiguously in the conventional notation of group theory that the latter will be used throughout the rest of the book. Although most chemists have at least a nodding acquaintance with this symbolism, familiarity with it will not be assumed and it will be brought in gradually as needed.

[15] In its emphasis on the the coupling of molecular orbitals by means of coordinates for nuclear motion, OCAMS resembles treatments based on the second-order Jahn-Teller effect [23, pp. 17–25], such as that of Bader [21] and developments ensuing from it [63, 64], in which a vibrational coordinate of appropriate symmetry couples occupied and unoccupied MOs of the same molecule and determines the manner of its decomposition.

The task of introducing the uninitiated reader to group theoretical concepts and terminology is, however, incidental to the main thrust of the following three chapters.[16] Their principal objective is to explore the relationships among orbital, configuration and state symmetry, molecular geometry, and energy. An understanding of these relations is a prerequisite of any attempt to use orbital symmetry arguments effectively as a rational, albeit qualitative, means of estimating differences in the energy – or free energy – of activation along different reaction paths. This, after all, is what the study of reaction mechanism is about.

1.6 References

[1] R.B. Woodward and R. Hoffmann: Angew. Chem. *81,* 797 (1969); Angew. Chem. Internat. Ed. (English) *8,* 789 (1969); *The Conservation of Orbital Symmetry.* Verlag Chemie, Weinheim and Academic Press, New York 1970.

[2] L.P. Hammett: *Physical Organic Chemistry.* McGraw-Hill, New York: *1st Ed.* 1940; *2nd Ed.* 1970.

[3] R.P. Bell: *Acid and Base Catalysis.* Oxford University Press, Oxford 1941.

[4] C.K. Ingold: *Structure and Mechanism in Organic Chemistry.* Cornell University Press, Ithaca: *1st Ed.* 1953; *2nd. Ed.* 1969.

[5] L. Melander: *Isotope Effects on Reaction Rates.* Ronald, New York 1970.

[6] E.A. Halevi: Secondary Isotope Effects. In: S.G. Cohen et al. (eds.) *Progress in Physical Organic Chemistry.* Wiley-Interscience, New York 1963, pp. 109–221.

[7] J.E. Leffler and E. Grunwald: *Rates and Equilibria of Organic Reactions.* Wiley, New York 1963.

[8] W. v.E. Doering and W.R. Roth: Tetrahedron *18,* 67 (1962); Angew. Chem. *18,* 27 (1962).

[9] See e.g. pp. 32–92 (1st Ed.) and 34–114 (2nd Ed.) of Reference [4] respectively.

[10] Nguyên Trong Anh: *Les Règles Woodward-Hoffmann.* Ediscience, Paris 1970; *Die Woodward-Hoffmann Regeln und ihre Anwendung.* Verlag Chemie, Weinheim 1972.

[11] T.L. Gilchrist and R.C. Storr: *Organic Reactions and Orbital Symmetry.* Cambridge University Press, Cambridge 1972.

[12] R.E. Lehr and A.P. Marchand: *Orbital Symmetry – A Problem Solving Approach.* Academic Press, New York and London 1972.

[13] A.V. Bellamy: *An Introduction to Conservation of Orbital Symmetry – A Programmed Text.* Longmans, London 1974.

[14] E. Wigner and E.E. Witmer: Z. Physik *51,* 859 (1928).

[15] F. Hund: Z. Physik *40*, 742 (1927); *ibid.* *42*, 93 (1927).

[16] J.E. Lennard-Jones: Trans. Faraday Soc. *51*, 668 (1929).

[17] R.S. Mulliken: Rev. Mod. Physics *4,* 1 (1932).

[18] A.D. Walsh: J. Chem. Soc. *1953,* 2260 and following papers.

[19] K.E. Shuler: J. Chem. Phys. *21,* 624 (1953).

[20] V.J. Griffing: J. Chem. Phys. *23,* 1015 (1955); V.J. Griffing and J.T. Vanderslice: *ibid.*, p.1039.

[16] This has been accomplished admirably by Hargittai and Hargittai [66] in their discursive overview of symmetry in chemistry.

[21] R.F.W. Bader: Canad. J. Chem. *40,* 1164 (1962).
[22] T.F. George and J. Ross: J. Chem. Phys. *55,* 3851 (1971).
[23] R.G. Pearson: *Symmetry Rules for Chemical Reactions.* Wiley, New York 1976.
[24] Nguyên Trong Anh: The Use of Aromaticity Rules, Frontier Orbitals and Correlation Diagrams: Some Difficulties and Unsolved Problems. In R. Daudel (ed.) *Quantum Theory and Chemical Reactions.* Reidel, Dordrecht 1980, pp. 177–89.
[25] I. Fleming: *Frontier Orbitals and Organic Chemical Reactions.* Wiley, London 1976.
[26] M.G. Evans and E. Warhurst: Trans. Faraday Soc. *34,* 614 (1938).
[27] M.G. Evans: Trans. Faraday Soc. *35,* 824 (1939).
[28] M.J.S. Dewar: Tetrahedron Suppl. *8,* 75 (1966); Angew. Chem. *83,* 859 (1971); Angew. Chem. Internat. Ed. (English) *10,* 761 (1971).
[29] E. Hückel: Z. Physik *70,* 204 (1931).
[30] See any textbook on MO theory, *e.g.*: a) A. Streitwieser: *Molecular Orbital Theory for Organic Chemists.* Wiley, New York 1961; b) L. Salem: *Molecular Orbital Theory of Conjugated Systems.* Benjamin, Reading 1966; c) E. Heilbronner and H. Bock: *Das HMO Modell und seine Anwendung, vol. 1.* Verlag Chemie, Weinheim 1968; d) H.E. Zimmerman: *Quantum Mechanics for Organic Chemists.* Academic Press, New York and London 1975.
[31] D.P. Craig: J. Chem. Soc. *1959,* 997.
[32] E. Heilbronner: Tetrahedron Lett. *1964,* 1928.
[33] H.E. Zimmerman: a) J. Amer. Chem. Soc. *88,* 1563,1566 (1966); b) Accts. Chem. Research *4,* 272 (1971); *ibid. 5,* 393 (1972). c) See also Reference [30d, pp. 37–41, 113–116, 157–158].
[34] See e.g. D.P. Craig: Aromaticity. In D. Ginsburg (ed.) *Non-Benzenoid Aromatic Compounds.* Interscience, New York 1959.
[35] J.H. Van Vleck and A. Sherman: Revs. Mod. Phys. *7,* 167 (1935).
[36] For a summary of the evidence on the structure and properties of cyclobutadiene, see W.T. Borden and E.R. Davidson: Ann. Revs. Phys. Chem. *30,* 134–149 (1979).
[37] K. Fukui, T. Yonezawa and H. Shingu: J. Chem. Phys. *20,* 722 (1952).
[38] K. Fukui, T. Yonezawa, C. Nagata and H. Shingu: J. Chem. Phys. *22,* 1433 (1954).
[39] K. Fukui: *Theory of Orientation and Stereoselection.* Springer, Heidelberg 1970.
[40] K. Fukui: Accts. Chem. Research *4,* 54 (1971) and papers cited therein.
[41] W.Th.A.M. van der Lugt and L.J. Oosterhoff: J. Amer. Chem. Soc. *91,* 6042 (1969).
[42] For a critical discussion of the applicability of orbital symmetry considerations to photochemical reactions, see J. Michl: Top. Curr. Chem. *46,* 1 (1974).
[43] San-Yan Chu: Tetrahedron *34,* 645 (1978).
[44] Various aspects of this popular approach are presented in G. Klopman (ed.) *Chemical Reactivity and Reaction Paths.* Wiley, New York 1974.
[45] M.J.S. Dewar and R.C. Dougherty: *The PMO Theory of Organic Chemistry.* Plenum, New York 1975.
[46] J.A. Berson and L. Salem: J. Amer. Chem. Soc. *94,* 8917 (1972); J.A. Berson: Accts. Chem. Research *5,* 406 (1971).
[47] S. David, O. Eisenstein, W.J. Hehre, L. Salem and R. Hoffmann: J. Amer. Chem. Soc. *95,* 3806 (1973).

[48] K.N. Houk: Accts. Chem. Research *8,* 361 (1975).
[49] O. Eisenstein, J.M. Lefour, Nguyên Trong Anh and R.F. Hudson: Tetrahedron *33,* 523 (1977).
[50] R.B. Woodward and R. Hoffmann: J. Amer. Chem. Soc. *87,* 2046 (1965).
[51] H.C Longuet-Higgins and E.W. Abrahamson: J. Amer. Chem. Soc. *87,* 2045 (1965).
[52] See e.g. D.M. Silver: J. Amer. Chem. Soc. *96,* 5959 (1974) for a discussion of this distinction and its role in the hierarchy of symmetry conservation rules.
[53] S. Meiboom and L.C. Snyder: J. Chem. Phys. *52,* 3857 (1970).
[54] D.M. Silver and M. Karplus: J. Amer. Chem. Soc. *97,* 2645 (1975).
[55] W.G. Dauben,L. Salem and N.J. Turro: Accts. Chem. Research *8,* 41 (1975).
[56] N.J. Turro and P. Lechtken: J. Amer. Chem. Soc. *95,* 264 (1973); P. Lechtken, R. Breslow, A.H. Schmidt and N.J. Turro: J. Amer. Chem. Soc. *95,* 3025 (1973).
[57] N.J. Turro and P. Lechtken: J. Amer. Chem. Soc. *94,* 2886 (1972).
[58] F.D. Mango and J.H. Schachtschneider: J. Amer. Chem. Soc. *89,* 2484 (1967); *94,* 2886 (1972).
[59] F. D. Mango: Coord. Chem. Rev. *15,* 109-205 (1975) and papers cited therein.
[60] See in particular: L. Cassar, P.E. Eaton and J. Halpern: J. Amer. Chem. Soc. *94,* 3515 (1970); J. Halpern: Accts. Chem. Research *3,* 386 (1970). For a full discussion, see Reference [23, p. 401ff.]
[61] E.A. Halevi: Helvet. Chim. Acta *58,* 2136 (1975).
[62] J. Katriel and E.A. Halevi: Theoret. Chim. Acta *40,* 1 (1975).
[63] L. Salem and J.S. Wright: J. Amer. Chem. Soc. *91,* 5947 (1969).
[64] R.G. Pearson: Accts. Chem. Research *4,* 152 (1971).
[65] E.A. Halevi: Angew. Chem. *88,* 664 (1976); Angew. Chem. Internat. Ed. (English) *15,* 593 (1976).
[66] I. Hargittai and M. Hargittai: *Symmetry through the Eyes of a Chemist.* Verlag Chemie, Weinheim 1986.

Part II

Symmetry and Energy

Chapter 2

Atoms and Atomic Orbitals

Prefatory note: This chapter and the others in Part II have been written in as non-mathematical a style as the author could manage. The theoretical validity of the various statements made without formal proof can be checked in any of the many available texts on Quantum Chemistry [1, 2, 3]. The initiate, who may be tempted to skip this and the following chapter, is urged to skim through them anyway. To paraphrase Shakespeare's Ulysses [8], "the author's drift" may be nothing to "strain at" but his "position" may not be altogether "familiar".

2.1 Is an Isolated Atom Spherically Symmetrical?

The intuitive answer to this question is: "Of course!". However, if by *atom* we mean "a nucleus and the electrons disributed around it", it depends on what we mean by "spherically symmetrical". The conventional picture of a many-electron atom is derived from that of the hydrogen atom: an electron in the central field of a positively charged nucleus. In its ground-state, the electron is in a spherically symmetric $1s$ orbital, so its charge density is spherically symmetric as well. By the same token, an alkali metal atom has spherical symmetry, because the lone valence electron can be regarded as moving in the central field of a *core*: the nucleus screened by a spherically symmetrical *closed shell*. Needless to say, a helium or neon atom, is which all of the electrons are in closed shells, also has spherical symmetry, as does a nitrogen with its closed half-shell. But consider a boron atom: its ground-state configuration is written $[1s^2 2s^2 2p^1]$, implying that one electron is in a p orbital, the other four being paired in $1s$ and $2s$.

In the classical model of the boron atom, depicted in Fig. 2.1, the $2p$ electron (e) behaves as if it is revolving about some axis (A) through the nucleus (n). Its motion is restricted to the plane (P) perpendicular to A, rather than to any of the infinite number of planes which, like P′, can be drawn to include it; its angular momentum is directed exactly along A and produces a magnetic moment μ antiparallel to A.

In order to determine the direction of A in space, we have to establish an external frame of reference, for example by exerting a magnetic field parallel to what we choose to define as the z axis. We then find that quantum theory limits the precision with which the plane of rotation can be fixed: No more than $h/2\pi$

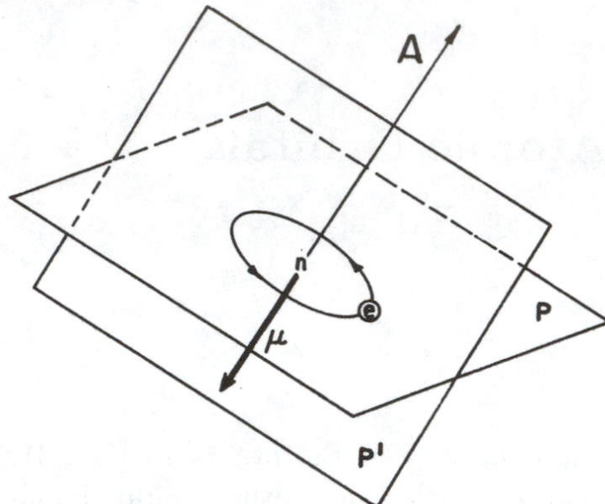

Figure 2.1. A $2p$ electron rotating about a nucleus

of the $2p$ electron's $\sqrt{2}h/2\pi$ units of angular momentum can be aligned either parallel or antiparallel to the field.

However, even the approximate specification of A is enough to guarantee that the distribution of the electron in space is not spherically symmetrical. The symmetry properties of the boron atom are determined by those of it single $2p$ electron, so when we say that the free boron atom is spherically symmetric we really mean that the *probability distribution* of the electron about the nucleus is spherically symmetrical because the direction of A is unknown.

In this familiar description, each of an atom's electrons is assigned separately to a *hydrogen-like* orbital: $1s$, $2s$, $2p$, etc. The screening of each electron by the others is treated approximately as if a spherical shell of negative charge were interposed between it and the nucleus. Thus, even in the case of atoms like boron, in which the distribution of electronic charge (as opposed to its *probability distribution*) is not spherically symmetrical, the potential energy of each electron is treated as if it were. Returning to the boron atom, we see that this assumption is reasonable as regards the $2p$ electron, which "sees" the spherically symmetrical core of the nucleus and four s electrons, but less so for the $2s$ electrons, which are screened from the nucleus not only by the $1s$ electrons and each other but by the p electron as well. The latter clearly provides a more effective shield against nuclear attraction when the $2s$ electron happens to be near the plane P than when it has moved away from it.

The departure from spherical symmetry is even more obvious in the case of the carbon atom. Its ground-state configuration is that obtained by adding a second p electron to its monovalent cation, which has the same electron configuration as the neutral boron atom. The potential energy of the $2p$ electrons, each of which is acted upon by the non-spherically symmetrical field of the other, cannot have spherical symmetry. It is generally treated as if it does, because it is only within this approximation that atomic orbitals can be legitimately regarded as s, p, d or f; and hydrogen-like AOs are so convenient. The quantitative deficiencies of the model are made up by antisymmetrizing the atomic wave function and by explicitly introducing orbital- and spin-angular momentum coupling, electron correlation, and relativistic effects.

2.2 Desymmetrization by an External Field

2.2.1 *p* Orbitals in a Magnetic Field

Now, however, a new difficulty arises: quantum theory tells us that our B atom has three distinct $2p$ states with the same energy and angular momentum, and – as we saw in the preceding section – they remain indistinguishable from each other as long as the potential energy is spherically symmetric. We have also seen that the plane in which the p electron is rotating can be characterized, if only approximately, by placing the atom in a magnetic field with which the magnetic moment of the electron can interact. The atom is "informed" in this way which direction we have specifed to be z, and the three p orbitals can now be distinguished from one another by the angle their axis of rotation (A) makes with the z axis and the sense of their rotation about it. We recall that A can align itself in one two ways: either (exactly) perpendicular to z, i.e. it lies somewhere in the plane containing the still unspecified x and y axes; or (nearly) parallel to z. In the former instance, μ is perpendicular to the field, so the energy of the p_0 orbital is the same as in its absence. The latter case can be subdivided: If the electron is rotating clockwise (p_{-1}) it is stabilized by the magnetic field; if counterclockwise (p_{+1}) – it is destabilized. The degeneracy of the triply degenerate p level has been split, the relative energies of the three states being dependent on the intensity of the magnetic field, as illustrated in Fig. 2.2.

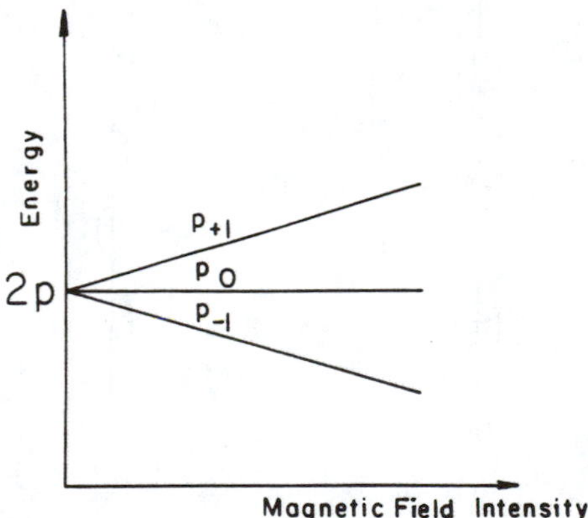

Figure 2.2. Splitting of an atomic p level by a magnetic field

All this would be too familiar to bear repetition, if it did not illustrate a paradox that we will encounter repeatedly: In order to characterize the three degenerate p orbitals that emerge as distinct solutions of the Schrödinger equation for the spherically symmetrical hydrogen-like atom, it is necessary to apply an external field, thus destroying the spherical symmetry that justifies giving the three orbitals the label p. The charge distribution is still symmetric to rotation about the z axis, because the atom has not been "told" how to distinguish x from y, and there are also an infinite number of mirror planes passing through the z axis. This set of *symmetry elements*, which confers *axial symmetry* on the atom, constitutes the *symmetry point group* labelled $\mathbf{C}_{\infty v}$ in the conventional *Schönfliess notation* [9, 10]. The paradox can be restated: In order to distinguish the three degenerate p orbitals, the spherical atom (symmetry point group $\mathbf{K}_h$) had to be *desymmetrized* to $\mathbf{C}_{\infty v}$, thus splitting the degeneracy of the p orbitals and *contaminating* their *pure p* character.

If the intensity of the magnetic field is now reduced to near zero, p_{+1}, p_0 and p_{-1} become very nearly degenerate, but not quite. A minimal field intensity must be maintained in order to define the z axis; as long as it does, the potential energy is not truly spherical but only effectively so. This distinction is important, because no observable of an atomic or molecular system can have higher symmetry than its potential energy. Thus, Fig. 2.3 would lead us to believe that the charge density of an electron in any of the three p orbitals is cylindrically symmetrical ($\mathbf{D}_{\infty h}$). In addition to the symmetry elements of $\mathbf{C}_{\infty v}$, it appears to have: a horizontal mirror plane, i.e. perpendicular to the z axis, and an infinite number of twofold rotational axes lying in it. The apparent cylindrical symmetry is, however, only achieved when the axial magnetic field becomes vanishingly small, so that the potential energy of the electron has become – *in effect* – spherically symmetric.

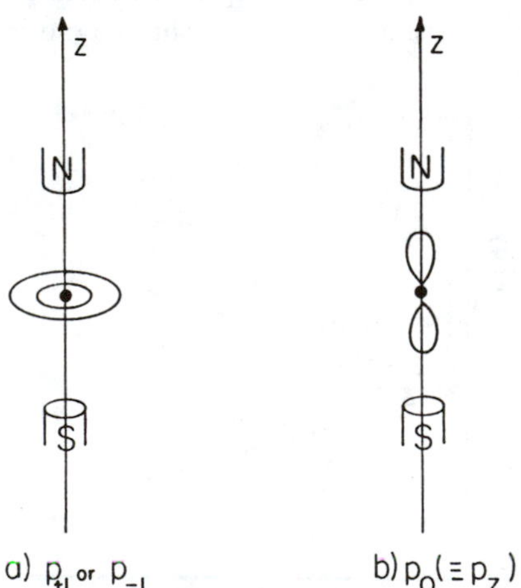

Figure 2.3a, b. Charge density of an electron in a vanishingly small magnetic field

The axial symmetry of a magnetic field, effective as it is in splitting the degeneracy of p (and also d, f, etc.) atomic orbitals, has several disadvantages from the point of view of chemical bonding. The limited mathematical sophistication of many experimentalists leaves them uncomfortable in the face of the complex wavefunctions p_{+1} and p_{-1}; if they must have any wavefunctions at all, let them at least be real. The latter, moreover, are better for depicting the directional properties of the electron density; the doughnut of charge shown in Fig. 2.3a is singularly uninformative about the bonding properties of the atom.

The textbook resolution of this dilemma is well known: Since p_{+1} and p_{-1} are degenerate, any orthogonal pair of linear combinations can serve as well. The pair invariably chosen:

$$p_x = (p_{-1} + p_{+1})/\sqrt{2}\ ;\ p_y = i\,(p_{-1} - p_{+1})/\sqrt{2} \tag{2.1}$$

is particularly attractive in that the charge density of an electron in each has the same shape as p_0, now rechristened p_z, but is cylindrically disposed about one of the other two cartesian axes. Equation 2.1 is formally unexceptionable, but a difficulty remains: How can an atom situated in a magnetic field parallel to z possibly tell which directions in space we have decided to label x and y? Furthermore, since our interest is centered on the directional properties of the electron density rather than on the chemically less important orbital magnetic moment, why resort to a magnetic field at all?

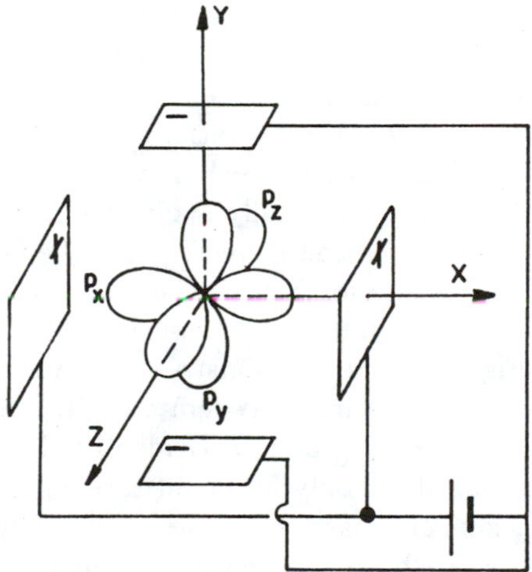

Figure 2.4. The three possible charge distributions of a p electron in a vanishingly small quadrupolar field

2.2.2 p Orbitals in a Quadrupolar Field

An electric field can serve to define a direction in space as well as a magnetic field, and its electrostatic interaction with the electron density makes it a good starting point for our exploration of the relation between symmetry and bonding. A simple dipolar field defines only one direction, so we begin with the slightly more elaborate arrangement schematically illustrated in Fig. 2.4, which produces a *quadrupolar field* that fixes all three cartesian axes in space. An electron in a p_x orbital is attracted by the positive poles towards which it is directed; that in a p_y orbital is repelled by the negative poles; the p_z orbital lies along an axis perpendicular to the field, so an electron in it is unaffected.

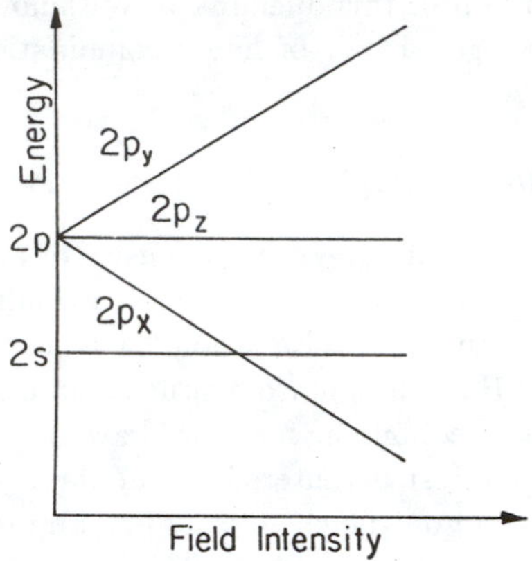

Figure 2.5. The effect of a quadrupolar field on the energy of 2*s* and 2*p* orbitals (schematic)

When the field intensity is vanishingly small – that is, when it is just strong enough to define the three axes – the potential energy is effectively spherical. The orbitals, though spatially distinct, remain virtually degenerate, and the charge density around each is cylindrically symmetric about its own axis. As shown in Fig. 2.5, the energy splitting increases with the field intensity.[1] The 2*s* orbital, included in the figure for completeness, is spherically symmetrical, so – like $2p_z$ – it ignores the field.

A comparison of Figs. 2.2 and 2.5 is instructive: At zero field intensity it is assumed in one case that the three degenerate orbitals are p_{+1}, p_0 and p_{-1}, and in the other that they are p_x, p_y and p_z. Evidently, the respective fields can be regarded as having "vanished" only for quantitative purposes; qualitatively, with regard to orbital shape and – as we shall see – orbital symmetry, it remains very much in evidence even when it is vanishingly small.

Let us place the nucleus at the center of the quadrupolar field illustrated in Fig. 2.6 and assume the electron to be located momentarily at a point p in the upper, anterior, right-hand octant, where x, y and z are all positive. Since the potential energy depends on the absolute values of x, y and z, there are seven other points in space that are energetically equivalent to p. The electron is taken from p to q by $C_2(z)$, a twofold (180°) rotation about the z axis; to r by $C_2(y)$, twofold rotation about y; and to s by $C_2(x)$. Inversion (i) of p through the origin, i, takes the electron to t. We can now move the electron from t to the three remaining equivalent points u, v and w by following the inversion by $C_2(z)$, $C_2(y)$ and $C_2(x)$ respectively. Alternatively, we could take it to these

[1] In practice, the energy will not vary linearly with the field, the non-linearity becoming more pronounced as the intensity increases.

points directly from p by reflection in the three perpendicular mirror planes, $\sigma(xy)$, $\sigma(zx)$ and $\sigma(xy)$ respectively.

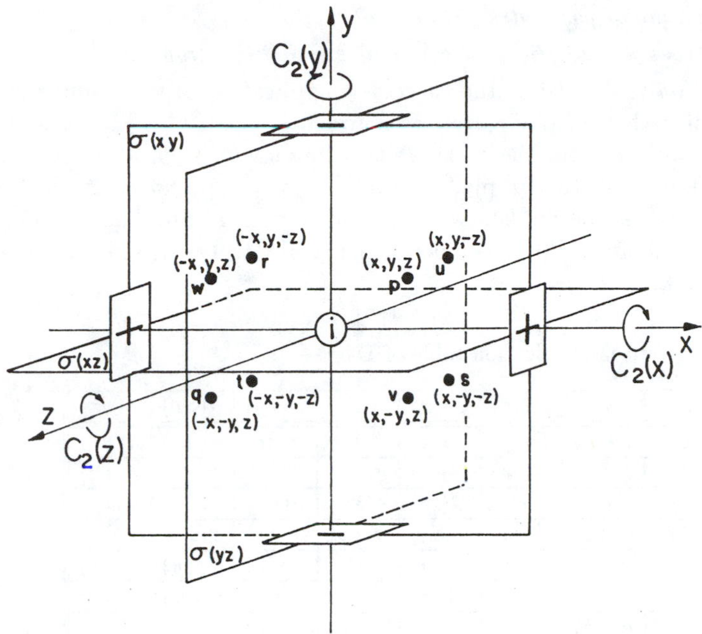

Figure 2.6. Energetically equivalent points in $\mathbf{D}_{2h}$ and its symmetry operations

The seven *symmetry operations (sym-ops)* just introduced are of three different kinds: (1) three twofold rotations, each of which leaves the sign of one cartesian coordinate unchanged and reverses the other two; (2) three reflections, each of which leaves the sign of two coordinates unchanged and reverses the third; and (3) inversion, which reverses all three coordinates. To these we add the operationally trivial but formally essential *identity* operation $(E)^2$, which does nothing and so leaves the electron at **p**, with x, y and z unchanged. These eight sym-ops[3] comprise $\mathbf{D}_{2h}$, a symmetry point group of *order* 8, that will appear repeatedly throughout the book in a variety of contexts.[4]

[2] From the German: *Einheit* = identity.

[3] In the formalism of group theory, the symmetry operations are said to be *elements* of the group. This usage of the term will be avoided because it can be confusing: *Symmetry elements*, such as rotational axes or mirror planes are conceptually distinct from the *operations* of rotation or reflection, which are the *elements* of the group. In $\mathbf{D}_{2h}$, each symmetry element is associated with a single sym-op, but – as will be seen in the following section – this will not always be the case.

[4] The reader whose familiarity with quantum- and/or group theory has made him or her impatient with the pace of this section is advised to skip to the next.

2.2.3 Digression: Some Elementary Group Theory

The eight sym-ops of $\mathbf{D}_{2h}$ are said to constitute a *group* because they comply with several *group postulates*[5] [9, 10]:

1. *There exists a multiplication rule under which the group is closed.*

Multiplication is defined as the successive application of two symmetry operations, reading their symbols from right to left. Thus, $i\, C_2(z) = \sigma(xy)$ means: "twofold rotation about the z axis *followed by* inversion produces the same result as reflection in the xy plane". Table 2.1 is the *group multiplication table* of $\mathbf{D}_{2h}$, which can be verified with the aid of of Fig. 2.6. The product of any two sym-ops is invariably a member of the group, so $\mathbf{D}_{2h}$ is indeed *closed under multiplication*.

Table 2.1. Group multiplication table of $\mathbf{D}_{2h}$

$\mathbf{D}_{2h}$	E	$C_2(z)$	$C_2(y)$	$C_2(x)$	i	$\sigma(xy)$	$\sigma(zx)$	$\sigma(yz)$
E	E	$C_2(z)$	$C_2(y)$	$C_2(x)$	i	$\sigma(xy)$	$\sigma(zx)$	$\sigma(yx)$
$C_2(z)$	$C_2(z)$	E	$C_2(x)$	$C_2(y)$	$\sigma(xy)$	i	$\sigma(yz)$	$\sigma(zx)$
$C_2(y)$	$C_2(y)$	$C_2(x)$	E	$C_2(z)$	$\sigma(zx)$	$\sigma(yz)$	i	$\sigma(xy)$
$C_2(x)$	$C_2(x)$	$C_2(y)$	$C_2(z)$	E	$\sigma(yz)$	$\sigma(zx)$	$\sigma(xy)$	i
i	i	$\sigma(xy)$	$\sigma(zx)$	$\sigma(yz)$	E	$C_2(z)$	$C_2(y)$	$C_2(x)$
$\sigma(xy)$	$\sigma(xy)$	i	$\sigma(yz)$	$\sigma(zx)$	$C_2(z)$	E	$C_2(x)$	$C_2(y)$
$\sigma(zx)$	$\sigma(zx)$	$\sigma(yz)$	i	$\sigma(xy)$	$C_2(y)$	$C_2(x)$	E	$C_2(z)$
$\sigma(yz)$	$\sigma(yz)$	$\sigma(zx)$	$\sigma(xy)$	i	$C_2(x)$	$C_2(y)$	$C_2(z)$	E

2. *The group contains an identity operation.*

The identity operation means "do nothing", so preceding or following it by another operation is the same as merely performing that operation.

3. *Every operation in the group has an inverse.*

The inverse of any sym-op simply undoes it, so preceding or following an operation by its inverse is equivalent to doing nothing, i.e. performing the identity operation.

4. *The associative law is obeyed.*

The product of three sym-ops $(c\,b\,a)$ is independent of the sequence in which the multiplication is carried out: Multiplying by c the product of b and a yields the same result as multiplying a by the product of c and b. Formally, $(c\,b)\,a = c\,(b\,a)$. For example:

$$\begin{aligned} [C_2(z)\, i]\; \sigma(zx) &= C_2(z)\; [i\, \sigma(zx)] \\ \sigma(xy)\; \sigma(zx) &= C_2(z)\; C_2(y) \\ C_2(x) &= C_2(x) \end{aligned}$$

In addition to conforming to the group postulates, as all groups must, $\mathbf{D}_{2h}$ obeys the *commutative rule*, according to which the result of performing any

[5] As worded here, the postulates apply specifically to symmetry point groups.

two symmetry operations in succession does not depend on the order in which they were carried out. A group in which every operation is its own inverse, is necessarily a *commutative (or Abelian) group*; this is clearly true of $\mathbf{D}_{2h}$, as is evident from the fact that E is the only symbol that appears along the diagonal in Table 2.1.

2.2.4 The Phase of an Orbital

Let us now assume that the intensity of the quadrupolar field in Fig. 2.5 happens to be that at which the energy of $2p_x$ is exactly equal to that of $2s$, so that the two orbitals have become degenerate. To be sure, such *accidental degeneracy* differs from the *essential degeneracy* of the three $2p$ orbitals in the free atom because it stems from the "accident" of the external field having attained a particular intensity. Nevertheless, like any degenerate pair of orbitals, these too can be equally well represented by any orthogonal pair of their linear combinations. In analogy with Equation 2.1 we can, in this special circumstance, replace $2s$ and $2p_x$ with the orthogonal pair of *hybrid* orbitals:

$$h_+^x = (2s + 2p_x)/\sqrt{2} \ ; \ h_-^x = (2s - 2p_x)/\sqrt{2} \tag{2.2}$$

We were justified in neglecting the relative phases of the orbitals up to this point in the discussion, because the sign of a pure p orbital, let alone that of an s orbital, is immaterial. In Fig. 2.4, for example, the positive lobe of $2p_x$ lies to the right of the nucleus and its negative lobe to the left, but the electron density – which varies as its square – is unchanged by any symmetry operation of $\mathbf{D}_{2h}$, whether or not it converts x to $-x$. The neglect of phase is no longer justified in the case of h_+^x and h_-^x, that are two distinct orbitals differing from one another in their directional properties.

However, it is not difficult to convince ourselves that there is very little to be gained from the *hybridization* expressed by Equation 2.2, as long as the potential energy of the atom has $\mathbf{D}_{2h}$ symmetry. Since p_x and $-p_x$ are two phases of the same orbital, the two hybrids – produced by an equal admixture of s into both – remain isoenergetic; Equation 2.2 has merely substituted one pair of degenerate orbitals for another. However, as will be seen shortly, the potential energy of the atom can be desymmetrized further, so that it makes an energetic distinction between x and $-x$. We therefore have to find a way of specifying the symmetry properties of the orbitals themselves, including phase, which for our purposes means specifying the signs of their different lobes.

2.2.5 Digression: A Bit More Group Theory

An s orbital remains unchanged under all of the symmetry operations of $\mathbf{D}_{2h}$, so – like the potential energy and the charge density of an electron occupying it – it is *totally symmetric* in $\mathbf{D}_{2h}$. A p_x orbital, on the other hand, changes sign under any symmetry operation that transforms x to $-x$; p_y and p_z behave analogously. The symmetry of a p orbital is therefore lower than that of the potential energy or charge density of an electron occupying it.

One way of expressing the symmetry properties of the three p orbitals would be to list the symmetry operations that leave each of them unchanged. p_x is untouched by E, $C_2(x)$, $\sigma(xy)$ and $\sigma(zx)$, all of which leave x invariant. It is easily confirmed with the aid of Table 2.1 that this set of four operations fulfils the Group Postulates, and thus constitutes the group conventionally named $\mathbf{C}_{2v}^{x}$[6]. The symmetry point groups $\mathbf{C}_{2v}^{y}$ and $\mathbf{C}_{2v}^{z}$, which leave the respective signs of y and z unchanged, could similarly be used to characterize the symmetry properties of p_y and p_z. Each of these three groups consists of four of the eight symmetry operation of $\mathbf{D}_{2h}$ and is thus a *subgroup* of it. Each includes: a C_2 axis, rotation about which reverses the sign of the two coordinates that are perpendicular to it; two mirror planes, reflection in each of which changes the sign of the coordinate normal to it; and – of course – the identity, which changes nothing. None of the four is capable of reversing the sign of the coordinate that is lined up along the rotational axis, so each p orbital retains its phase under all of the operations included in its own subgroup.

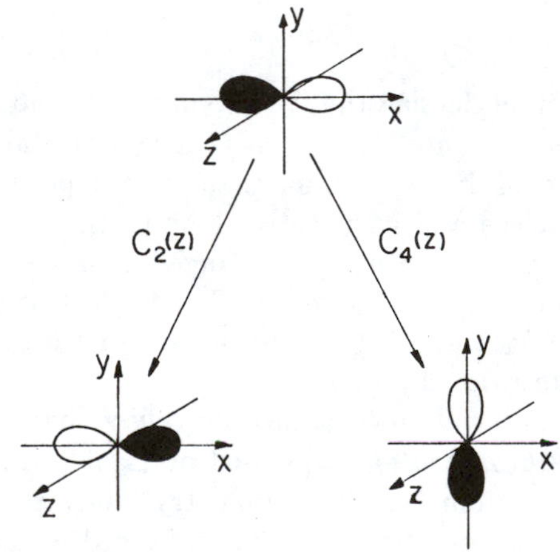

Figure 2.7. The effect of $C_2(z)$ and $C_4(z)$ on p_x

The characterization of orbitals by subgroup, though legitimate, is unattractive for at least two reasons. First of all, dealing with each of the orbitals of the same atom within the framework of a different symmetry point group is cumbersome. Secondly, the procedure obscures an important distinction, illustrated in Fig. 2.7, between symmetry operations that are excluded from a particular subgroup for quite different reasons: $C_2(z)$, for example, is excluded from $\mathbf{C}_{2v}^{x}$ because rotating the p_x orbital by $180°(\frac{2\pi}{2})$ reverses its phase; $C_4(z)$ is excluded because it converts p_x into p_y. The former, a twofold rotation that leaves the potential energy of an electron in that orbital unchanged, is included

6 $\mathbf{C}_2$ denotes the twofold rotational axis, v indicates the presence of (two) mirror planes that intersect at the (*vertical*) rotational axis, and x specifies the direction of the axis in space.

in $\mathbf{D}_{2h}$, the symmetry point group that characterizes the potential energy of an electron in a quadrupolar field. The latter, a rotation about z by $90°(\frac{2\pi}{4})$, interconverts two orbitals that are directionally and energetically non-equivalent, so it is excluded not only from the subgroup $\mathbf{C}_{2v}^{x}$ but from the parent group $\mathbf{D}_{2h}$ as well.

The way to have our cake and eat it is to retain all of the symmetry elements of $\mathbf{D}_{2h}$ and specify under each its *character* with respect to the coordinate or orbital under consideration. In contrast to $C_4(z)$, none of these eight sym-ops can convert one cartesian coordinate to another; at most, it can change its sign. Its character can therefore only take on one of two values: 1 and −1. The latter, A in the conventional notation adopted in Chapter 1, signifies that the symmetry operation reverses the sign of the coordinate or the phase of the orbital; the former, equivalent to S, indicates that it does not. The row of eight digits, each either 1 or −1, which describe the behaviour of an orbital under the eight symmetry operations of $\mathbf{D}_{2h}$, is called its *symmetry species* or *irreducible representation*, commonly abbreviated to *irrep*. An s orbital is unchanged under all of the sym-ops in the group, so it belongs to the *totally symmetric representation*, A_g[7]: all of its eight characters are 1. In the irreps of p_x, p_y and p_z, both 1 and −1 appear four times, differently ordered in each. A basic theorem in group theory requires $\mathbf{D}_{2h}$ to have exactly eight irreducible representations. These are listed in Table 2.2 as an 8×8 array, called the *Character Table* of the group.

The irreducible representations are labeled in the Schönfliess notation, which is explained in all books on group theory. Those in Table 2.2 can be easily understood as follows: The irreps labelled A are symmetric to twofold rotation about all three twofold axes. Those labelled B are symmetric to rotation about one axis and antisymmetric to rotation about the other two; the subscripts specify the unique axis, 1, 2, and 3 respectively referring to z, y and x. Symmetry with respect to inversion is indicated by g and antisymmetry by u.[8] Symmetric or antisymmetric behavior with respect to reflection in the mirror planes is implicit but unambiguous. It is clear from Table 2.1 that reflection in a mirror plane, that reverses the sign of the cartesian coordinate perpendicular to it, is equivalent to the sequence: inversion, that reverses all three coordinates, followed (or preceded) by a twofold rotation about the perpendicular axis, that restores the original sign to the two in-plane coordinates.

The cartesian coordinates conventionally appear at the right of the Table beside their respective representations. Each is symmetric with respect to rotation about the axis parallel to it and changes sign under inversion, so z, y and x transform as B_{1u}, B_{2u} and B_{3u} respectively. It follows that any binary product of the cartesian coordinates must be symmetric to inversion. It is clear that x^2, y^2 and z^2 have to come under A_g, since the charge density of an electron, that transforms like the square of the orbital, is totally symmetric in $\mathbf{D}_{2h}$. The

[7] A purist would say "2s transforms as A_g", or "2s is a basis for A_g". Loosely speaking, "2s *has* the representation A_g" and "2s *belongs to* A_g" mean the same thing.

[8] From the German: *gerade* = even, *ungerade* = odd.

Table 2.2. Character table of $\mathbf{D}_{2h}$

$\mathbf{D}_{2h}$	E	$C_2(z)$	$C_2(y)$	$C_2(x)$	i	$\sigma(xy)$	$\sigma(zx)$	$\sigma(yz)$	
A_g	1	1	1	1	1	1	1	1	x^2, y^2, z^2
B_{1g}	1	1	−1	−1	1	1	−1	−1	xy, R_z
B_{2g}	1	−1	1	−1	1	−1	1	−1	zx, R_y
B_{3g}	1	−1	−1	1	1	−1	−1	1	yz, R_x
A_u	1	1	1	1	−1	−1	−1	−1	
B_{1u}	1	1	−1	−1	−1	−1	1	1	z
B_{2u}	1	−1	1	−1	−1	1	−1	1	y
B_{3u}	1	−1	−1	1	−1	1	1	−1	x

$A \otimes A = A$

$B_i \otimes B_i = A \; ; \; B_i \otimes B_j = B_k$

$g \otimes g = u \otimes u = g \; ; \; g \otimes u = u$

representations of the cross-products, xy, zx and yz, are easily obtained by multiplying the characters of the rows representing the two coordinates, column by column. Multiplying two irreps in this way, character by character, yields their *direct product*, which is itself an irrep; thus $B_{3u} \otimes B_{2u} = B_{1g}$, $B_{1u} \otimes B_{3g} = B_{2u}$, and so on.[9]

A concise set of rules for obtaining the direct product appears at the bottom of Table 2.2; they confirm that multiplying any irrep by A_g leaves it unchanged and that the product of any irrep with itself yields A_g. Also tabulated for completeness are the coordinates of rotation (R_x, R_y and R_z) about the three cartesian axes. Their symmetry properties, along with those of vibrational and reaction coordinates, will be dealt with in subsequent chapters.

A second kind of product is the *scalar product*, obtained by multiplying the characters sym-op by sym-op, summing the products over the operations of the group, and normalizing by division by the *order* of the group (g), the number of sym-ops comprising it. For example:

$$B_{2u} \cdot B_{3u} = (1 + 1 - 1 - 1 + 1 + 1 - 1 - 1)/8 = 0 \quad (2.3)$$

$$B_{2u} \cdot B_{2u} = (1 + 1 + 1 + 1 + 1 + 1 + 1 + 1)/8 = 1 \quad (2.4)$$

Equations 2.3 and 2.4 illustrate the analogy between irreducible representations and mutually perpendicular vectors. The former states that they are *orthogonal* to one another: the scalar product of any two irreps – like that between two orthogonal vectors – is zero. The latter indicates that they are *normalized*: the scalar product of any irrep with itself is unity.

Let us return to the hybrid orbitals defined by Equation 2.2, in a quadrupolar field exactly strong enough to make s and p_x degenerate (Fig. 2.5). It is

[9] For example:

	$\mathbf{D}_{2h}$	E	$C_2(z)$	$C_2(y)$	$C_2(x)$	i	$\sigma(xy)$	$\sigma(zx)$	$\sigma(yz)$	
	B_{2u} :	1	−1	1	−1	−1	1	−1	1	y
$\otimes$	B_{3u} :	1	−1	−1	1	−1	1	1	−1	x
=	B_{1g} :	1	1	−1	−1	1	1	−1	−1	xy

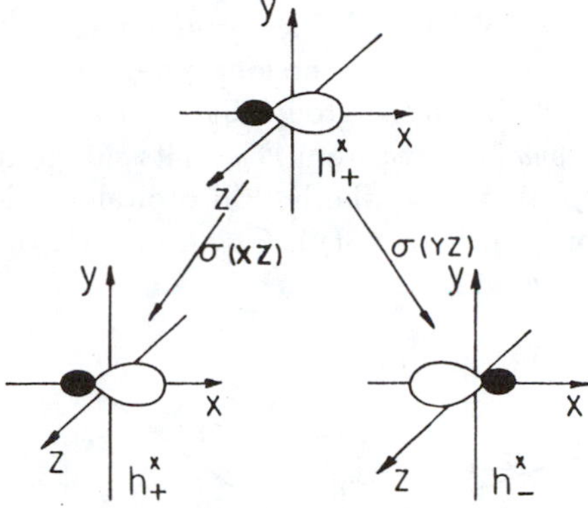

Figure 2.8. Transformation of an *sp* hybrid under two symmetry operations of $\mathbf{D}_{2h}$

clear from Fig. 2.8 that neither h^x_+ nor h^x_- can be assigned to an irreducible representation of $\mathbf{D}_{2h}$ because, whereas some of its symmetry operations leave h^x_+ and h^x_- unchanged, others interconvert them. The hybrids, and the charge-density of electrons occupying them as well, show a directional preference for $+x$ or $-x$ that is inconsistent with the inherent symmetry of the quadrupolar field. A pair of degenerate orbitals that behaves in this way is said to belong to a *reducible representation*, because it will always be possible to find a pair of linear combinations that transform as irreps of the group. In the present case, we merely regenerate the original hydrogenlike orbitals, $s(a_g)$ and $p_x(b_{3u})$:[10]

$$(h^x_+ + h^x_-)/\sqrt{2} = 2s \; ; \; (h^x_+ - h^x_-)/\sqrt{2} = 2p_x \; . \tag{2.5}$$

2.2.6 Hybridization

If hybridization of orbitals has turned out to be a sterile device in the context of the preceding section, it was because the quadrupolar field cannot distinguish between *sp* hybrids that are oppositely directed. Needless to say, hybridization is much too important a concept in chemical bonding to be abandoned. What makes it important, however, is the enhanced electrostatic attraction of an electron in a suitably directed hybrid orbital of one atom towards the nucleus of another. This effect can be simulated by putting our isolated model atom in a dipolar field lined up along the x axis, in which the potential increases with x. An electron in that atom will be stabilized in regions of positive potential ($x > 0$) and destabilized where x – and the potential – are negative. We do this in two ways:

In the first, illustrated on the right hand side of Fig. 2.9, we maintain the strong quadrupolar field of Fig. 2.5 in which $2s$ and $2p_x$ have become degenerate,

[10] The irreps of orbitals and coordinates will henceforth be labeled with lower-case symbols, e.g. $\gamma(2s) = a_g$; capital letters will be reserved for labeling the representations themselves and for specifying the state and configurational symmetry of molecules.

and superimpose on it a dipolar field of increasing intensity, thus stabilizing h_+^x relative to h_-^x. Now, since h_+^x and h_-^x are no longer degenerate, no sym-ops which interconvert them are included in the group characterizing the perturbed atom. The system has been *desymmetrized* from $\mathbf{D}_{2h}$ to its subgroup $\mathbf{C}_{2v}^x$, that retains only those sym-ops $\mathbf{D}_{2h}$ that leave the hybrid orbitals unchanged $\{E, C_2(x), \sigma(xy), \sigma(zx)\}$, and from which $\{i, C_2(y), C_2(z), \sigma(yz)\}$, that interconvert h_+^x and h_-^x, have been eliminated.

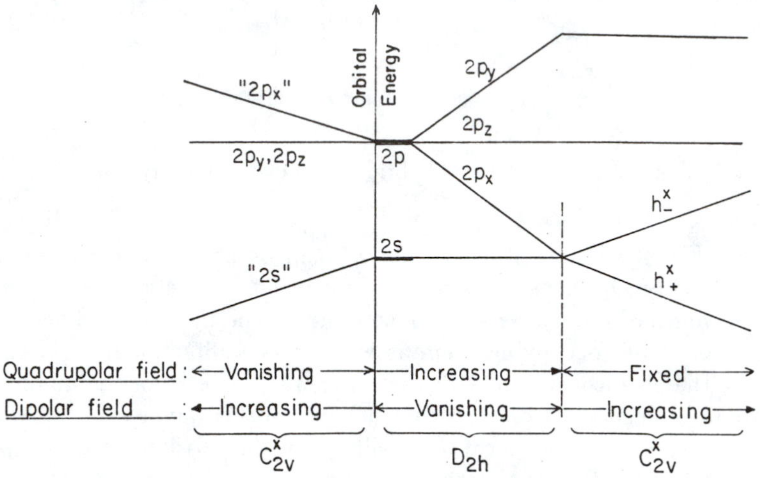

Figure 2.9. Splitting and interaction of $2s$ and $2p_x$ orbitals in variously superimposed electrostatic fields

At this point we consult Table 2.2 and see that B_{3u} is the only irrep in which -1 is the character of each of the four symmetry operations in the eecond, excluded set. Conversely, only B_{3u} – along with A_g, the totally symmetric representation – has 1 as the character of each of the sym-ops retained in $\mathbf{C}_{2v}^x$. This latter fact is expressed by the statement that the subgroup $\mathbf{C}_{2v}^x$ is the *kernel* of the irrep B_{3u} of the parent group $\mathbf{D_{2h}}$. We note further that B_{3u} is the representation of the coordinate x, and realize that, after the system has been perturbed by a dipolar field along the x axis, the energy of an electron in a p_x orbital can remain unaffected only under sym-ops that do not convert x to $-x$. The perturbation, which has the representation B_{3u}, has evidently reduced the symmetry of the system to that of its kernel, i.e. from $\mathbf{D_{2h}}$ to its subgroup $\mathbf{C}_{2v}^x$.

Alternatively, we retain only the vanishingly small quadrupolar field that is required to define the cartesian axes and, by imposing $\mathbf{D}_{2h}$ symmetry, to establish $2p_x$, $2p_y$ and $2p_z$ as distinct orbitals. Now, the imposition of an infinitesimal dipolar field introduces the energetic distinction between x and $-x$ by reducing the symmetry further to $\mathbf{C}_{2v}^x$. The effect of increasing the strength of the dipolar field on orbital energy is shown on the left side of Fig. 2.9: $2p_y$ and $2p_z$ are unaffected and remain virtually degenerate, whereas $2s$ and $2p_x$ interact, becoming more and more effectively hybridized as the field intensity increases.

Table 2.3. Character table of $\mathbf{C}_{2v}^{x}$

$\mathbf{C}_{2v}^{x}$	E	$C_2(x)$	$\sigma(xy)$	$\sigma(zx)$	
A_1	1	1	1	1	$x,\ x^2, y^2, z^2$
A_2	1	1	-1	-1	$yz,\ R_x$
B_1	1	-1	1	-1	$y,\ xy,\ R_z$
B_2	1	-1	-1	1	$z,\ zx,\ R_y$

$A \otimes A = B \otimes B = A\ ;\ A \otimes B = B$
$1 \otimes 1 = 2 \otimes 2 = 1\ ;\ 1 \otimes 2 = 2$

Some of $2p_x$ is admixed into $2s$, giving it the favorable directional properties that stabilize it in the field; orthogonality of the two sp hybrids then requires the $2p_x$ orbital to accept a destabilizing admixture of $2s$. Denoting the extent of the admixture by the field-dependent parameter λ, we write:

$$\text{“}2s\text{”} = (2s + \lambda 2p_x)/\sqrt{1+\lambda^2}\ ;\ \text{“}2p_x\text{”} = (\lambda 2s - 2p_x)/\sqrt{1+\lambda^2} \tag{2.6}$$

When the dipolar field is strong enough to effect an equal mixture of $2s$ and $2p_x$ ($\lambda = 1$),“$2s$” becomes identical with h_+^x, “$2p_x$” with h_-^x, and Equation 2.6 with Equation 2.2.

2.2.7 The Formal Expression of Desymmetrization

The left side of Fig. 2.9 illustrates a common phenomenon: the interaction of two initially orthogonal orbitals under the influence of a field that mixes them. There are two equivalent ways of rationalizing this behavior formally in terms of orbital symmetry:

2.2.7.1 I. Desymmetrization to a Subgroup

The dipolar field has reduced the symmetry of the system from $\mathbf{D}_{2h}$ to $\mathbf{C}_{2v}^{x}$, the Character Table of which is displayed in Table 2.3. It is obtained from Table 2.2 by striking out the columns of characters under those symmetry operations of $\mathbf{D}_{2h}$ which are not included in $\mathbf{C}_{2v}^{x}$. It has already been noted that the characters of all of the sym-ops retained in $\mathbf{C}_{2v}^{x}$ are the same in B_{3u} as in the totally symmetric representation (A_g) of $\mathbf{D}_{2h}$, $\mathbf{C}_{2v}^{x}$ being the kernel of B_{3u} in the parent group. Both of these irreps of $\mathbf{D}_{2h}$ *correlate* with the totally symmetric representation, A_1, of $\mathbf{C}_{2v}^{x}$; as a result, x and x^2, which transform in $\mathbf{D}_{2h}$ as B_{3u} and A_g respectively, appear together in A_1 of $\mathbf{C}_{2v}^{x}$. A comparison of Tables 2.2 and 2.3 will confirm that the other six irreps of $\mathbf{D}_{2h}$ coalesce pairwise to the remaining three irreducible representations of $\mathbf{C}_{2v}^{x}$: A_u and B_{3g} going to A_2; B_{2u} and B_{1g} to B_1; and B_{1u} and B_{2g} to B_2.

Evidently, the s and p_x orbitals are both totally symmetric in $\mathbf{C}_{2v}^{x}$ and are capable of interacting with one another. The mixing is accomplished by the dipolar field, which – aligned along the x axis and therefore totally symmetric in $\mathbf{C}_{2v}^{x}$ – increases λ from 0 to 1, smoothly correlating $2s$ with h_{+}^{x} and $2p_x$ with h_{-}^{x}.

2.2.7.2 II. Desymmetrization by a Perturbation

We observe the mechanism of orbital mixing in finer focus, recognizing that $2s$ and $2p_x$ are mutually orthogonal in spherical symmetry as they are in a quadrupolar field – vanishing or substantial. Their orthogonality is summed up by the mathematical statement: "The intergral of their product over all values of x, y and z vanishes". A sufficient condition is:

$$\int_{-\infty}^{\infty} 2s\, 2p_x\, dx = 0 \tag{2.7}$$

This is so because s is everywhere positive, whereas p_x is positive to the right of the nucleus ($x > 0$) but negative to its left ($x < 0$). Their product at any point to the right of the origin exactly cancels their product an equal distance to the left, because the value of p_x at $-x$ is equal and opposite to its value at x. Clearly, a symmetry operation that converts x to $-x$ can have no effect on the energy. In the dipolar field, the potential energy of an electron in p_x is lowered on the right and raised on the left by a quantity proportional to x, say by λx. *Perturbation theory*[11] then tells us that the energy of the lower of these two orbitals (s) is decreased and that of the upper (p_x) is increased by an amount that is inversely proportional to their energy difference and directly proportional to the square of integral:

$$\int_{-\infty}^{\infty} 2s\, \lambda x\, 2p_x\, dx \neq 0 \tag{2.8}$$

This integral does not vanish, because the product of $2p_x$ with x transforms as x^2, and so – unlike $2p_x$ itself – it does not change sign under any symmetry operation that converts x to $-x$. In order for Equation 2.6 to hold in any commutative symmetry point group, it is necessary that:

$$\gamma(1^{st} orbital) \otimes \gamma(perturbation) \otimes \gamma(2^{nd} orbital) = \gamma_I \tag{2.9}$$

where γ_I is the totally symmetric – or *identity* – representation of the group. Equation 2.9 holds trivially in our example, since $2s$, λx and $2p_x$ are all totally symmetric (a_1) in $\mathbf{C}_{2v}^{x}$. More significantly, it is also true in $\mathbf{D}_{2h}$, where $\gamma(2s) = a_g$, $\gamma(\lambda x) = b_{3u}$ and $\gamma(2p_x) = b_{3u}$:

$$a_g \otimes b_{3u} \otimes b_{3u} = a_g \tag{2.10}$$

The important point to note is that the question, whether a particular pair of orbitals will be mixed under a given perturbation, can be posed and answered

[11] Equation 2.6 is derived from *first order perturbation theory*, that normally holds only for small perturbations. However, when symmetry is broken by a small perturbation, it remains no less broken as the perturbation gets larger. Therefore, qualitative symmetry arguments based on first order theory also hold for much larger perturbations, for which first order theory no longer yields reliable quantitative results.

in the point group that describes the unperturbed system. Thus, still staying in $\mathbf{D}_{2h}$, we can ask, for example: "What type of field is capable of mixing $2s$ and $2p_z$?"; the question is formally expressed by the equation:

$$a_g \otimes \gamma(\text{field}) \otimes b_{1u} = a_g \tag{2.11}$$

Since character-by-character multiplication is commutative, and the direct product of any irrep of $\mathbf{D}_{2h}$ with A_g leaves it unchanged, it follows that $\gamma(\text{field}) = b_{1u}$: the field must be set up parallel to the z axis. We also know that such a field will reduce the symmetry to $\mathbf{C}^z_{2v}$, because – as a glance at Table 2.2 will show – it is the kernel subgroup of B_{1u}[12].

As an additional example, we might inquire what field – if any – is capable of hybridizing p_x and p_y. Turning to Table 2.2 once more, we see that the formal requirement is:

$$b_{3u} \otimes \gamma(\text{field}) \otimes b_{2u} = a_g \tag{2.12}$$

$$\gamma(\text{field}) = b_{3u} \otimes b_{2u} = b_{1g} \tag{2.13}$$

Since no cartesian coordinate has the representation b_{1g}, no external dipolar field can induce $2p_x$ and $2p_y$ to interact. Begging the question of the precise nature of the perturbation called for, the requirement can be read as follows: "In order to mix a p_x and a p_y orbital, the potential energy of the system must be desymmetrized from $\mathbf{D}_{2h}$ to the subgroup which is the kernel of B_{1g}, *viz.* $\mathbf{C}^z_{2h}$, comprising $\{E,\ C_2(z), i,\ \sigma(xy)\}$." In other words, in order to hybridize these two orbitals effectively, the symmetry must be lowered in a way that destroys the individual identity of x and y. Reflection symmetry in the xy plane can remain, but now that x and y are no longer defined individually, $\sigma(xy)$ is more reasonably referred to as σ_h, the *horizontal* mirror plane, perpendicular to the remaining (*vertical*) rotational axis, $C_2(z)$.

It will become obvious in subsequent chapters that when the approach outlined above under **I** is extended to chemical reactions it will lead naturally to the correlation procedures of Woodward and Hoffmann and of Longuet-Higgins and Abrahamson (WH-LHA). The more flexible, if somewhat more elaborate, approach **II** points the way to OCAMS. The presentation just completed should suffice to convince the reader that the two are fully equivalent, the choice of one or the other being a matter of convenience. In subsequent applications, we will have recourse to both points of view, always making sure that they are adopted in a mutually consistent manner. [18]

[12] Its Character Table appears in Appendix A, along with those of the other common symmetry point groups. In standard sets of Character Tables the principal rotational axis is conventionally chosen to be z.

2.3 Something About *d* Orbitals

The discussion to this point has been limited to *s* and *p* orbitals. They, and the elementary group theoretical concepts and procedures introduced so far in this chapter, should suffice for dealing with nearly all of the illustrative organic reactions that will follow. It has already been mentioned that qualitative orbital symmetry arguments have played a minor role in the analysis of inorganic reaction mechanisms, presumably because their reliablity had been authoritatively questioned. The writer is convinced that they can be used to much better effect than they have in the past, provided that the symmetry of *d* orbitals is properly taken into account.[13] The higher symmetry of many inorganic molecules, those involving transition metals in particular, makes it necessary to extend our exposition of elementary group theory to include *non-commutative groups* and *degenerate irreducible representations*, which will turn out to be useful when considering reactions involving highly symmetrical organic molecules as well.[14] As in our previous excursions into group theory, there will be no pretension to rigor or completeness; only those concepts that are essential to an understanding of the following chapters will be introduced, and they too through examples rather than by formal precept.

2.3.1 Splitting *d* Orbitals by an External Field

The degenerate orbitals of the 3*d* level split in an axial magnetic field much as do the 2*p* orbitals in Fig. 2.2. Their five-fold degeneracy is removed completely: two orbitals (d_{+1} and d_{+2}) are destabilized, two (d_{-1} and d_{-2}) are stabilized, and the fifth (d_0) is unaffected. As in the case of p_0, we have to conclude that the magnetic moment of the last lies somewhere in the xy plane; its orbital motion keeps taking it by indeterminate paths through the region of the z axis, where its charge density is consequently maximal. Again, the behavior in a magnetic field does not contain much directly useful chemical information.

This is about as far as the analogy between 2*p* and 3*d* atomic orbitals can be taken. The quadrupolar field of Fig. 2.4, which conveniently gives the former three the same shape and aligns each of them along one of the three cartesian axes, cannot do the same for the latter set of five. Only four of them can take the familiar form depicted in Fig. 2.10 as c–f, the fifth (g) retaining the elongated form which is also taken by (d_0) in the magnetic field. We can, of course, draw two more identically shaped "clover leafs" (a and b in Fig. 2.10) bringing the total up to six *d* orbitals of similar shape: The lobes of the first three (a–c) lie along two cartesian axes in a plane perpendicular to the third; d_{xy} (f) lies in the same plane as $d_{x^2-y^2}$ (c), but is rotated by 45°, so that its lobes lie midway

[13] The extension to *f* orbitals offers nothing new in principle. They do not occur in many useful examples, so they will not be considered in this book.

[14] The organic chemist who has had enough can skip to the next chapter with little loss; he or she has probably already done so.

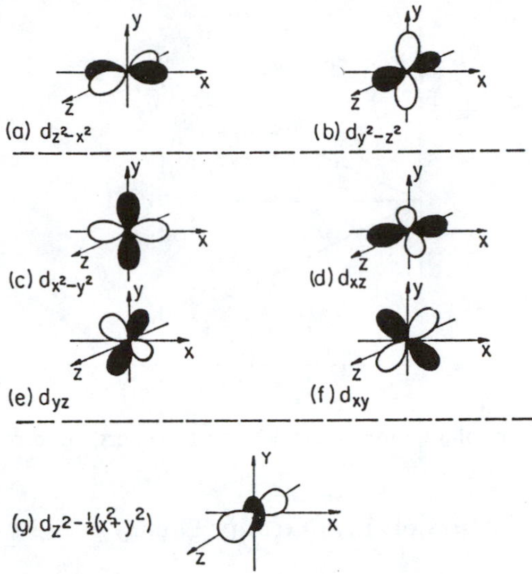

Figure 2.10a–g. The d orbitals. **(a–f)**: Geometrically equivalent set of six; **(c–g)**: orthogonal set of five

between the x and y axes; d_{xz} (d) and d_{yz} (e) bear the same relation to $d_{z^2-x^2}$ (a) and $d_{y^2-z^2}$ (b) respectively.

These six geometrically identical AOs evidently have the same energy as long as the quadrupolar field is vanishingly small, but only five of them can be independent. We can get rid of the redundancy by taking orthogonal combinations of two of them, conventionally those shown in a and b of Fig. 2.10. The symmetry properties of the orbitals are the same as those of their indices, so we simply combine the latter:

$$d_+ = (d_{z^2-x^2} + d_{y^2-z^2}) = d_{y^2-x^2}\ ;\ d_- = (d_{z^2-x^2} - d_{y^2-z^2}) = d_{2z^2-x^2-y^2} \quad (2.14)$$

The positive combination is not a new orbital but merely the negative phase of $d_{x^2-y^2}$. The negative combination, usually abbreviated to d_{z^2} after its dominant term, reproduces Fig. 2.10 g, the fifth independent d AO.

The irreps of the five independent d orbitals in $\mathbf{D}_{2h}$ are easily read from Table 2.2. All of them are symmetric to inversion: d_{z^2} and $d_{x^2-y^2}$ are totally symmetric (a_g); d_{xy}, d_{xz} and d_{yz} transform as b_{1g}, b_{2g} and b_{3g} respectively. Like the p orbitals in Fig. 2.5, the d orbitals should be stabilized by the greater proximity of their lobes to the positive than to the negative poles, whereas the three orbitals that are symmetrically disposed towards them should remain unaffected. The expected behavior is illustrated schematically in Fig. 2.11. The quadrupolar field is incapable of splitting the degeneracy completely, three orbitals remaining isoenergetic in it. A more elaborate arrangement of poles would have to be dreamt up to split the remaining threefold degeneracy by means of an external electrostatic field; the exercise hardly seems worth while.

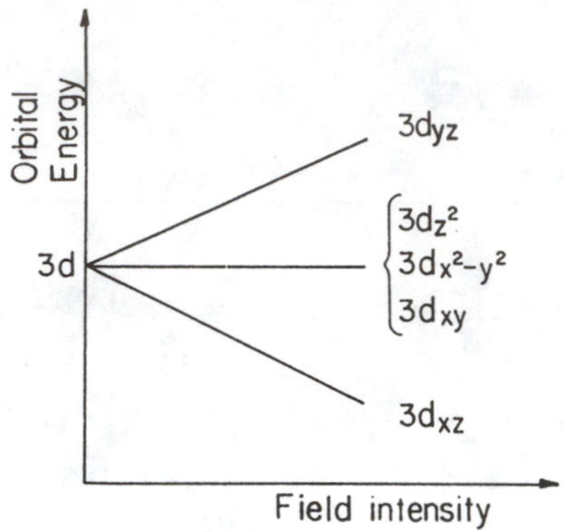

Figure 2.11. The effect of a quadrupolar field on the energy of *d* orbitals (schematic)

2.3.2 A Further Excursion into Group Theory

The familiar splitting of the *d* level in square-planar, tetrahedral or octahedral molecules and complex ions are less conveniently discussed in terms of a quadrupolar field, because their symmetry is higher than $\mathbf{D}_{2h}$. The symmetry point groups involved: $\mathbf{D}_{4h}$, $\mathbf{T}_d$ and $\mathbf{O}_h$, are *non-commutative*, by which is meant that the product of two symmetry operations may depend on the order in which they are carried out. The relevant properties of non-commutative symmetry point groups are illustrated below with the smallest of the three, $\mathbf{D}_{4h}$. It contains just twice the number of sym-ops as its subgroup $\mathbf{D}_{2h}$.[15] The relation between the two groups can be visualized by looking ahead to Fig. 2.12 and comparing the complex ion *trans*-$[NiF_2Cl_2]^{-2}$, which has $\mathbf{D}_{2h}$ symmetry, with $[NiF_4]^{-2}$, which is truly square-planar ($\mathbf{D}_{4h}$).

As its name implies, $\mathbf{D}_{4h}$ contains a fourfold rotation about the principal axis, conventionally specified to be z. It has been noted in connection with Fig. 2.7, that this operation converts p_x to p_y; at the same time it converts p_y to $-p_x$ and leaves p_z untouched. If it is followed by $\sigma(yz)$, which is one of the sym-ops included in $\mathbf{D}_{2h}$, p_y in not affected and the positive sign is restored to p_x. The net result, interchange of p_x and p_y without changing the sign of either, can be achieved at once by $\sigma_d(+)$, reflection in a plane that includes the z axis and the diagonal between the x and y axes along which x and y are either both positive or both negative. Formally: $\sigma(yz)C_4(z) = \sigma_d(+)$. Let us note first that taking the product in the reverse order is equivalent to $\sigma_d(-)$, reflection in the second diagonal plane, where x and y are equal but of opposite sign. The sym-ops $\sigma(yz)$ and $C_4(z)$ obviously do not commute, and $\mathbf{D}_{4h}$, which includes them both, is necessarily a *non-commutative* group.

[15] $\mathbf{D}_{4h}$ is itself a subgroup of $\mathbf{O}_h$; so is $\mathbf{T}_d$.

$\mathbf{D}_{4h}$ can be generated by taking the *semidirect product*[16] of $\mathbf{D}_{2h}$ with any subgroup of $\mathbf{D}_{4h}$ that is not a subgroup of $\mathbf{D}_{2h}$ as well. For example, multiplying each of the eight sym-ops of $\mathbf{D}_{2h}$ by E and $\sigma_d(+)$, the two elements of $\mathbf{C}_s^{d+}$, we obtain the parent group as their semidirect product: $\mathbf{C}_s^{d+} \wedge \mathbf{D}_{2h} = \mathbf{D}_{4h}$. As can be seen in Table 2.4, multiplication by E regenerates the eight operations of $\mathbf{D}_{2h}$, whereas multiplying them by $\sigma_d(+)$ produces eight new ones, raising the order of $\mathbf{D}_{4h}$ to 16.

Table 2.4. $\mathbf{D}_{4h}$ as the semidirect product of $\mathbf{C}_s^{d+}$ and $\mathbf{D}_{2h}$

E	$C_2(z)$	$C_2'(y)$	$C_2'(x)$	i	$\sigma_h(xy)$	$\sigma_v(zx)$	$\sigma_v(yz)$
$\sigma_d(+)$	$\sigma_d(-)$	S_4	$-S_4$	$C_2''(-)$	$C_2''(+)$	$-C_4$	C_4

Several points are worth noting about Table 2.4:

1. The new symmetry operations appear in pairs: a) a clockwise and a counterclockwise rotation by 90°;[17] b) reflection in the two symmetrically disposed diagonal mirror planes; c) two new twofold rotations about the diagonals; d) two S_4 operations, equivalent to clockwise and counterclockwise 90° rotations followed by reflection in the xy-plane.

2. $C_2(x)$ and $C_2(y)$, are no longer rotations about axes lying along energetically distinct directions, so they too can be considered to be a pair. They are relabeled C_2', to distinguish them from rotation about the diagonals, labeled C_2''; the unprimed C_2 is retained for rotation by 180° about the principal symmetry axis, z.

3. Similarly, reflection in either of the two original – now energetically equivalent – *vertical* mirror planes is labeled σ_v to distinguish it from σ_d, reflection in the new *diagonal* mirror planes.

4. Four of the sym-ops of $\mathbf{D}_{2h}^z$ remain "unpartnered" in $\mathbf{D}_{4h}$: E; C_2, for which the label z is no longer necessary; i; and $\sigma(xy)$, relabeled σ_h (*horizontal*) because the xy plane is perpendicular to the principal (*vertical*) axis.

5. The fourfold rotations, $C_4^{(\pm)}$, are inverse to one another, rather than being self-inverse; so are the corresponding *improper rotations* $S_4^{(\pm)}$. As a result, the Group Multiplication Table for $\mathbf{D}_{4h}$, unlike that of $\mathbf{D}_{2h}$ (Table 2.1), cannot have all E-s along the diagonal.

2.3.2.1 Classes and Degenerate Representations

We have noted that as a result of the energetic indistinguishability of x and y in $\mathbf{D}_{4h}$, twelve of its symmetry operations come in closely related pairs; they are

[16] It is the *direct product* if all sym-ops of the first subgroup commute with all those of the second. (e.g. $(\mathbf{C}_{4v}^z \times \mathbf{C}_s^{xy} = \mathbf{D}_{4h})$.

[17] Note that the two rotations (symmetry operations) are associated with the same rotational axis (symmetry element); see footnote 3.

said to fall into six *classes*[18]. Each of the four remaining operations is in a class by itself, as it is in $\mathbf{D}_{2h}$. Taking the cartesian coordinates as our basis, we see that each of the latter four operations treats them differently: E leaves all three as they were, C_2 changes the signs of x and y, σ_h reverses that of z, and i inverts all three. In no case, however, is one cartesian coordinate exchanged for another. The sym-ops in each of the other six classes also affects them in a characteristic manner: $C_4^{(\pm)}$ interchanges x and y, reversing the sign of one of them; $S_4^{(\pm)}$ does the same, but changes the sign of z as well. σ_v reverses either x or y, but not both; C_2' does the same and also changes the sign of z. σ_d interchanges x and y, reversing both or neither; C_2'' does the same, and additionally changes the sign of z.

It is easy to characterize the transformations of a p_z orbital under the operations of $\mathbf{D}_{4h}$: as in $\mathbf{D}_{2h}$, its character is 1 under those sym-ops which leave its phase unchanged and -1 under those that reverse it. It is symmetric to C_4 so its irrep has the label A, it is antisymmetric to C_2' and C_2'' so its suffix is 2 rather than 1, and to inversion so it is u rather than g: the full label of its irrep is therefore A_{2u}. The d_{z^2} orbital transforms like z^2, and is clearly totally symmetric; like s, it belongs to A_{1g}. Consultation of Fig. 2.10 allows us to characterize $d_{x^2-y^2}$ and d_{xy} without difficulty: both change sign under C_4 and are invariant to inversion, so they belong to B_g representations. The former, symmetric to C_2' and antisymmetric to C_2'' belongs to B_{1g}; the latter, in which the two are reversed, to B_{2g}. It is noteworthy that group theory requires the character of each of these orbitals to be the same for different operations of the same class.

It is less easy to characterize the remaining four atomic orbitals in the valence shell. The p_x and p_y orbitals are interchanged under the twelve "paired" symmetry operations, as are d_{zx} and d_{yz}, because they remain *degenerate* when the potential energy has $\mathbf{D}_{4h}$ symmetry, i.e. there is no energetic distinction between the x and y axes. A full description of the symmetry properties of either pair of orbitals requires every sym-op to be represented by a two-dimensional matrix which differs for different operations in the same class.[19]

A less complete, but still remarkably perspicuous, description is afforded by the *character* of the matrix[20] like the rows of 1 and -1 that characterize the representations referred to in the preceding paragraph – and can be regarded as one-dimensional matrices – they are the same for all (in $\mathbf{D}_{4h}$, both) operations in the same class. In evaluating the character of a given sym-op with respect to a particular pair of degenerate orbitals in $\mathbf{D}_{4h}$, we disregard those that are interchanged and only consider orbitals that are returned to themselves with or

[18] The formal definition of *class*, which need not concern us here, specifies that preceding any given sym-op by any operation of the group and then following it by the inverse of that operation will either regenerate the original sym-op or produce another member of the same class. When this procedure, called a *similarity transformation*, is applied to any sym-op of a commutative group, it invariably regenerates the original operation; every sym-op of a commutative group is thus in a class by itself.

[19] This point will be addressed in Section 4.4.3 in somewhat greater detail.

[20] The *character* is formally defined as the *trace* of the matrix that represents the sym-op: the sum of its diagonal elements.

without a change of sign; each one that retains its sign contributes +1 to the character of the operation, one that reverses it contributes −1.[21]

Neither x nor y changes its sign under E or σ_h, whereas both do under C_2 and i, so the character of the first two operations with respect to (x, y) is 2 and that of the second two is −2. All of the other, paired, sym-ops have the character 0, because they either interchange x and y or – when they do not – reverse one but not the other. Similarly, when the products (zx, yz) are chosen as the basis, the character of E and i is seen to be 2, that of C_2 and σ_h to be −2, and that of all of the others to be 0.

The representations of the two pairs of orbitals, $[p_x, p_y]$ and $[d_{zx}, d_{yz}]$, though not one-dimensional, are *irreducible*, because no combination of the two orbitals can be found that will keep them separate under all operations of the group.[22] The two-dimensional irreps of $\mathbf{D}_{4h}$, conventionally labelled by a suitably subscripted letter E,[23] are displayed in Table 2.5.

Table 2.5. The irreducible representations of degenerate p and d orbitals in $\mathbf{D}_{4h}$

$\mathbf{D}_{4h}$	E	$2C_4$	C_2	$2C'_2$	$2C''_2$	i	$2S_4$	σ_h	$2\sigma_v$	$2\sigma_d$	
E_u	2	0	−2	0	0	−2	0	2	0	0	(p_x, p_y)
E_g	2	0	−2	0	0	2	0	−2	0	0	(d_{zx}, d_{yz})

The complete Character Table of $\mathbf{D}_{4h}$ can be found in Appendix A. A point worth noting is that the number of irreducible representations, here 10, is the same as the number of classes of sym-ops in the group. The reader might care to check this point in the Character Tables of some of the other listed symmetry point groups, such as the cubic groups, $\mathbf{T}_d$ and $\mathbf{O}_h$, which have three-dimensional irreps (labelled T) because the x, y and z axes are energetically equivalent, i.e. *triply degenerate*, in them.

2.3.2.2 Invariant Subgroups, Kernels and Co-Kernels

Familiarity with these topics, which are given short shrift in most textbooks of group theory for chemists, will prove useful in subsequent chapters. They are most easily explained with the help of molecular examples, so – for logical consistency with the earlier sections of this chapter – the reader is asked to think of each of the illustrative molecules as a central atom or ion in the perturbing field set up by the surrounding ions.

---

[21] In other non-commutative groups, the cartesian coordinates are not merely interchanged, but may be mixed by some of the symmetry operations. For example, a threefold rotation about the z axis, which is a sym-op of $\mathbf{C}^z_{3v}$, mixes x and y inextricably: $x \rightarrow (-x + \sqrt{3}y)/2$; $y \rightarrow (-y - \sqrt{3}x)/2$. Nevertheless, sym-ops in the same class have the same character in these groups as well.

[22] It will be recalled that the analogous *reduction* of the pair of hybrids h^x_+ and h^x_- to the one-dimensional representations of $\mathbf{D}_{2h}$ was possible (cf. Equation 2.3); it always is in a commutative group, where all of the irreps are necessarily one dimensional.

[23] Use of the same letter to symbolize doubly-degenerate representations and the identity operation is regrettable, but should be no more confusing than that of "σ" to represent both an axially symmetric MO and a mirror plane.

The fact that the symmetry operations of a non-commutative group fall into classes leads to the recognition that its subroups are of two different kinds: *invariant subgroups*, that consist entirely of whole classes of the original group [19], and *non-invariant subroups*, one or more elements of which are members of a class of the parent group that is not included in the subgroup *in toto*.

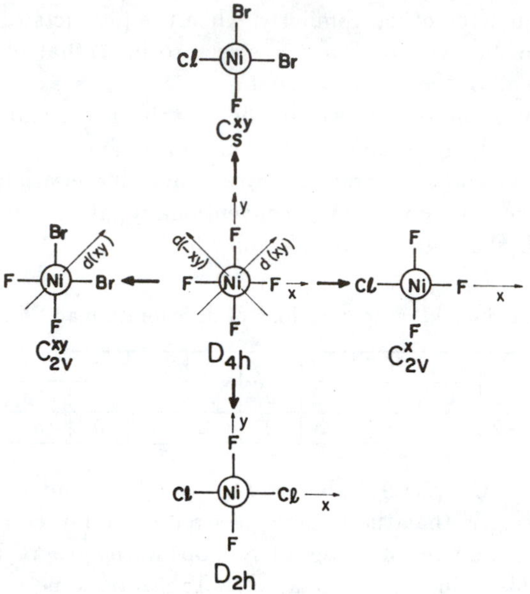

Figure 2.12. Substitutional desymmetrization of $[NiF_4]^{-2}$ from $\mathbf{D}_{4h}$ to several of its subgroups

Substitution of different halide ions for F^- in the complex, square-planar complex ion $[NiF_4]^{-2}$, shown in the center of Fig. 2.12, produces various nearly, but not quite, square-planar ions, each of which is of lower symmetry than $\mathbf{D}_{4h}$. None of them has any symmetry elements that are not present in $[NiF_4]^{-2}$, so each belongs to some subgroup of $\mathbf{D}_{4h}$. Replacement of two *trans*-situated F^- ions by Cl^- reduces the symmetry to $\mathbf{D}_{2h}$. The replacement of flourine by the less electronegative chlorine stabilizes an electron in p_y(or d_{yz}) of the central Ni-ion relative to one in p_x (or d_{zx}); the analogy with an atom in a quadrupolar field (Fig. 2.4) is obvious. A comparison of the Character Tables of the two groups shows that $\mathbf{D}_{2h}$ is an invariant subgroup of $\mathbf{D}_{4h}$, as can also be deduced from Table 2.4: the sym-ops retained in $\mathbf{D}_{2h}$ are either in separate classes of $\mathbf{D}_{4h}$ (E, C_2, i, σ_h), or else paired in C_2' and σ_v.

Replacement of a single fluoride ion by chloride reduces the symmetry of the system to $\mathbf{C}_{2v}$. Two of its elements, $C_2(x)$ and $\sigma(zx)$ are retained but $C_2(y)$, which is paired with the former in $\mathbf{D}_{4h}$ as one of two equivalent C_2' operations, and $\sigma(yz)$, which is in the same class (σ_v) as the latter, are excluded from the subgroup; evidently it is not invariant. The same can be said about the symmetry of *cis*-$[NiF_2Br_2]^{-2}$, which retains only one of the two C_2'' axes and one of the two diagonal mirror planes. Desymmetrization to $[NiFClBr_2]^{-2}$ leaves reflection in the molecular plane, $\sigma(xy)$, as the only symmetry element except for the ubiquitous E, but since σ_h – like E – is in a class by itself in the parent group, the subgroup comprising these two elements, $\mathbf{C}_s^{xy}$, is an invariant subgroup of $\mathbf{D}_{4h}$. It might be noted that the still more highly substituted complex, $[NiFClBrI]^{-2}$, is no less symmetrical, since it too retains the horizontal mirror plane.

The kernel of an irreducible representation was defined in the preceding section as the subgroup comprising those elements that have the character 1 rather than -1 in it. Now that we have encountered irreps of higher dimension, the definition of *kernel* has to modified so that it includes the previous definition as a special case: *The kernel of an irreducible representation is the set of elements that have the same character as the identity in that representation.* The more restrictive original definition holds not only for commutative groups but for one-dimensional representations of non-commutative groups as well. Thus, B_{1g} is the only irrep of $\mathbf{D}_{4h}$ in which the eight sym-ops retained in $\mathbf{D}_{2h}$ have the character 1 and the other eight have the character -1; clearly, $\mathbf{D}_{2h}$, is the kernel of B_{1g}. It hardly needs saying that the kernel of the totally symmetric representation A_{1g} is the full parent group, $\mathbf{D}_{4h}$. As formulated above, the definition also applies to $\mathbf{C}_s^{xy}$, which is the kernel of the two-dimensional irrep E_u, since σ_h is the only operation of $\mathbf{D}_{4h}$ that has the same character (2) as the identity in that representation.

Let us now take note of the fact that $\mathbf{C}_s^{xy}$, the kernel of E_u, and $\mathbf{D}_{2h}$, the kernel of B_{1g}, are invariant subgroups.[24] In contrast, $\mathbf{C}_{2v}$ – the non-invariant subgroup characterizing the mono- and *cis*-disubstituted ions in Fig. 2.13 – is not the kernel of any irrep of $\mathbf{D}_{4h}$; it is the *co-kernel* of E_u. A co-kernel [20, 22, 21, 23] of a representation is the subgroup retaining, in addition to the kernel, selected symmetry operations from classes of the parent group that are not included in full. Thus, $\mathbf{C}_{2v}$ has, in addition to the two symmetry elements of the kernel (E, σ_{xy}), another mirror plane and a twofold rotational axis at the line of intersection of the two planes. The additional symmetry elements can be chosen in one of several ways: If the mirror plane selected is one of the two vertical planes, say σ_{zx}, the rotational axis is necessarily C_x, which was one the two vertical axes. This particular choice, labelled $\mathbf{C}_{2v}^x$ in the figure, is appropriate for characterizing $[NiF_3Cl]^{-2}$, whereas for *cis*-$[NiF_2Cl_2]^{-2}$, one of the diagonal rotational axes (C_2'') and the corresponding mirror plane are retained instead.

One final point: $[NiFClBr_2]^{-2}$ need not be made directly by trisubstitution in $[NiF_4]^{-2}$, but might be produced from it stepwise, either by replacing two adjacent fluoride ions of $[NiF_3Cl]^{-2}$ by bromide or one F^- of *cis*-$[NiF_2Br_2]^{-2}$ by Cl^-. Invariably, $\mathbf{C}_s^{xy}$ is the kernel of that irrep of $\mathbf{C}_{2v}$ which is symmetric

[24] The same invariant subgroup may, however, be the kernel of more than one irrep.

to reflection in the molecular plane,[25] regardless of what direction in space is chosen to define the C_2 axis.

Desymmetrization to the co-kernel subgroups will become important in connection with the mechanism of reactions involving transition metal complexes as well as of highly symmetric organic molecules, such as cubane or tetrahedrane. A *Table of Kernels and Co-Kernels* is, therefore, included as Appendix B, but reference to it will be deferred until the the ideas developed in this chapter are extended beyond single atoms, first to diatomic and then to polyatomic molecules.

2.4 References

[1] J.W. Linnett: *Wave Mechanics and Valency.* Methuen, London 1960. A delightfully simple introduction to an inherently difficult subject.

[2] R. McWeeney: *Coulson's Valence, Third Edition.* Oxford University Press, Oxford 1975. The modern version of a didactic classic.

[3] Several of the more recent Quantum Chemistry texts of intermediate difficulty are listed below: [4, 5, 6, 7].

[4] H.F. Hameka: *Quantum Theory of the Chemical Bond.* Haffner, New York 1975.

[5] J.P. Lowe: *Quantum Chemistry.* Academic Press, New York 1978.

[6] R.L. Flurry, Jr.: *Quantum Chemistry, an Introduction.* Prentice-Hall, Englewood-Cliffs 1980.

[7] I.N. Levine: *Quantum Chemistry.* Allyn and Bacon, Y. New York 1983.

[8] William Shakespeare: *Troilus and Cressida*, Act 3, Scene 3.

[9] Most of the Group Theory required for this book can be found in Reference [5, Chap. 13] or Reference [6, Chap. 13].

[10] Several of the many available presentations of Group Theory for chemists are listed below in rough order of increasing difficulty: [11, 12, 13, 14, 15, 16, 17].

[11] H.H. Jaffe and M. Orchin: *Symmetry in Chemistry.* Wiley, New York 1965.

[12] S.F. Kettle: *Symmetry and Structure.* Wiley, Chichester 1985.

[13] F.A. Cotton: *Chemical Applications of Group Theory.* Wiley, New York 1963.

[14] D.S. Schonland: *Molecular Symmetry.* Van Nostrand, London 1965.

[15] B.E. Douglas and C.A. Hollingsworth: *Symmetry in Bonding and Spectra.* Academic Press, London 1985.

[16] L.H. Hall: *Group Theory and Symmetry in Chemistry.* McGraw-Hill, New York 1969.

[17] R.M. Hochstrasser: *The Molecular Aspects of Symmetry.* Benjamin, New York 1966.

[18] See References [1, 50, 51, 61] and – in particular – [62] in *Chapter 1*.

[19] E.P. Wigner: *Group Theory.* Academic Press, New York 1959, p.67.

[20] M.A. Melvin: Rev. Mod. Phys. *28*, 18 (1956).

[21] W.A. Bingel: Acta Phys. Acad. Sci. Hungar. *51*, 13 (1956).

[25] Whether it is b_1 or b_2, depends on the choice of an axis convention, which should always be specified.

[22] P. Murray-Rust, H.-B. Bürgi and J.D. Dunitz: Acta. Cryst. *A35*, 703 (1979). A lucid exposition of the distinction between *kernel* and *co-kernel* in terms of distortional desymmetrization of a molecule in a crystalline environment. The *Table of Kernels and Co-kernels* in Appendix B was adapted from this paper.

[23] E.A. Halevi: J. Chem. Research (S), *1985*, 206. A brief discussion of the application of kernels and co-kernels to the description of substitutional desymmetrization in the context of chirality and prochirality.

Chapter 3

Diatomic Molecules and Their Molecular Orbitals

Just as the electronic configuration of an atom is built up by stepwise population – electron by electron – of hydrogen-like atomic orbitals, that of a diatomic molecule is constructed by successively filling the molecular orbitals derived from the hydrogen molecule ion, H_2^+ [1].

3.1 The Hydrogen Molecule Ion

3.1.1 The Symmetry of H_2^+

The first thing to note about Fig. 3.1 is that the two hydrogen nuclei define an internuclear axis, which is conventionally labelled z; the x and y axes remain undefined, but the xy plane can be – with care. In principle, either the center of nuclear mass or the center of nuclear charge can be chosen to serve as the origin of our coordinate system. When both H_A^+ and H_B^+ are protons, either criterion fixes the origin at **i**, the point midway between them. However, when H_A^+ is a proton and H_B^+ is a deuteron or *vice versa*, the center of charge remains at **i** but the center of mass, to which rotational and vibrational motion have to be referred, has moved closer to the heavier nucleus.

Since we are primarily concerned with electronic properties, we implicitly adopt the *Born-Oppenheimer approximation*, according to which the electronic wave functions, the electronic energy and the charge density are all calculated as if the nuclei were clamped a fixed distance, R_{AB}, from each other, prohibiting nuclear motion and thus rendering the nuclear mass irrelevant. Only after the electronic properties of the "frozen" molecule have been evaluated as a function of R_{AB} and the equilibrium value of the internuclear distance (R_e) ascertained, are the rotational and vibrational motions taken up explicitly. Therefore, in temporary disregard of a possible ponderal distinction between H_A^+ and H_B^+, we set the origin at **i**, constraining the xy plane to pass through it.

Assume that the electron is momentarily situated at an arbitrary point **p**. Its potential energy is uniquely determined by two distances, r_A and r_B. Neither of these distances is altered when the electron is rotated about the z axis through any angle ϕ, say to point **p′**, as long as the perpendicular distance to the axis remains constant ($\mathsf{r}_{p'} = \mathsf{r}_p$). One particular rotation of this kind, specifically by $\phi = 180°$, brings the electron to **q**, which can also be related to **p** by reflection in the plane (not drawn in Fig. 3.1) which includes the z axis

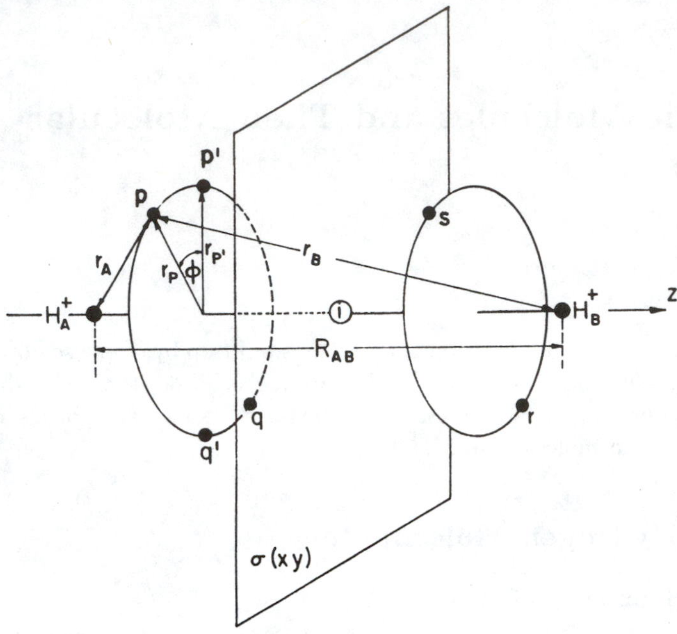

Figure 3.1. Cylindrical symmetry ($\mathbf{D}_{\infty h}$) of the hydrogen molecule ion

and is perpendicular to the original direction of r_p. An infinite number of such *vertical* mirror planes exist; one such is that which interrelates $\mathbf{p}'$ and $\mathbf{q}'$. All of the symmetry operations of $\mathbf{C}_{\infty v}$, which were enumerated in Section 2.2.1, leave r_A and r_B unchanged, so H_2^+ has at least axial symmetry. However, the potential energy of an electron at $\mathbf{r}$ is also the same as that at $\mathbf{p}$, which is related to $\mathbf{r}$ by inversion through $\mathbf{i}$ or by a twofold rotation about one of an infinite number of axes lying in the *horizontal* (xy) plane. Nor is it different at $\mathbf{s}$, which can be reached from $\mathbf{p}$ by a twofold rotation about another axis in the xy plane, or simply by reflection in that plane. These additional sym-ops, which are not included in $\mathbf{C}_{\infty v}$, interchange r_A and r_B as they interconvert z and $-z$; since the potential energy is invariant to all of them, H_2^+ is not merely axially symmetric, but has cylindrical ($\mathbf{D}_{\infty h}$) symmetry. Clearly, $\mathbf{C}_{\infty v}$ is a subgroup of $\mathbf{D}_{\infty h}$, bearing the same relation to it that $\mathbf{C}_{2v}$ does to $\mathbf{D}_{2h}$ (Section 2.2.5).

At infinite nuclear separation, the electron has to choose between one nucleus and the other, producing either ($H_A + H_B^+$) or ($H_A^+ + H_B$). In either case, it is localized on one of the nuclei in what is effectively an isolated atom, the electron being too far from the the second nucleus for its attraction to have an appreciable effect on the energy. However, in order to define the z axis, the bare nucleus – say H_B^+ – must be sufficiently close to H_A for its attraction to be perceptible, if only as a minute perturbation of the energy of the bound electron. Therefore, although $\mathsf{r}_B >> \mathsf{r}_A$, the cylindrical symmetry of the molecule, i.e. its invariance to symmetry operations that interchange r_A and r_B, ensures that its true stationary state is one in which the two nuclei are afforded equivalent status. The electron may be associated with a single atom, slightly perturbed by a remote bare nucleus, but we have no way of knowing which nucleus is which.[1]

3.1.2 The Molecular Orbitals of H_2^+

When H_A^+ and H_B^+ have come close enough to establish cylindrical symmetry, the $1s$ orbitals centered on the two nuclei have to be combined to MOs that transform according to the symmetry operations of $\mathbf{D}_{\infty h}$, that convert $1s(A)$ to $\pm 1s(B)$. We therefore have to combine the two AOs into MOs that belong to irreducible representations of $\mathbf{D}_{\infty h}$. This requirement is satisfied by the two *linear combinations of atomic orbitals* (*LCAO*s):

$$1s\sigma_g = (1s(A) + 1s(B))/\sqrt{2} \;;\; 1s\sigma_u = (1s(A) - 1s(B))/\sqrt{2} \tag{3.1}$$

Axial symmetry, i.e. total symmetry with respect to all of the sym-ops of $\mathbf{C}_{\infty v}$, is denoted by σ. The positive combination, being symmetric with respect to inversion (g) as well, is also totally symmetric in $\mathbf{D}_{\infty h}$; the negative combination changes sign whenever $1s(A)$ and $1s(B)$ are interchanged, so it is antisymmetric (u) with respect to inversion and to all of the other sym-ops of $\mathbf{D}_{\infty h}$ that are not included in its subgroup, $\mathbf{C}_{\infty v}$.

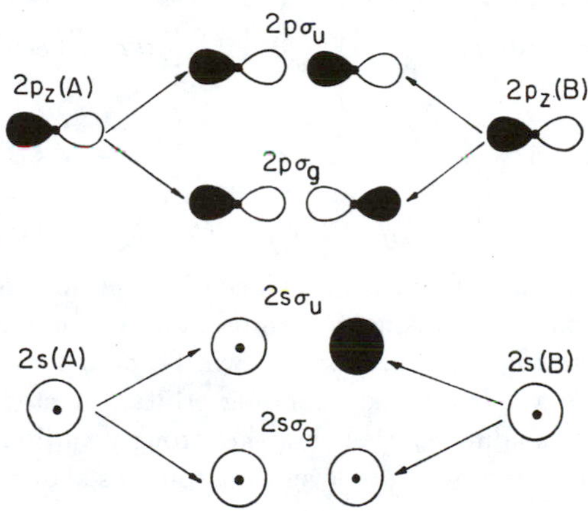

Figure 3.2. Construction of the $2s\sigma$ and $2p\sigma$ MOs of H_2^+

We can construct similar pairs of MOs from the $2s$ orbitals:

$$2s\sigma_g = (2s(A) + 2s(B))/\sqrt{2} \;;\; 2s\sigma_u = (2s(A) - 2s(B))/\sqrt{2} \tag{3.2}$$

and, taking proper account of orbital phase, from the $2p_z$ orbitals[2]:

[1] The time-dependence of an electron jump from an atom to an approaching nucleus is beyond the scope of this book.

[2] Footnote 13 of Chapter 1 explains the choice of sign.

$$2p\sigma_g = (2p_z(A) - 2p_z(B))/\sqrt{2}\ ;\ 2p\sigma_u = (2p_x(A) + 2p_z(B))/\sqrt{2} \qquad (3.3)$$

As in the $1s\sigma$ orbitals, mutual orthogonality of the two in each of the pairs shown in Fig. 3.2 is assured by the fact that one is g and the other is u. The four orbitals are degenerate at infinite R_{AB} but the degeneracy is split as the nuclei approach one another; the g orbitals become bonding and the u orbitals antibonding. Furthermore, the two former have the same irreducible representation (σ_g) in $\mathbf{D}_{\infty h}$, so – from the considerations outlined in Section 2.2.6 – they should mix, as should the two antibonding σ_u orbitals, one of each pair being stabilized and the other destabilized.[3]

The interaction of $2s\sigma_g$ and $2p\sigma_g$, for example, necessarily implies hybridization between the $2s$ and $2p_z$ orbitals of both atoms. This being so, there can be no objection to carrying out the sp-hybridization in the separate atoms, and then allowing the initially degenerate hybrid AOs to interact as R_{AB} decreases. Predictably, the result is the same as before: two interacting σ_g and two interacting σ_u orbitals. Now, however, the way in which the quadruply degenerate energy level splits is more instructive.

The hybrids are familiar from Section 2.2.4:

$$h_+^z(A) = (2s(A) + 2p_z(A))/\sqrt{2}\ ;\ h_-^z(A) = (2s(A) - 2p_z(A))/\sqrt{2} \qquad (3.4)$$

$$h_+^z(B) = (2s(B) + 2p_z(B))/\sqrt{2}\ ;\ h_-^z(B) = (2s(B) - 2p_z(B))/\sqrt{2} \qquad (3.5)$$

and their *symmetry adaptation* to $\mathbf{D}_{\infty h}$ produces the molecular orbitals illustrated in Fig. 3.3:

$$\sigma_g = (h_+^z(A) + h_-^z(B))/\sqrt{2}\ ;\ \sigma_u' = (h_+^z(A) - h_-^z(B))/\sqrt{2} \qquad (3.6)$$

$$\sigma_g' = (h_-^z(A) + h_+^z(B))/\sqrt{2}\ ;\ \sigma_u = (h_-^z(A) - h_+^z(B))/\sqrt{2} \qquad (3.7)$$

The distinction between $2s\sigma$ and $2p\sigma$ has been lost, all four MOs being virtually degenerate as long as R_{AB} is nearly infinite, but as the nuclei approach one another, the most strongly bonding orbital will be be σ_g, whereas σ_g' (primed because it lies above σ_g), is a weakly bonding orbital directed outwards. Similarly, the weakly antibonding σ_u lies below the strongly antibonding σ_u'. As R_{AB} decreases, the quadruply degenerate level thus splits as shown schematically in Fig. 3.3: $\sigma_g < \sigma_g' < \sigma_u < \sigma_u'$.

Degenerate with $2p_z$ at $R_{AB} \approx \infty$ are the remaining $2p$ orbitals, two centered on each nucleus. Having defined only the z axis, we cannot distinguish between them in terms of their symmetry properties with respect to x and y. We could, of course, adopt the viewpoint of Section 2.2.1, and looking along the internuclear axis from positive z towards the origin, assign an electron rotating around it clockwise to $2p_{-1}$ and one rotating counterclockwise to $2p_{+1}$. Then we might split the degeneracy between them by setting up a hypothetical magnetic

[3] In principle, $1s\sigma_g$ also interacts with $2s\sigma_g$ and $2p\sigma_g$ and $1s\sigma_u$ with $2s\sigma_u$ and $2p\sigma_u$. The $1s$ orbitals lie so low below $2s$ and $2p$, however, that these interactions can be neglected for most practical purposes.

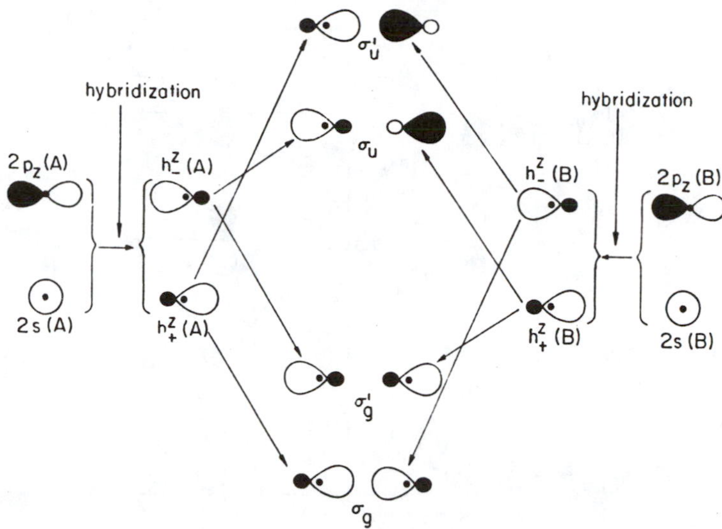

Figure 3.3. Hybridization and symmetry adaptation of $2s\sigma$ and $2p\sigma$ orbitals

field parallel to the axis, thereby reducing the overall symmetry from cylindrical $\mathbf{D}_{\infty h}$ to axial $\mathbf{C}_{\infty v}$, and proceed to construct π orbitals as appropriate linear combinations of $2p_{-1}(A)$, $2p_{-1}(B)$, $2p_{+1}(A)$ and $2p_{+1}(B)$. This is in fact standard practice, but our hypothetical field need not be magnetic. Instead, harking back to Section 2.2.2, we can hypothesize a quadrupolar field at right angles to the z axis, so as to fix common x and y axes for both atoms, and allow for the construction of the π MOs as orthogonal linear combinations of $2p_x(A)$, $2p_x(B)$, $2p_y(A)$ and $2p_y(B)$.

The charge density of an electron in any of the molecular orbitals so constructed has to be totally symmetric in $\mathbf{D}_{2h}$, so the MOs themselves must be either symmetric or antisymmetric to each of its sym-ops. They are depicted in Fig. 3.4, their explicit forms being:

$$2p\pi_u^x = (2p_x(A) + 2p_x(B))/\sqrt{2}\ ;\ 2p\pi_g^x = (2p_x(A) - 2p_x(B))/\sqrt{2} \tag{3.8}$$

$$2p\pi_u^y = (2p_y(A) + 2p_y(B))/\sqrt{2}\ ;\ 2p\pi_g^y = (2p_y(A) - 2p_xyB))/\sqrt{2} \tag{3.9}$$

The π_u orbitals, which are degenerate when the potential energy has strictly cylindrical symmetry,[4] are converted in a quadrupolar field to two orthogonal orbitals that remain virtually degenerate when the field intensity is near zero. It is easily confirmed with the Character Table for $\mathbf{D}_{2h}$ that the respective representations of $2p\pi_u^x$ and $2p\pi_u^y$ are b_{3u} and b_{2u}; similarly, the irreps of $2p\pi_g^x$ and $2p\pi_g^y$ are b_{3g} and b_{2g} respectively. We complete the list by noting that the representation of σ_g in $\mathbf{D}_{2h}$ is a_g and that of σ_u is b_{1u}.

[4] They belong to the same two-dimensional irrep (Π_u) of $\mathbf{D}_{\infty h}$.

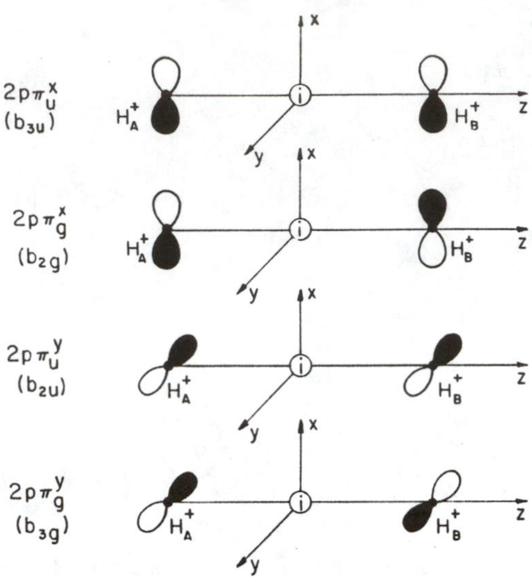

Figure 3.4. The π orbitals of H_2^+

3.2 Homonuclear Diatomic Molecules

The logical next step should be the construction of an orbital correlation diagram for H_2^+, which we would obtain by ignoring internuclear repulsion and decreasing R_{AB} to zero, eventuallly uniting the two H^+ nuclei to He^{2+}. However, there are adequate reasons for skipping it and moving on to many-electron diatomics. In the one-electron atoms H and He^+, an energy level with main quantum number n is n^2-fold degenerate, so a clutter of correlation lines would necessarily terminate at every atomic energy level of He^+ above $n = 1$. Moreover, H_2^+ does not have much interesting chemistry, its principal distinction being the utility of its molecular orbitals as models for those of many-electron diatomic molecules. Its correlation diagram even has to be modified for the neutral hydrogen molecule; the $2s$ and $2p$ orbitals are not degenerate in the united helium atom because the $1s$-electron that is invariably present in the lower excited states of He, screens the nucleus more effectively from a $2p$ than from a $2s$ electron.

We therefore proceed directly to the homonuclear diatomic molecule, X_2, in which X represents a second-period element like oxygen or nitrogen. These molecules are not only interesting in themselves, but provide a natural bridge to linear polyatomic organic molecules like acetylene and, from there, to ethylene and beyond.

3.2.1 Mulliken's Orbital Correlation Diagram

Fig. 3.5 displays Mulliken's [2] generalized *orbital correlation diagram* for homonuclear diatomic molecules, which has been of seminal importance for the elucidation of the electronic structure of molecules.[5] Only the atomic levels $n = 1, 2$ have been included on its right; on the left, the orbitals of the united atom have been extended sufficiently that all of the necessary correlation lines can be drawn. The energetic spacing of the AOs is purely schematic, their order on the left being that common to silicon and sulfur, the united atoms corresponding to N_2 and O_2. The abcissa, R_{AB}, is – of course – quite non-linear.

Let us read the diagram from right to left. With the two X atoms separated, we imagine a very weak quadrupolar field perpendicular to their line of centers,

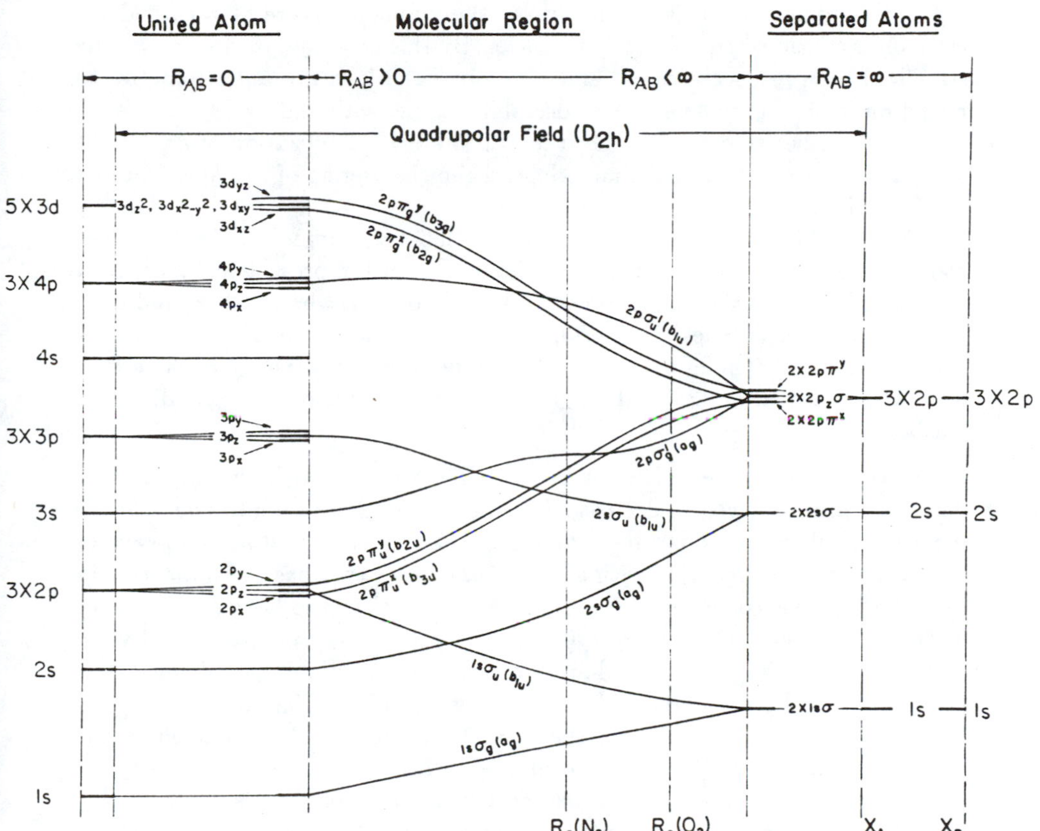

Figure 3.5. Schematic orbital correlation diagram for homonuclear diatomic molecules

[5] It has been said [3] "that it might well be on the walls of chemistry buildings...beside the Mendeléef periodic table...Just as the latter affords an understanding of the structure of atoms so does the former afford an understanding of the structure of molecules".

which has already been defined to be the z axis. The field establishes the other cartesian axes by an infinitesimal stabilization of p_x and destabilization of p_y, an effect that is exaggerated in Fig. 3.5 for the sake of clarity. The potential energy of an electron has been desymmetrized by the field from $\mathbf{D}_{\infty h}$ to $\mathbf{D}_{2h}$, so we can borrow the symmetric and antisymmetric MOs of H_2^+, recognizing that the bonding and antibonding orbitals in each pair (Equations 3.1-3.3, 3.8, 3.9) remain virtually degenerate at large R_{AB}. Moreover, since $2s$ and $2p$ do not have the same energy in many-electron atoms, sp-hybridization is not assumed to take place until the nuclei have come within bonding range of one another.

As the internuclear distance gets shorter, the distinction between bonding and antibonding orbitals makes itself felt. The familiar criteria for bonding are:

1. An MO is bonding if the electron density can be large between the nuclei and antibonding when it is excluded from the xy plane, where $\mathsf{r}_A = \mathsf{r}_B$ and the electron is attracted equally to both nuclei. By this criterion, σ_g and π_u will be stabilized as R_{AB} decreases, σ_u and π_g – both of which are antisymmetric to reflection in the xy plane – will be destabilized, i.e. antibonding.

2. The relative extent of the stabilization or destabilization of an MO at any particular internuclear distance depends on the overlap of the AOs centered on X_A and X_B.

3. Two MOs with the same symmetry label will interact, the lower becoming more stable at the expense of the upper. In the present instance, $2s\sigma_g$ and $2p\sigma_g$, belong to the same representation, a_g, of $\mathbf{D}_{2h}$, whereas $2s\sigma_u$ and $2p\sigma_u$ both belong to b_{1u}. These four orbitals, which are best represented by Equations 3.2 and 3.3 at infinite separation, are described increasingly well as R_{AB} decreases by Equations 3.6 and 3.7, in which they are regarded as bonding and antibonding combinations of sp hybrids.

Since internuclear repulsion is being deliberately ignored, the internuclear distance can be allowed to decrease until the two nucleii fuse into one. The X_2 molecule gradually becomes more and more like the united atom, and each of its MOs correlates smoothly with an AO of the latter that has the same irrep in $\mathbf{D}_{2h}$. The σ_g AOs correlate with atomic s orbitals, that also have a_g symmetry in $\mathbf{D}_{2h}$; σ_u goes to $p_z(b_{1u})$, π_u^x to $p_x(b_{3u})$, π_u^y to $p_y(b_{2u})$, π_g^x to $d_{xz}(b_{2g})$, and π_g^y to $d_{yz}(b_{3g})$. The appropriate united-atom orbitals are thus uniquely defined, but not all of them have counterparts on the right side of the diagram. The unused $3p_x$ and $3p_y$ AOs correlate with the $3p\pi_u$ MOs that are not included in the diagram, $4s$ with σ_g'' – the next σ_g orbital – and so on.

The remaining threefold degeneracy in the $3d$ level sounds a cautionary note. It was conceded in Section 2.3.1 that the quadrupolar field, of which we have made repeated use, cannot split this degeneracy.[6] The electron density of an electron in $d_{x^2-y^2}$ or d_{z^2}, both of which are totally symmetric in $\mathbf{D}_{2h}$, and also of one in d_{xy}, that is not, are all symmetrically disposed in the field. Evidently, the formal reduction of symmetry from $\mathbf{D}_{\infty h}$ to $\mathbf{D}_{2h}$ will prove to be an unreliable device whenever the occupied MOs include those which – like $3d\delta$

[6] To first order.

– are constructed from higher AOs of the X atoms,[7] but can be used to good advantage when they are limited to combinations of $1s$, $2s$ and $2p$ orbitals, as they are in the present context.

3.2.2 The Symmetry of Electron Configurations

Fig. 3.5 was drawn in complete disregard of nuclear repulsion, that prevents the two atoms from merging in practice. The internuclear distance achieves its equilibrium value (R_e) when the electronic energy, which – at the present level of approximation – is the sum of the orbital energies, each counted once for every occupying electron, just balances the internuclear (and interelectronic) repulsion, which is introduced separately as a parameter. In general, R_e varies from state to state, but our molecules of interest are still strongly so bonded in their low-lying electronic states that R_e is not very different in them from its value in the ground-state. If the correlation lines in Fig. 3.5 are drawn so as to reproduce the correct energetic order of MOs in N_2 and O_2 at their ground-state equilibrium distances, we can assume with some confidence that the order will remain the same in their lower excited states.

The ground-state configuration of N_2, with its 14 electrons, is read from Fig. 3.5 to be:

$$[1s\sigma_g^2\ 1s\sigma_u^2\ 2s\sigma_g^2\ 2s\sigma_u^2\ 2p\pi_u^{x\ 2}\ 2p\pi_u^{y\ 2}\ 2p\sigma_g'^{\ 2}]$$

Neglecting the core of $1s$ electrons, recognizing that $2p\pi_u^x$ and $2p\pi_u^y$ are isoenergetic in the unperturbed molecule, and condensing the notation, it can be rewritten: $[\sigma_g^2\ \sigma_u^2\ \pi_u^4\ \sigma_g'^{\ 2}]$. It then follows that the lowest excited configuration is generated by exciting an electron from σ_g' – which is a nearly non-bonding lone-pair orbital – to an antibonding π_g-orbital, producing the configuration: $[\sigma_g^2\ \sigma_u^2\ \pi_u^4\ \sigma_g'\ \pi_g]$, whereas the more highly excited $\pi \longrightarrow \pi^*$ configuration is written: $[\sigma_g^2\ \sigma_u^2\ \pi_u^3\ \sigma_g'^{\ 2}\ \pi_g]$ and so on. At the eqililibrium bond length of O_2, which is larger than that of N_2, σ_g' lies below π_u, so its ground-configuration is $[\sigma_g^2\ \sigma_u^2\ \sigma_g'^{\ 2}\pi_u^4\ \pi_g^2]$.

The notation just introduced is rather more than a convenient shorthand for specifying which orbitals are occupied and by how many electrons. It expresses the fact that the MO approximation to the molecular wave function is a product of one-electron wave functions, i.e. orbitals, each taken to a power equal to the number of electrons occupying it.[8] We recall that the irreducible representation of a product of coordinates is the direct product of their irreps; extending the same idea to the product of orbitals, we see that the irrep of an electron configuration is simply the direct product of the irreps of its occupied

[7] Such MOs have to be considered in connection with the higher excited states of simple diatomics, as well as in the ground-states of transition metal dimers.

[8] The state wave function is not simply the orbital product; it also includes the spin and has to be made antisymmetric with respect to the interchange of any two electrons in order to comply with the *Pauli exclusion principle*. This type of permutational antisymmetrization will become important when electron spin is taken into account in Part IV, but we need not concern ourselves with it for the present.

orbitals.[9] The irrep of the ground-state configuration of N_2 is:

$$\Gamma_1(N_2) = \gamma^2(\sigma_g) \otimes \gamma^2(\sigma_u) \otimes \gamma^2(\pi_u^x) \otimes \gamma^2(\pi_u^y) \otimes \gamma^2(\sigma_g')$$

Two more points should be noted before going on to characterize the configurations fully by symmetry species:

1. All of the electrons in a closed shell configuration, in which all orbitals of the same energy are fully occupied, are paired, so the net electron spin is necessarily zero ($S = 0$) and its multiplicity ($2S + 1$) makes it a singlet. The irrep of every doubly-occupied MO appears multiplied by itself in the irrep of the configuration, so – since the product of any irrep with itself is totally symmetric – a closed shell singlet configuration is necessarily totally symmetric.[10] The irreducible representations of open shell configurations, of whatever multiplicity, will require more detailed consideration.

2. From the strict molecular orbital point of view, the configuration, which identifies the occupied orbitals of a molecule, determines its energy and uniquely defines its state. Equating the electronic ground-state with the lowest electron configuration is a good approximation for many, but not all, closed shell ground-states. It is considerably less reliable for open shell states – whether ground- or excited – in which one or more of the orbitals is singly-occupied; these have to be dealt with more carefully.

3.2.3 State Symmetry in $\mathbf{D}_{\infty h}$ and $\mathbf{D}_{2h}$

Wigner and Witmer's [5] characterization of the electronic states of diatomic molecules marked the triumphant entry of group theory into the field of molecular structure. It was focused primarily on considerations of angular momentum, which we have managed to sidestep almost completely thus far in the book by artificially reducing spherical or cylindrical symmetry to $\mathbf{D}_{2h}$.[11] Thus, the σ and π orbitals of homonuclear diatomic molecules were assigned in the foregoing paragraphs to the irreps of $\mathbf{D}_{2h}$. These molecules are cylindrically symmetrical, and their configurational and state symmetry is ordinarily described accordingly. It therefore seems advisable to suggest a simple general procedure for interrelating the irreps of the continuous, non-commutative group $\mathbf{D}_{\infty\mathbf{h}}$ and those of its discrete, commutative subgroup $\mathbf{D}_{2h}$.

In the Character Table of $\mathbf{D}_{2h}$ we find four representations that are symmetric under (twofold) rotation about the z axis, and can thus be related to Σ.[12]

---

[9] This is straightforward for commutative groups like $\mathbf{D}_{2h}$. It is also true for $\mathbf{D}_{\infty h}$ and other non-commutative groups, but the direct product of two-dimensional representations like E_g an E_u is no longer a single irrep. For our immediate purposes, the loss of generality resulting from desymmetrization from $\mathbf{D}_{\infty h}$ to $\mathbf{D}_{2h}$ is more than offset by the gain in simplicity.

[10] The restriction to commutative groups is being maintained temporarily but it is not a neccesary condition; closed shell configurations are totally symmetric in non-commutative groups as well.

[11] This is the pedestrian equivalent of Hurley's [6] subduction of a simply subducible group by a commutative subgroup.

[12] We postpone facing the complication that it can also be related to Δ.

Two of these, A_g and B_{1g}, are assigned to Σ_g and the remaining two, A_u and B_{1u}, to Σ_u on the basis of their inversion symmetry. The distinction between the members of each pair must therefore rest on their behavior with respect to symmetry elements other than $C_2(z)$ or i. Choosing reflection symmetry with respect to *both* the zx and yz planes as the additional criterion, we obtain:

$$A_g \Longleftrightarrow \Sigma_g^+ \,;\, B_{1g} \Longleftrightarrow \Sigma_g^- \,;\, A_u \Longleftrightarrow \Sigma_u^- \,;\, B_{1u} \Longleftrightarrow \Sigma_u^+$$

The ground-configuration of N_2 is a closed shell singlet, so its designation can only be ${}^1\Sigma_g^+$, in which the upper left index 1 specifies it to have singlet spin-multiplicity. In the first excited configuration of N_2, σ_g' and π_g are singly-occupied. The $\mathbf{D}_{2h}$ representation of the former is a_g, whereas the latter is either b_{2g} or b_{3g}, depending on whether the electron is in π_g^x or π_g^y. We recall that the direct product of any representation with a_g is itself, so the configuration is either B_{2g} or B_{3g}. It has also been noted that in cylindrical symmetry – or even in $\mathbf{D}_{4h}$ – C_4 interconverts p_x and p_y (cf. Fig. 2.7), which must therefore be assigned to the same two-dimensional irrep, Π_u. It follows that B_{3u} and B_{2u} of $\mathbf{D}_{2h}$ *map onto* Π_u of $\mathbf{D}_{\infty h}$; this is simply the converse of the by now familiar statement, that when a cylindrically symmetric molecule is desymmetrized to $\mathbf{D}_{2h}$, for example by the imposition of an external quadrupolar field, a degenerate pair of π_u orbitals split to $\pi_x(b_{3u})$ and $\pi_y(b_{2u})$. Similarly, B_{2g} and B_{3g}, likewise interconvertible in $\mathbf{D}_{\infty h}$, map onto Π_g. The first excited configuration of N_2 therefore corresponds to two states, a singlet (${}^1\Pi_g$) and a triplet (${}^3\Pi_g$). Both have the same orbital occupancy and – within the simple molecular orbital approximation – the same energy. However, it is a well known empirical fact [7, pp. 196–198, 240–245]; [8, pp. 28–32], amply confirmed by theory when *inter-electronic repulsion* is taken into account [9], that the lowest triplet generally lies below the singlet with the same electron configuration.[13]

States in which two electrons can occupy different π orbitals must be handled with more care. From the chemical viewpoint, the most interesting example is the oxygen molecule, dioxygen, the essential feature of which is the presence of two π_g electrons. We can conceive of three variations on this configurational theme: $[...\pi_g^{x\ 2}]$, $[...\pi_g^{y\ 2}]$ and $[...\pi_g^x\ \pi_g^y]$. Assuming the presence of a substantial quadrupolar field, that establishes $\mathbf{D}_{2h}$ symmetry and stabilizes π_g^x relative to π_g^y, we recognize the first two to be totally symmetric closed shell singlets, and relabel them ${}^1A_g$; the third has an open shell, so it corresponds to a singlet (${}^1B_{1g}$) and to a triplet (${}^3B_{1g}$) – that is the most stable of all. The four states, ordered accordingly,[14] appear at the right of Fig. 3.6.

We now reduce the intensity of the quadrupolar field to zero. The energy of the triplet, which is unaffected by the field, stays constant, and the rules given above suggest the correlation: ${}^3B_{1g} \Longleftrightarrow {}^3\Sigma_g^-$. These rules, which assume a set of simple one-to-one correlations between the irreps of $\mathbf{D}_{2h}$ and those of $\mathbf{D}_{\infty h}$,

[13] This is strictly true when a single open shell configuration suffices to characterize both states fully.

[14] It is assumed that the external field is not so intense as to bring the lowest singlet below the triplet.

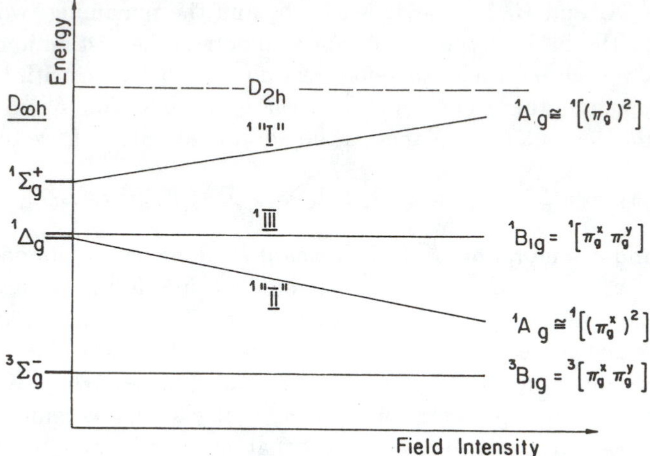

Figure 3.6. The lower states of O_2 in a quadrupolar field

break down when they are applied to the three singlet states. The difficulty stems from the fact that, now that the x and y axes are no longer defined, a perfectly legitimate sym-op, rotation by 90°, interconverts π_g^x and π_g^y. As a result, the two closed shell configurations, in each of which one π orbital is doubly-occupied, have become degenerate, so it would be quite wrong to map both of them onto the same one-dimensional irrep, Σ_g^+.

We therefore take linear combinations in the usual manner:

$$^1\mathrm{I} = (\,(\pi_g^x)^2 + (\pi_g^y)^2\,)/\sqrt{2}\;;\; ^1\mathrm{II} = (\,(\pi_g^x)^2 - (\pi_g^y)^2\,)/\sqrt{2} \tag{3.10}$$

Only the first of these two combinations, which transform as $z^2(x^2 + y^2)$ and $z^2(x^2 - y^2)$ respectively,[15] is totally symmetric in $\mathbf{D}_{\infty h}$ as it is in $\mathbf{D}_{2h}$. The second, viewed along the z axis, is reminiscent of the atomic $d_{x^2-y^2}$-orbital cf. Fig. 2.10 c). Rotation about the internuclear axis (z) by 45°, also a valid sym-op of $\mathbf{D}_{\infty_h}$, converts it to the open shell singlet configuration $[...\pi_g^x\ \pi_g^y]$, labelled ^{1}III in Fig. 3.6, which is the analog of d_{xy}. (cf. Fig. 2.10f.).

Evidently, of the three singlets, only ^{1}I can be designated $^1\Sigma_g^+$; ^{1}II and ^{1}III are necessarily two components of a degenerate representation, which includes x^2-y^2 and xy, and allows ^{1}II and ^{1}III to interconvert under the sym-ops of $\mathbf{D}_{\infty h}$. The two distinct $\mathbf{D}_{2h}$ configurations, one closed shell and the other open shell, thus coalesce in $\mathbf{D}_{\infty h}$ and become components of a single degenerate state; the similarity of these MOs to atomic d orbitals prompted labeling the configuration comprising them Δ_g. Empirical evidence identifies $^1\Delta_g$ as the more stable form of singlet dioxygen [10, pp. 345–346], allowing completion of the *state correlation diagram* in Fig. 3.6.

[15] Consult the $\mathbf{D}_{2h}$ Character Table: π_g^x transforms like $xz(b_{2g})$ and π_g^y like $yz(b_{3g})$, so their squares transform like x^2z^2 and y^2z^2 respectively.

Table 3.1. Correlation Table. $\mathbf{D}_{\infty h} \Longleftrightarrow \mathbf{D}_{2h}$

$\mathbf{D}_{\infty h}$	$\mathbf{D}_{2h}$
...	
Δ_u	B_{1u}, A_u
Δ_g	B_{1g}, A_g
Π_u	B_{3u}, B_{2u}
Π_g	B_{3g}, B_{2g}
Σ_u^-	A_u
Σ_g^-	B_{1g}
Σ_u^+	B_{1u}
Σ_g^+	A_g

The $\mathbf{D}_{\infty h} \Longleftrightarrow \mathbf{D}_{2h}$ correlations that have just been worked out, together with several others that can be rationalized along similar lines, are summarized in Table 3.1[16]. The assignment of symmetry labels to the three singlets and three triplets that are consistent with the $\pi \rightarrow \pi^*$ configuration of N_2 is left as an exercise.

At this point, a few words are called for about the mechanism of the interaction induced by the quadrupolar field between the two closed shell singlets of the cylindrically symmetrical molecule. We read Fig. 3.6 from left to right. After the two components of ${}^1\Delta_g$ have been nominally split to ^{1}II and ^{1}III, the totally symmetric field (in $\mathbf{D}_{2h}$) mixes ^{1}I and ^{1}II, the degree of mixing depending on

[16] The reader who has not skipped Section 2.3 is advised to look at Table 3.1 in conjuction with the Character Table of $\mathbf{D}_{\infty h}$ in Appendix A. Several additional points will then emerge: a) Each of its two-dimensional irreps splits to two of $\mathbf{D}_{2h}$; b) there is an infinite number of two-dimensional representations, indicated in the table by "...", each of which splits similarly; c) $\mathbf{D}_{2h}$ is the co-kernel of Δ_g (not its kernel, which is $\mathbf{C}_i$), as can be seen from the fact that Δ_g maps onto the totally symmetric representation of $\mathbf{D}_{2h}$, indicating that the quadrupolar field can be regarded as a perturbation which has the symmetry species Δ_g in $\mathbf{D}_{\infty h}$.

the field intensity. If Λ represents a field-dependent mixing parameter:

$$^1\text{``II''} = (^1\text{II} + \Lambda^1\text{I})/\sqrt{1+\Lambda^2} \;\; ; \;\; ^1\text{``I''} = (-\Lambda^1\text{II} + ^1\text{I})/\sqrt{1+\Lambda^2} \tag{3.11}$$

The similarity between Equations 3.11 and 2.6, and the resemblance of Fig. 3.6 to the left side of Fig. 2.9 should not obscure the very real difference between them. The phenomenon that was described in Section 2.5 is the interaction between two orbitals – specifically, the hybridization of two AOs centered on the same nucleus – under the influence of an external field. The formally similar Equations 3.1–3.9 describe the interaction of AOs centered on two different atoms in response to the simultaneous attraction of one electron to two nuclei, resulting in the orbital correlations illustrated in Fig. 3.5. In contrast to both of the above, Fig. 3.6 is a state correlation diagram, in which two electron configurations of the same symmetry species (in $\mathbf{D}_{2h}$) undergo *configuration interaction (CI)* as a result of the correlated motion of the electrons[17]. Neither "I" nor "II" is a *pure* configuration; the inclusion of some I in "II" stabilizes it, whereas the contribution of II to "I" is destabilizing. The interaction is possible in the quadrupolar field because:

$$\Gamma(\text{I}) \otimes \Gamma(field) \otimes \Gamma(\text{II}) = \Gamma_I \; . \tag{3.12}$$

As in the case of the analogous Equation 2.9, Equation 3.12 can be regarded in either of two ways: In $\mathbf{D}_{2h}$, $\Gamma(\text{I})$, $\Gamma(field)$ and $\Gamma(\text{II})$ are all totally symmetric (A_g), and so is their triple direct product. Alternatively, Equation 3.12 is applied in $\mathbf{D}_{\infty h}$, the point group of the unperturbed molecule, in which neither $\Gamma(field)$ nor $\Gamma(\text{II})$ is totally symmetric; both have Δ_g symmetry, but their product fulfils Equation 3.12 nonetheless[18]. The latter interpretation is less restrictive, because it does not assume a particular perturbing field *a priori*, but rather deduces the irrep of that symmetry-breaking perturbation which is capable of inducing two originally non-interacting states to interact.

3.3 Heteronuclear Diatomic Molecules

The axially symmetric ($\mathbf{C}_{\infty v}$) heteronuclear molecule, CO, has the same united atom, silicon, as the cylindrically symmetric ($\mathbf{D}_{2h}$) homonuclear molecule N_2. A comparison of its orbital correlation diagram (Fig. 3.7) with that of N_2 (Fig. 3.5) is instructive. As before, we split the degeneracy between x and y by assuming the presence of a hypothetical quadrupolar field perpendicular to the z axis. Rotational symmetry about z – the only axis present – is reduced from C_∞ to C_2; the inversion center and $\sigma(xy)$ are absent, leaving E, $C_2(z)$, $\sigma(zx)$ and $\sigma(yz)$.

[17] Configurations with the same irrep are mixed in the absence of an external field by coulomb repulsion between electrons; it thus takes into account electron correlation, which is ignored in the simple MO picture

[18] Direct products of degenerate representations comprise more than one irrep. For example: $\Delta_g \otimes \Delta_g = \Sigma_g^+ \oplus \Sigma_g^- \oplus \Delta_g$, so the "equals" sign (=) in Equation 3.12 has to be replaced by "includes" ($\ni$)

The symmetry point group of the molecule in the quadrupolar field is thus $\mathbf{C}_{2v}$, which bears the same relation to $\mathbf{C}_{\infty v}$ as $\mathbf{D}_{2h}$ does to $\mathbf{D}_{\infty h}$. The reduction of symmetry from $\mathbf{D}_{\infty h}$ to $\mathbf{C}_{\infty v}$ – or from $\mathbf{D}_{2h}$ to $\mathbf{C}_{2v}$ – is a result of a dipolar field along the z axis. The greater electronegativity of oxygen than carbon is the natural equivalent of the artificial dipolar field introduced in Section 2.2.6, the effect of which on an isolated atom was illustrated in Fig. 2.9.

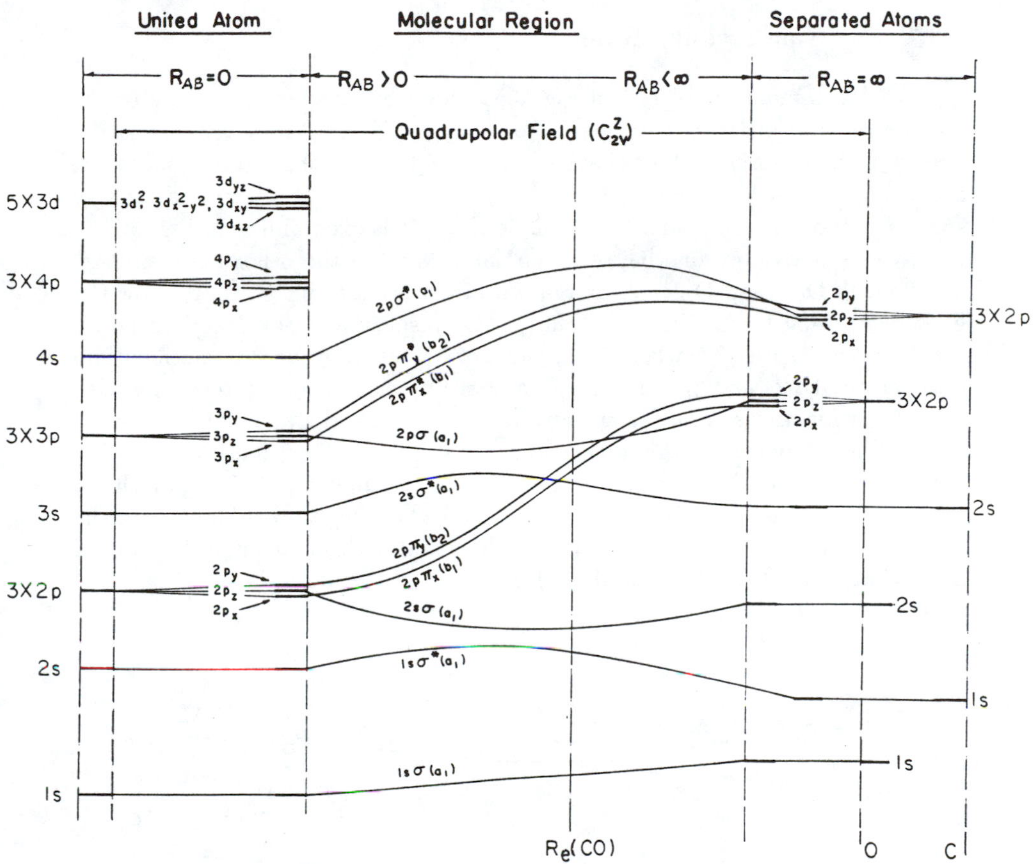

Figure 3.7. Schematic orbital correlation diagram of CO

The orbital levels of the oxygen atom on the right of Fig. 3.7 are set lower than the corresponding ones for carbon, in deference to the former's higher nuclear charge. The distinction between σ and π remains, as does that between π_x and π_y, but that between g and u is lost. The most grievous loss of all is that of the horizontal mirror plane, $\sigma(xy)$, which establishes a nodal plane in the σ_u and π_g orbitals of homonuclear diatomic molecules, and thus compels them to be antibonding. In contrast, each antibonding orbital of a heteronuclear diatomic molecule has the same irrep as its bonding counterpart, and so has to be distinguished from it by an asterisk. This symbol indicates the presence of a nodal surface perpendicular to the internuclear axis, that affects the orbital energy strongly, but is unrelated to orbital symmetry. As a consequence, an antibonding σ^* orbital can correlate with either an s or a p_z AO of the united atom, and a π^* orbital is not constrained to go to an atomic d orbital, but can correlate with p_x (or p_y) instead, if such an AO happens to be available at a lower energy.

3.3.1 The Non-Crossing Rule

Fig. 3.7 looks much simpler than Fig. 3.5. The orbitals span only three of the four representations of $\mathbf{C}_{2v}$ compared to six of $\mathbf{D}_{2h}$'s eight, so fewer correlation lines cross one another. This behavior is a consequence of the *non-crossing rule*, according to which correlation lines between orbitals of the same symmetry species cannot cross one another [4, p. 66 ff.][19]. It is exemplified in Fig. 3.7 by two avoided crossings, one between $1s\sigma^*$ and $2s\sigma$ and the other between $2s\sigma^*$ and $2p\sigma$. In $\mathbf{D}_{2h}$ (and $\mathbf{D}_{\infty h}$) the g and u orbitals cannot mix, so they follow the lines connecting them with the s and p AOs respectively of the united atom. At the point in Fig. 3.5 where $1s\sigma_u$ and $2s\sigma_g$ cross, they become accidentally degenerate like $2s$ and $2p_x$ in Fig. 2.5, and are prevented from interacting with each other for precisely the same reason. Unlike them, the analogous $1s\sigma^*$ and $2s\sigma$ in Fig. 3.7 belong to the same symmetry species in $\mathbf{C}_{2v}$ (and $\mathbf{C}_{\infty v}$) and are obliged to interact, like $2s$ and $2p_x$ in Fig. 2.9. Reading Fig. 3.7 from right to left, $1s\sigma^*$ and $2s\sigma$ are seen to approach one another, but they do not cross; the correlation lines appear to repel one another, and each proceeds to the united atom that seemed to be the original destination of the other.

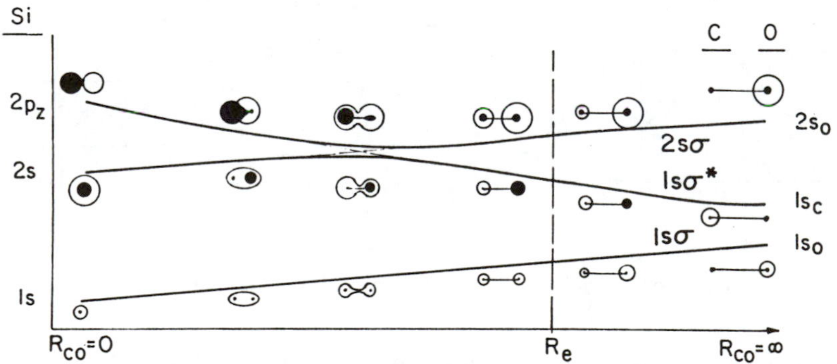

Figure 3.8. The avoided crossing between $1s\sigma^*$ and $2s\sigma$

A perhaps overfanciful pictorial representation of this avoided crossing is portrayed in Fig. 3.8. If $1s\sigma^*$ and $2s\sigma$ were of different symmetry species, they

[19] The non-crossing rule was originally deduced for correlation lines between *states* with the same irrep. However, as explained in the preceding section, the formalism governing the interaction of orbitals and of configurations is fully analogous, so the non-crossing rule can be applied to either type of correlation diagram, as the occasion demands.

would have continued along the dashed lines and crossed. Instead, well before the intended crossing, they begin to mix, the lower orbital accepting a stabilizing admixture of the upper and becoming less antibonding. The upper, by default, has to incorporate more and more of $-1s\sigma^*$ ($1s\sigma^*$ in its negative phase), becoming increasingly less bonding and eventually antibonding. In the region of R_e, $1s\sigma^*$ and $2s\sigma$ are still far apart. The positive and negative lobes are well separated, not by a plane of symmetry but by a curved nodal surface; the same is true of the other σ^* orbitals. The π^* orbitals too have much higher energies than the corresponding π orbitals, thanks to the nodal surface in each. Thus, though π and π^* have the same symmetry label, there is no doubt about which is bonding and which is antibonding. Of course, π and σ orbitals cannot mix in the axial potential of CO any more than in the cylindrical potential of N_2, so the π and σ correlation lines cross each other with impunity in Fig. 3.7, just as they do in Fig. 3.5.

Table 3.2. Correlation Table: $\mathbf{D}_{\infty h} \Longleftrightarrow \mathbf{C}_{\infty v} \Longleftrightarrow \mathbf{C}_{2v}$

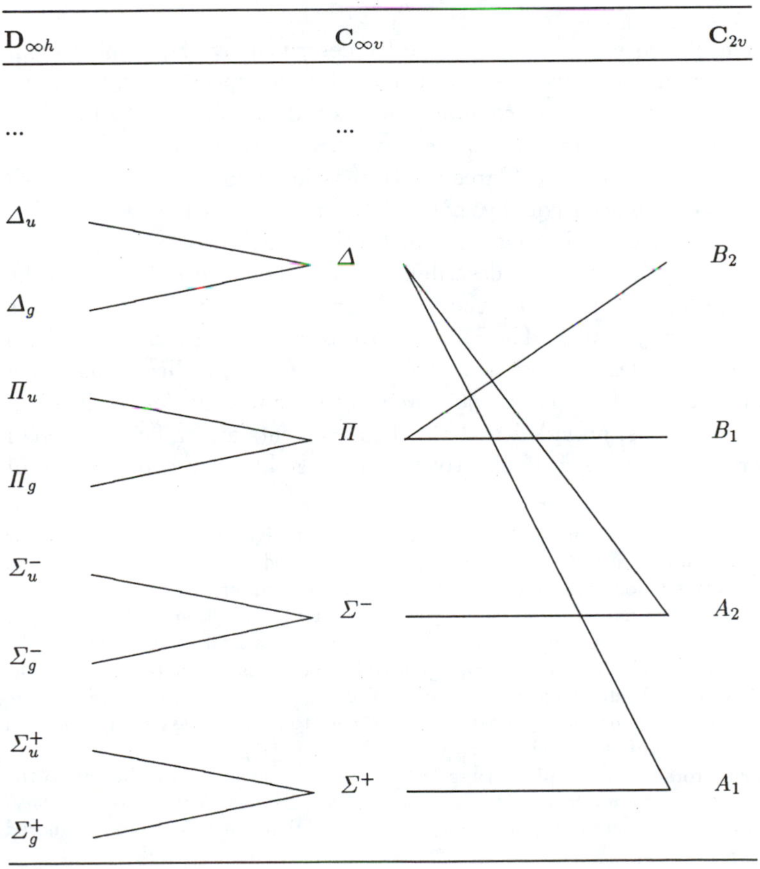

$\mathbf{D}_{\infty h}$	$\mathbf{C}_{\infty v}$	$\mathbf{C}_{2v}$
...	...	
Δ_u		
	Δ	B_2
Δ_g		
Π_u		
	Π	B_1
Π_g		
Σ_u^-		
	Σ^-	A_2
Σ_g^-		
Σ_u^+		
	Σ^+	A_1
Σ_g^+		

3.3.2 Configuration and State Correlation

The state symmetries of heteronuclear diatomic molecules are simpler than those of homonuclear diatomics; the g and u states of $\mathbf{D}_{\infty h}$ simply coalesce in $\mathbf{C}_{\infty v}$. Table 3.2 summarizes the correlations of the irreps of $\mathbf{D}_{\infty h}$ with those of its subgroup $\mathbf{C}_{\infty v}$ and between the latter and those of $\mathbf{C}_{2v}$[20].

The ground-state configuration of CO can be read off from Fig. 3.7 to be $[2s\sigma^2\ 2s\sigma^{*2}\ 2p\pi^4\ 2p\sigma^2]$, so it can only be $^1\Sigma^+$. Its first excited configuration, $[2s\sigma^2\ 2s\sigma^{*2}\ 2p\pi^4\ 2p\sigma\ 2p\pi^*]$, gives rise to $^3\Pi$ and above it to $^1\Pi$, each of which splits into a B_1 and a B_2 component in $\mathbf{C}_{2v}^z$. The $\pi \to \pi^*$ states are easily confirmed to be Σ^+, Σ^- and Δ, a singlet and a triplet of each.

3.4 Symmetry Coordinates

3.4.1 Homonuclear Diatomics

An X_2 molecule has been set up in Fig. 3.9, where – as has been our wont – the x and y directions are fixed with a hypothetical quadrupolar field, so $\mathbf{D}_{2h}$ labels can be used.

The motion of two nuclei in space is fully described by the displacement of each of them along the three cartesian axes. Having specified the symmetry of the system to be $\mathbf{D}_{2h}$, we combine these six displacements to form six linearly independent *symmetry coordinates*, each transforming as one of the irreducible representations of $\mathbf{D}_{2h}$. Three are translational coordinates, in which the two atoms are displaced equally, parallel to one of the cartesian axes, so the original XX distance and orientation of the internuclear axis in space are retained. When the pairs of arrows describing each of them are subjected to the symmetry operations of the group, they are seen to fall into the three irreps which characterize x, y, and z. The atoms move parallel to the cartesian axes in the other three modes of motion as well, but in opposite directions. Such motions parallel to x and y – i.e. at right angles to the internuclear axis – can be regarded as rotations, provided that the displacements are sufficiently small that the internuclear distance is effectively unchanged.[21] The combination of

[20] There is a trivial but sometimes bothersome notational ambiguity associated with $\mathbf{C}_{2v}$: B_1 and B_2 can be confused if the axis convention is not specified clearly. When the twofold rotational axis is aligned along z, as was done throughout this chapter, the conventional $\mathbf{C}_{2v}$ Character Table in Appendix A lists B_1 and B_2 as symmetric to reflection in the zx and yz planes respectively. When the C_2 axis lies along x or y, these two irreps will necessarily be defined differently (cf. Table 2.3). A convenient mnemonic device is to list the labels of the coordinates (i, j, k) in cyclically permuted alphabetical order ($ijk = xyz, zxy$ or yzx); then, if the C_2 axis lies along i, functions symmetric with respect to $\sigma(ij)$ are assigned to b_1, and those symmetric to $\sigma(ik)$ are assigned to b_2.

[21] As the molecule rotates, the displacements are assumed to remain perpendicular to the internuclear axis. Inertia does in fact tend to increase the XX distance, leading to *centrifugal distortion* and thus to interaction between rotation and vibration; this effect can be ignored in the present context.

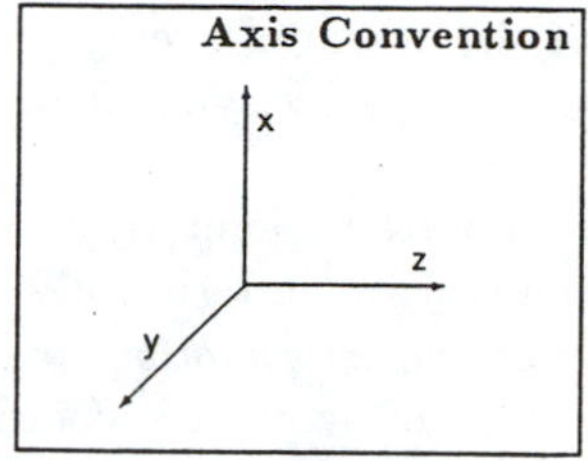

Symmetry Coordinates:

Translations:	X—X	X—X	X—X
	$T_x(b_{3u})$	$T_y(b_{2u})$	$T_z(b_{1u})$
Rotations	X—X	X—X	X—X
and Vibration:	$R_y(b_{2g})$	$R_x(b_{3g})$	$\xi(a_g)$

Example of Composite Motion:

X—X

$$b_{1u} \oplus a_g \oplus b_{3u} \oplus b_{2g}$$

X—X		X—X		X—X		X—X
$T_z(b_{1u})$	$\oplus$	$\xi(a_g)$	$\oplus$	$T_x(b_{3u})$	$\oplus$	$[-R_y](b_{2g})$

Figure 3.9. Symmetry coordinates of a homonuclear diatomic molecule ($\mathbf{D}_{2h}$) (Consult Table 3.1 to obtain the $\mathbf{D}_{\infty h}$ labels)

displacements along x that produces rotation about y is easily seen to have the irrep b_{2g} and rotation about x to transform as b_{3g}.

In contrast to the first five symmetry coordinates, the sixth (ξ) represents an increase or – in the opposite phase – a decrease of the internuclear distance and a consequent change in the potential energy. At the equilibrium internuclear distance, R_e, the simultaneous attraction of the electrons for both nuclei is just balanced by the internuclear (and interelectronic) repulsion. Therefore, displacement along ξ, which changes R from its optimum value, is opposed by a restoring force, causing the bond length to oscillate about R_e. The two arrows that represent motion along ξ transform under the sym-ops of $\mathbf{D}_{2h}$ (or of $\mathbf{D}_{\infty h}$) either into themselves or into one another, so the molecule retains its full symmetry, and the potential energy – which varies periodically as the molecule vibrates – remains totally symmetric. This is the first point that Fig. 3.9 is intended to make.

A no less important point to note is that the six symmetry coordinates comprise a *complete set*, in terms of which any arbitrary molecular motion can be described. A *composite motion* like the one in Fig. 3.9 can be constructed by a superposition of symmetry coordinates with suitably chosen phase and amplitude, and is therefore assigned to a *reducible representation*, the *direct sum* of its component irreps. It is easy to see that the motion of a single atom also belongs to a reducible representation: Displacement of the left-hand X atom parallel to x is clearly a superposition of T_x and the negative phase of R_y, so it belongs to $b_{3u} \oplus b_{2g}$, whereas that of the right-hand atom along z, composed of T_z and ξ, transforms as $b_{1u} \oplus a_g$

3.4.2 Heteronuclear Diatomics

When the atoms are not identical – and it is immaterial whether the dissymmetry is substantial, as in HCl, or minimal, as in $^{35}Cl^{37}Cl$ – the symmetry of the diatomic molecule is $\mathbf{C}^z_{\infty v}$, reduced in our hypothetical quadrupolar field to $\mathbf{C}^z_{2v}$. Their symmetry labels are obtainable from those in Fig. 3.9 by inspection, by comparing Tables 3.1 and 3.2, or by consulting the *Correlation Table* for $\mathbf{D}_{2h}$ in Appendix C. Most simply, we consult the Character Table of $\mathbf{C}^z_{2v}$: The three translations transform as a_1, b_1 and b_2 and the rotations about x and y as b_2 and b_1 respectively. There is no molecular rotation about z in a linear molecule, where the nuclei lie on the internuclear axis;[22] as in the case of the homonuclear molecule, the vibration is necessarily totally symmetric (a_1), because the molecular symmetry remains $\mathbf{C}^z_{2v}$.

3.5 References

[1] References [1–7] of Chapter 2 all go on to deal with diatomic molecules.
[2] R.S. Mulliken: Rev. Mod. Physics *4*, 1 (1932).
[3] J.H Van Vleck and A. Sherman: Rev. Mod. Physics *7*, 167 (1935), cited by Coulson [4, p. 93].
[4] C.A. Coulson: *Valence.* Oxford Univ. Press, Oxford 1952.
[5] E. Wigner and E.E. Witmer: Z. Physik *51*, 859 (1928).
[6] A.C. Hurley: Chem. Phys. Letters *91*, 163 (1982).
[7] J.W. Calvert and J.N. Pitts, Jr.: *Photochemistry.* Wiley, New York 1966.
[8] N.J. Turro: *Modern Molecular Photochemistry.* Benjamin-Cummings, Menlo Park 1978.
[9] J. Katriel and R. Pauncz: Adv. Quantum Chem. *10*, 143 (1977).
[10] G. Herzberg: *Spectra of Diatomic Molecules.* Van Nostrand, Princeton, NJ 1950.

[22] One can, of course, imagine inducing rotation of electrons about the molecular axis; this, does not produce molecular rotation, but amounts to excitation to a higher electronic state.

Chapter 4

Formation and Deformation of Polyatomic Molecules

Diatomic molecules are necessarily linear, but a triatomic molecule can be either linear like CO_2 and HCN or bent like SO_2 and H_2O. Mulliken's correlation diagram procedure was extended to tri- and tetraatomic molecules by Walsh [1], who promulgated a set of simple but remarkably viable [2, 3, 4] rules for predicting whether or not a molecule will remain linear, from the effect of the departure from linearity on the energy of its occupied molecular orbitals.

For our present purpose, which is to consider ways of following a transition from one stable molecular system – the reactant(s), to another having the same atomic composition – the product(s), it will generally be sufficient to know that N_2O, for example, is linear and O_3 is bent, without delving too deeply into the whys and wherefores thereof. The discussion of *Walsh Diagrams* in this Chapter will therefore be limited to two simple linear systems, one tri- and the other tetra-atomic. These will suffice for our exploration of the interrelations among orbital symmetry, molecular geometry and energy, in which vibrational deformation and chemical reaction will be regarded to be two closely related means of symmetry breaking. The extension to inherently non-linear molecules will follow naturally.

4.1 Triatomic Molecules

The formation of an HXH molecule, in which X is an atom of a second row element like carbon or oxygen, can be thought of in a variety of ways: The attachment of a hydrogen atom to an XH radical, of a proton to XH^- or of a hydride ion to XH^+; the addition of two H atoms to X; the insertion of X into an H_2 molecule, *etc*... Adopting the molecular orbital viewpoint, we begin by conceptually bringing two protons up to an X^{N+} ion, forming the skeleton of XH_2, and then adding $N+2$ electrons to the nuclear frame one at a time, successively filling its MOs. N is conventionally taken to be the nuclear charge (Z) of X less the number of inner shell electrons. For second-row atoms, with which we will be mostly concerned, $N = Z - 2$, the two $1s$ electrons having been included with the nucleus in the *atomic core*; the four valence-shell atomic orbitals ($2s$ and $3\times 2p$) comprise the *minimal basis set* needed to represent the contribution of each such atom to the MOs of any molecule which includes it.

4.1.1 Molecular Orbitals and Walsh Diagrams

If the three atoms are placed on a line, with the X nucleus between the two H nucleii, $\mathbf{D}_{\infty h}$ symmetry will have been established. An electron within range of the trinuclear system will distinguish the internuclear z axis from any other direction in space, but will not be able to tell z from $-z$, because it cannot "know" whether it is closer to the H nucleus on the left than to the one on the right. As before, we can communicate the whereabouts of the x and y axes to any electron in the vicinity by imposing, at right angles to z, an external quadrupolar field that is barely strong enough to split the degeneracy between the $2p_x$ and $2p_y$ orbitals of X; $+x$ and and $-x$ are equivalent once more, as are $\pm y$. The $\mathbf{D}_{\infty h}$ symmetry of the potential energy of the electron in the field of the nuclear frame has been artificially reduced to $\mathbf{D}_{2h}$.

Hybridization of $2s$ with $2p_z$ produces h_z^+ and h_z^-, one of which can form a bonding and an antibonding orbital with the $1s$ AO of the H atom towards which it is directed. The resultant bond orbitals are *symmetry-adapted* to $\mathbf{D}_{\infty h}$ or, adequately for our immediate purposes, to $\mathbf{D}_{2h}$.[1] The set of six MOs on the left side of Fig. 4.1 comprise all of those that can be constructed from the valence-shell orbitals of the three atoms: $[2s, 3{\times}2p](\mathrm{X}), 2{\times}1s(\mathrm{H})$. The MOs, appropriately labeled, are stacked in order of increasing energy; we begin with $2\sigma_g$, the totally symmetric bonding orbital, reserving the label $1\sigma_g$ for the omitted inner-shell $1s$ orbital of X.

Following Walsh [1], we rationalize the fact that molecules like H_2O and CH_2 – as well as their analogs H_2S and CF_2 – are non-linear in their ground-states, in terms of the variation of orbital energies with the HXH angle. Avoiding intuitive arguments, which are often obfuscatory and occasionaly misleading, we go directly to the extreme right of Fig. 4.1, where the two XH bonds are at right angles to one another. Within our minimal basis set of AOs, these can only be constructed from two pure $2p$ orbitals, one pointing towards each of the H nuclei, in the $(x, -z)$ and $(-x, -z)$ directions respectively. They evidently form weaker XH bonds than could be formed with more effectively directed hybrid orbitals, so both their symmetric and antisymmetric bonding combinations are higher – and their antibonding counterparts lower – than in the linear system. HH repulsion destabilizes both antisymmetric combinations and increases the separation of each from the corresponding symmetric MO. The $2p_y$ orbital remains non-bonding, so – in our crude qualitative approximation – it is unaffected. $2p_x$ and $2p_y$ have been incorporated fully into the XH orbitals, and – as long as the bond angle is 90° – $2s$ is not involved in XH-bonding. The latter could have been placed above both σ orbitals or below them, rather than between the two; it will soon be clear that the qualitative result would have remained the same.

We now recognize that the symmetry point group of the bent molecule is $\mathbf{C}_{2v}^x$, all four operations of which leave X in place. Two of them, E and $\sigma_v(zx)$, also leave the H atoms in place and the other two, $C_2(x)$ and $\sigma_d(yz)$, exchange

[1] Formally, we are classifying the MOs according to the commutative subgroup $\mathbf{D}_{2h}$ of the molecular symmetry point group $\mathbf{D}_{\infty h}$.

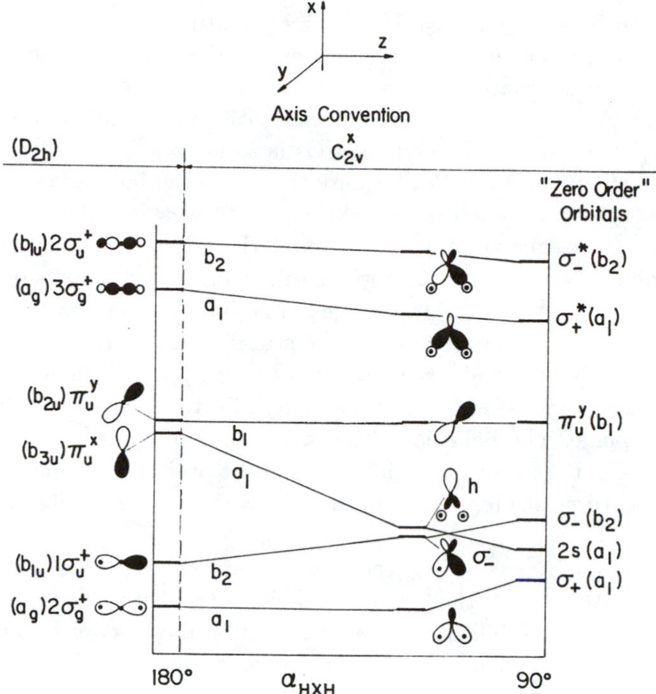

Figure 4.1. Schematic Walsh diagram for HXH molecules

them, but the potential energy of any electron in the vicinity is unaffected by the interchange. Labeling the *zero-order* orbitals with the aid of Table 2.3, we see that the two lowest have the same irrep, a_1, and would be expected to interact. Whether the lower of the two is $2s$ or σ_+, it is stabilized by an admixture of the upper, becoming – in effect – the in-phase combination of two localized sp^2-hybridized XH-bonding orbitals.[2] The same interaction turns the upper into a lone-pair sp^2 hybrid directed along the positive x axis. The latter is destabilized by unfavorable overlap with the XH-bonding orbital but not as much as the lower is stabilized, because the symmetric XH-antibonding orbital (σ_+^*) is mixed into it constructively, pushing it down slightly and being itself raised in the process. A similarly weak interaction occurs between the two well-separated b_2 orbitals.

Turning to the left side of Fig. 4.1, we recall that even the slightest decrease of the angle α_{HXH} below 180° reduces the symmetry from $\mathbf{D}_{\infty h}$ to $\mathbf{C}_{2v}^x$. The nuclear frame suffices to define the plane occupied by the three atoms as zx, so we can dispense with our hypothetical quadrupolar field. The irreps of the orbitals on the left in $\mathbf{C}_{2v}$ are easily identified, and correlation lines are drawn across the diagram, connecting MOs with the same symmetry label while obeying the *non-crossing rule*. Walsh Diagrams are often drawn curved for aesthetic

[2] As a result α_{HXH} becomes larger than 90°, approaching the more realistic value of 120°.

appeal or – with better justification – if the curvature is known to be borne out by computation; when they are based on qualitative arguments of the sort just invoked, straight correlation lines are all that are warranted. In the case of our present example, the most that can be said with confidence about the effect of bending on the energy of the various MOs is as follows:

Bending should stabilize $2\sigma_g$ because there is no nodal plane to prevent HH bonding, but the conversion of sp hybrids to the weaker sp^2 hybrids is destabilizing. In the absence of information about the relative magnitude of the two opposing effects, we leave the energy of this orbital unchanged. Both HH antibonding and the hybridization change destabilize $1\sigma_u$ on bending. The most pronounced change in the diagram is the stabilization of the p orbital (π_u^x) as it aqcuires s character and becomes more like an sp^2 hybrid. In contrast, π_u^y remains a pure p orbital and is not markedly affected by bending the molecule in its nodal plane. The behavior of $3\sigma_g$ and $2\sigma_u$, which are unoccupied in the ground-states of the molecules of interest, can be predicted similarly. The gross qualitative features of Fig. 4.1 are borne out by detailed quantitative investigation. [3]

In the ground-state of H_2O, the four lowest orbitals are doubly-occupied. Fig. 4.1 attributes the fact that the water molecule is bent in its ground- and first excited states to the stabilization of π_u^x, which is doubly-occupied in both states. The most stable closed shell singlet of methylene would also be expected to be bent, with electron configuration $[a_1^2\ b_1^2\ a_1^2]$, as indeed it is ($\alpha_{\mathrm{HCH}} = 105°$). [5] This is not its ground-state, because the first open shell triplet lies ≈10 kcal/mol below it. The conventional explanation, which has been shown to be something of an oversimplification, [6] is that lodging two electrons with parallel spin separately in two MOs of slightly different energy relieves their mutual repulsion sufficiently to outweigh the lower orbital energy of the closed shell singlet. Even so, linear geometry, in which $2p_x$ and $2p_y$ become degenerate, is not achieved. Instead, a compromise between the two effects is reached at the rather wide HCH angle of 136°; the configuration is $[a_1^2\ b_1^2\ a_1\ b_1]$ and the state-label is accordingly $^3B_1$.

4.1.2 Symmetry Coordinates of a Symmetric Triatomic Linear Molecule

The symmetry of nuclear displacements arises most commonly in connection with vibrational spectroscopy of polyatomic molecules [7]. Let us compare the nuclear displacements of a symmetric linear triatomic, XYX, illustrated in Fig. 4.2, with those shown in Fig. 3.9 for a homonuclear diatomic molecule, which also has cylindrical symmetry. It was pointed out that in the latter case there is no way of reducing the symmetry of the potential energy of X_2 below $\mathbf{D}_{\infty h}$ by nuclear motion; in the case of the triatomic molecule, there is.

Since each of its three atoms can move parallel to any one of the three coordinate axes, we can construct nine symmetry coordinates. Assuming the molecule to be linear, and artificially reducing its symmetry from $\mathbf{D}_{\infty h}$ to $\mathbf{D}_{2h}$ as

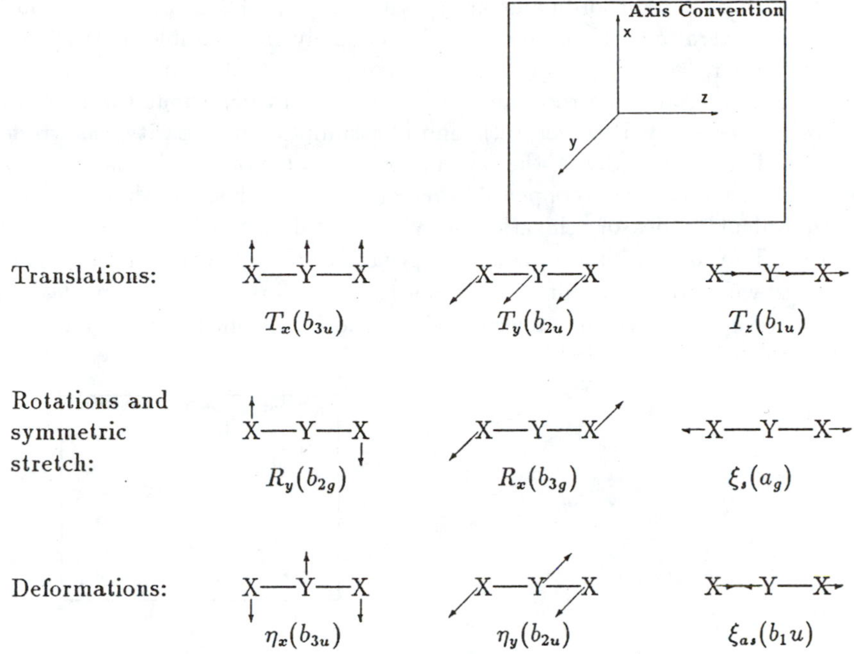

Figure 4.2. Symmetry coordinates of a linear XYX molecule ($\mathbf{D}_{2h}$). (Consult Table 3.1 to obtain the $\mathbf{D}_{\infty h}$ labels)

before, we construct and characterize them in Fig. 4.2. The three translations, in which all of the atoms are displaced to an equal extent parallel to one of the axes, again transform like the cartesian coordinates. The two rotational coordinates and the totally symmetric stretching coordinate (ξ_s) are also fully analogous to those of X_2 because the central Y atom remains in place in all three, and the two X atoms undergo equal and opposite displacements, just as in Fig. 3.9. Like ξ_s, the three new symmetry coordinates all involve changes in the internuclear distances and thus affect the potential energy. In them, the two equivalent X atoms are displaced equally in one direction while the central Y atom moves in the other. Two, η_x and η_y, are a degenerate pair of bending coordinates in $\mathbf{D}_{\infty h}$, that split in $\mathbf{D}_{2h}$ to b_{3u} and b_{2u}. The third (ξ_{as}) is an antisymmetric stretching coordinate, σ_u^+ in $\mathbf{D}_{\infty h}$ and b_{1u} in $\mathbf{D}_{2h}$.

As its suffix implies, ξ_s leaves the symmetry of the molecule unchanged, but motion along any of the other three vibrational symmetry coordinates reduces the symmetry of the molecule to the subgroup that is the kernel of its representation. Thus[3]:

$$\eta_x(b_{3u}) \Rightarrow \mathbf{C}_{2v}^x \;\; ; \;\; \eta_y(b_{2u}) \Rightarrow \mathbf{C}_{2v}^y \;\; ; \;\; \xi_{as}(b_{1u}) \Rightarrow \mathbf{C}_{2v}^z$$

[3] ξ_{as} does not destroy the axial symmetry of the molecule, which remains in $\mathbf{C}_{\infty v}$, represented here by $\mathbf{C}_{2v}$ thanks to our hypothetical quadrupolar field. The bending modes genuinely do reduce the symmetry all the way down from $\mathbf{D}_{\infty h}$ to $\mathbf{C}_{2v}$.

Note that these three coordinates are specified in Fig. 4.2 to be *deformations* rather than *vibrations*. If the molecule is genuinely more stable in the linear than in the bent geometry, as is CO_2 for example, each of them is opposed by a restoring force, and the resultant motion is a vibration about the common equilibrium geometry. However, if the initial assumption of linearity was wrong, because – like H_2O or SO_2 – the presumptive linear molecule is stabilized by bending, neither η_x nor η_y is opposed by a restoring force; instead, they represent two equivalent modes of relaxation to a more stable, albeit less symmetric, structure. The convention commonly adopted is to align linear molecules along the z-axis; whenever the assumption of linearity is incorrect, dispacement along either η_x or η_y will take the molecule to its true $\mathbf{C}_{2v}$ geometry, with x or y as the twofold rotational axis.[4]

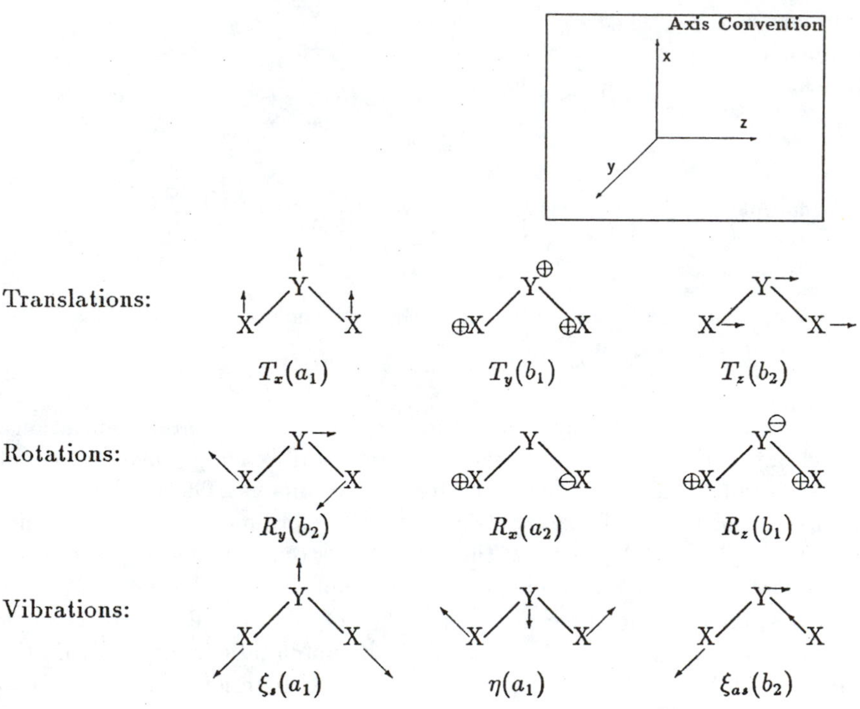

Figure 4.3. Symmetry coordinates of a non-linear XYX molecule ($\mathbf{C}_{2v}^x$). ($\oplus$ and $\ominus$ represent motion along $+y$ and $-y$ respectively)

4.1.3 Symmetry Coordinates of a Symmetric Non-Linear Triatomic

The symmetry coordinates of the non-linear molecule, relabelled with the aid of Table 2.3, are shown in Fig. 4.3. It has been produced by bending the linear

[4] Any superposition of both will do the same, putting the molecule into a plane intermediate between yz and zx.

molecule along η_x into $\mathbf{C}_{2v}^x$. A new rotational coordinate has appeared; linear molecule has been replaced by rotation about z , which is no longer the common internuclear axis. Two of the vibrational coordinates are totally symmetric in $\mathbf{C}_{2v}^x$; the third describes an in-plane vibration, throughout which the molecule remains symmetric with respect to $\sigma(zx)$ but loses both $C_2(x)$ and $\sigma(xy)$, regaining them momentarily as it passes back and forth through the symmetrical geometry. None of the three vibrational coordinates retains $C_2(x)$ or $\sigma(xy)$ alone: the former is retained in R_x whereas T_y and R_z preserve the latter; but translations and rotations, which do not change internuclear distances, have no effect on the potential energy.[5]

4.2 Linear Tetraatomics and Their Deformation

For the tetraatomic system HXXH, representing both the linear acetylene and the non-linear hydrogen peroxide, we expect to be able to construct twelve symmetry coordinates. Three of them are translational, whereas two of the remaining nine in the linear conformation and three in the non-linear one are reserved for rotations. Linear tetraatomics thus have seven vibrational coordinates, motion along which changes the potential energy, whereas their non-linear counterparts have six. Those of the linear HXXH molecule are shown in Fig. 4.4 with the subgroup into which each is taken, if only momentarily, by the displacement.

Displacing the atoms along the symmetric XH- and XX-stretching coordinates does not reduce the symmetry of the molecule,[6] Like its counterpart in Fig. 4.3, ξ_{as} reduces the molecular symmetry to $\mathbf{C}_{2v}$ (to $\mathbf{C}_{\infty v}$ if the hypothetical quadrupolar field is not imposed), in which z is the unique rotational axis. One pair of symmetric bending coordinates, labelled π_u in cylindrical symmetry, again takes the system into two equivalent, differently oriented, *cis* conformations. The principal new feature, which is absent in triatomic molecules, is the appearance of two additional bending coordinates; these too are degenerate in cylindrical symmetry, but have the irrep π_g rather than π_u. The symmetry elements retained in each of the two ensuing *trans* conformations are easily confirmed with the aid of Table 2.2 to be the identity (E), a center of inversion (i), one twofold rotational axis ($C_2(x)$ in b_{3g} and $C_2(y)$ in b_{2g}) and the mirror plane perpendicular to it ($\sigma(yz)$ or $\sigma(zx)$). The four sym-ops retained constitute either $\mathbf{C}_{2h}^x$ or $\mathbf{C}_{2h}^y$, the respective kernels of b_{3g} and b_{2g}.[7]

---

[5] In principle, the nine symmetry coodinates could here too have been constructed from nuclear displacements parallel to the cartesian axes. They would still span the same irreps: $[3\times a_1 \oplus a_2 \oplus 2\times b_1 \oplus 3\times b_2]$, but would no longer be separated into translational, rotational, and vibrational symmetry coordinates.

[6] The kernel of the totally symmetric representation of a group is the full group, which is one its two *trivial subgroups*, the other consisting only of the identity operation.

[7] Considered as subgroups of $\mathbf{D}_{\infty h}$, $\mathbf{C}_{2h}^x$ and $\mathbf{C}_{2h}^y$ are the *co-kernels* of the two orthogonal components of Π_g; its kernel is $\mathbf{C}_i$, the subgroup comprising the two sym-ops common to both: E and i (Appendix B).

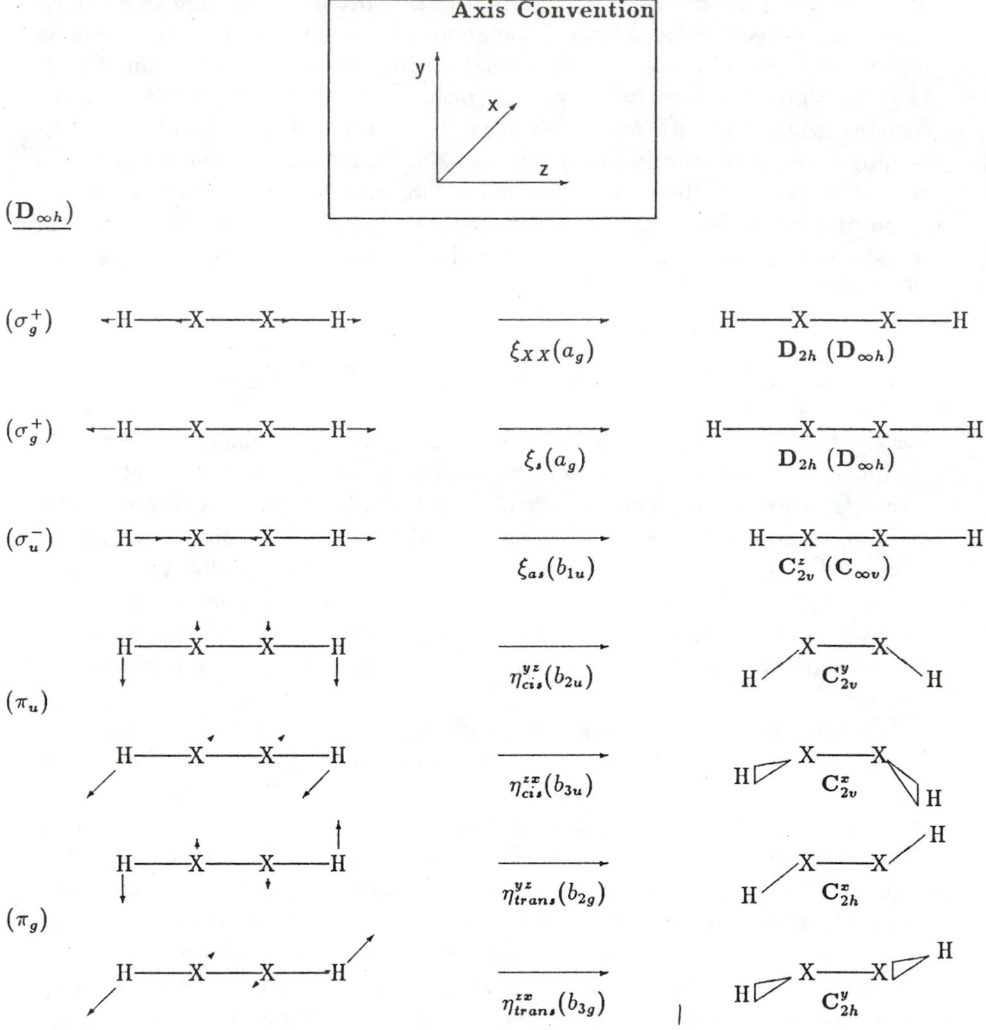

Figure 4.4. Symmetry coordinates of a linear HXXH molecule and the ensuing subgroups of $\mathbf{D}_{2h}$ ($\mathbf{D}_{\infty h}$)

In order to determine whether HXXH will remain a linear molecule or, if not, whether it will relax preferentially to assume a *cis* or a *trans* geometry, two Walsh Diagrams must be constructed. First, however, the valence-shell bond orbitals have to be set up and symmetry-adapted, and the resulting combinations with the same symmetry label allowed to interact, like the *zero-order* orbitals of bent HXH in Fig. 4.1. The ordering of the symmetry-adapted bond orbitals on the left side of Fig. 4.5 is almost self-evident: XH bonds are stronger than XX σ bonds, which are stronger than π bonds. The corresponding antibonding orbitals are stacked in reverse order, because the more strongly bonding a given

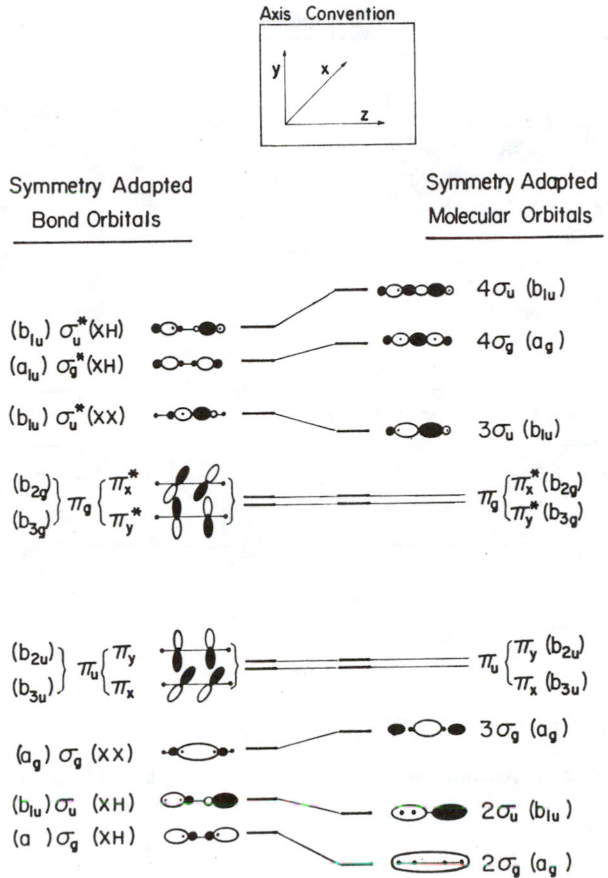

Figure 4.5. Valence-shell molecular orbitals of linear HXXH

MO is, the more highly destabilized is its antibonding counterpart. The symmetric member of any pair of symmetry-adapted combinations is placed below its antisymmetric partner, which is bisected by a nodal plane. As usual, the MOs are labelled according to their irreps in both $\mathbf{D}_{2h}$ and $\mathbf{D}_{\infty h}$.

The main interactions that have to be considered before the molecule is distorted away from linearity are between the two totally symmetric bonding σ orbitals, σ_g(XH) and σ_g(XX), and between the the two corresponding antibonding orbitals, σ_u^*(XH) and σ_u^*(XX). Next, we recognize that σ_g(XX) can also interact with σ_g^*(XH) as can σ_u(XH) with σ_u^*(XX). The latter two interactions are less effective than the former two because the interacting localized orbitals are farther apart in energy. The resultant σ MOs are no longer *pure* XH or XX orbitals, nor – except for the lowest and the highest – can they be firmly categorized as bonding or antibonding. They are therefore simply labeled σ_g or σ_u and numbered in order of increasing energy, the labels $1\sigma_g$ and $1\sigma_u$ having been appropriated by the two combinations of the $1s$ AOs of X, which were left

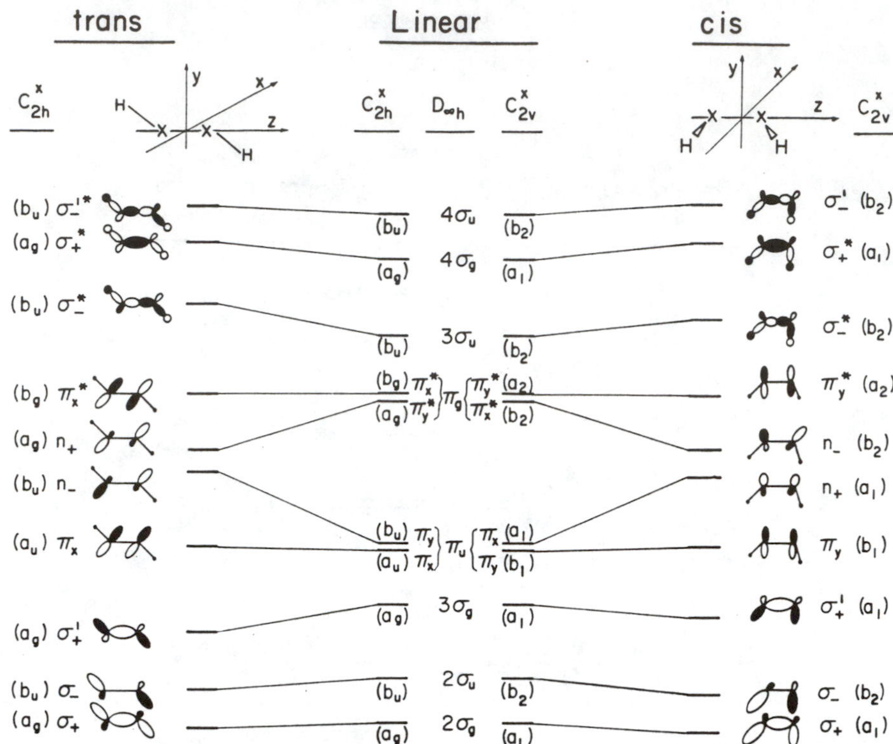

Figure 4.6. Walsh diagrams for *cis*- and *trans*-bending of HXXH

out of the diagram. We note further that each σ_g orbital has an even number (0, 2 or 4) and each σ_u orbital an odd number (1, 3 or 5) of nodal surfaces; in the latter case, one such surface is necessarily the bisecting mirror plane, $\sigma(xy)$. The localized π and π^* orbitals interact with no others, so they constitute MOs just as they are.

The molecule is now bent out of linearity in two different ways, along either η_{cis}^{zx} or η_{trans}^{yz} of Fig. 4.4, retaining $C_2(x)$ as the rotational axis in both cases. The first deformation takes the molecule to a *cis* conformation, with the molecular plane at right angles to that of the diagram; the second takes it to a *trans* conformation in the plane of the diagram. The MOs of linear HXXH have been transferred from Fig. 4.5 to the center of Fig. 4.6, and the Walsh Diagrams for its deformation to *cis* and *trans* appear to its right and left respectively. The symmetry labels of the MOs in $\mathbf{C}_{2v}^x$ can be read from Table 2.3; those for $\mathbf{C}_{2h}^x$ can be confirmed by consulting the Character Table for $\mathbf{C}_{2h}^z$ in Appendix A, taking note of the need to interchange x and z wherever they appear.

The most important effect of bending the XH bonds away from linearity is to convert one π orbital and its antibonding partner into combinations of lone-pair hybrid orbitals, the former going up in energy and the latter going down. In addition, the consequent desymmetrization allows each of these two orbitals to interact with the σ MOs that have the same irrep in the resultant subgroup, pushing down the one immediately beneath it and pushing up the one lying just above it. Since the interaction is stronger between orbitals that are close in energy than between those that are farther apart, the lowering of $3\sigma_g$ by π_y^* on the left of Fig. 4.6 should be more pronounced than that of $2\sigma_u$ by π_x^* on the right. Similarly, $3\sigma_u$ should be raised higher by π_y on the left than $4\sigma_g$ by π_x^* on the right. Numerous lesser interactions have to be taken into account in any quantitative assessment of the preferred mode of symmetry breaking [3], but we have gone as far as is profitable at the level of qualitative discourse.[8]

The ground-state configuration of acetylene is $[2\sigma_g^2\, 2\sigma_u^2\, 3\sigma_g^2\, \pi_u^4]$, so the destabilization which either π_x or π_y suffers on bending is sufficient inducement for it to remain linear. Going on to H_2O_2, we note that if it were linear in the ground-state, its configuration would be $[2\sigma_g^2\, 2\sigma_u^2\, 3\sigma_g^2\, \pi_u^4\, \pi_g^4]$. Bending destabilizes one π-orbital and stabilizes its antibonding partner, which is also doubly occupied, so the preference of H_2O_2 for a bent conformation must be ascribed to stabilization of the occupied σ MOs. Of these, $3\sigma_g$ is closest in energy to the π orbitals, and – being essentially an OO-bonding orbital – should interact strongly with π_y (or π_x) once bending has removed the orthogonality between them. Inspection of Fig. 4.6 leads to the prediction that H_2O_2 should be bent *trans* rather than *cis*. Experiment bears this expectation out, the molecule actually taking up a *gauche* conformation, retaining C_2 as its only non-trivial symmetry element, in which it is slightly more stable than in the exact *trans* geometry. [9]

To summarize our brief discussion of Walsh Diagrams, let us put the concepts and methods on which we have been relying into perspective. It is gratifying to be able to obtain all of the MOs of a molecule that can be constructed from its minimal basis set of AOs, to rank them in terms of energy, and to be able to make an educated guess about its preferred ground-state geometry. However, we need only attempt to predict the conformations of the lowest excited singlet and triplet of acetylene on the basis of Fig. 4.6 and then check them against the results of experiment or of detailed computation [3], to realize that qualitative arguments should not be relied upon too heavily:

1. One π^*-orbital goes down strongly in each bending mode, as expected, but its bonding partner goes up very little – if at all.
2. Neither the singlet nor the triplet is derived from a $\pi_x \rightarrow \pi_x^*$ or $\pi_y \rightarrow \pi_y^*$ excitation, as one might suppose, but rather from $\pi_x \rightarrow \pi_y^*$ or $\pi_y \rightarrow \pi_x^*$.
3. The excited states of both spin states are bent, as expected, but take up different conformations: the singlet is indeed *trans* but the triplet is *cis*.

Despite their obvious quantitative deficiencies, the ideas developed in this section are instructive, and the methods based on them are useful – in their proper qualitative context:

[8] For example, one might argue that *through-space* interaction should stabilize n_+ in the *cis* conformation but not in the *trans*; the former, however, is destabilized by *through-bond* interaction with σ'_+ [8]. There is little to be gained by qualitative speculation about the quantitative balance between two opposing effects, so we leave the separation of n_+ and n_- the same on both sides of Fig. 4.6.

1. They provide a means of specifying just how a particular nuclear displacement reduces symmetry below that of the original symmetry point group.

2. They allow the identification of those MOs on the same side of a correlation diagram which – although they have different symmetry labels in the point group of the undistorted molecule – map onto the same irrep of the subgroup that characterizes the distorted molecule, and thus become newly able to interact.

3. As will be seen in the following section, they allow the identification of MOs situated on opposite sides of a correlation diagram that cannot be brought into correlation with one another unless the symmetry is reduced, and make it possible to specify the irrep of the distortion that takes the system into a subgroup in which they can correlate.

4.3 Dimerization of Methylene and Its Reversal

4.3.1 The Dimerization of Methylene

It is a short step from acetylene to ethylene, but instead of constructing a correlation diagram for addition of H_2 across the acetylenic triple bond, let us consider the dimerization of two coplanar methylene molecules. For consistency with the axis convention of Figs. 4.1–4.4, the methylene molecules on the the left side of Fig. 4.7 are placed in in the zx plane and allowed to approach each other along their common C_2 axis (x), leaving the p_y orbitals free for π bonding. The symmetry of the field exerted on the electrons by the nuclear frame is $\mathbf{D}_{2h}$; no hypothetical external field need be postulated.

The four combinations of the CH-bonding orbitals at the bottom of the diagram are of different symmetry species, each of which also characterizes one of the CH-antibonding orbitals at the top. The symmetric and antisymmetric combinations of sp^2 hybrids, which are degenerate at infinite separation, become the strongly bonding σ and antibonding σ^* orbitals of ethylene . As the bond is formed, each of these interacts with the CH orbitals that have the same irrep: the former mixes strongly with σ_{++} and weakly with σ^*_{++} and the latter stabilizes σ_{+-} less than it destabilizes σ^*_{+-}, which is closer to it in energy. Nevertheless, all eight CH-orbitals retain their identity across the diagram as long as the CH bonds stay in the zx plane and remain equivalent with respect to all of the symmetry elements of $\mathbf{D}_{2h}$. The positive and negative combinations of carbon $2p_y$ AOs also split, becoming π and π^* as the CH_2 fragments come together to form ethylene.

The dashed half-arrows in Fig. 4.7 represent electrons present during the approach of two triplet methylenes. The full half-arrows (enclosed in a dashed circle to indicate their absence in the triplet dimerization) represent electrons present during the approach of two singlets.

Omitting the occupied CH orbitals, because they retain their symmetry labels across the diagram, the ground-state configuration of ethylene is adequately characterized as $[\sigma(a_g)^2\,\pi(b_{2u})^2]$. Now, suppose that each of the methylenes is

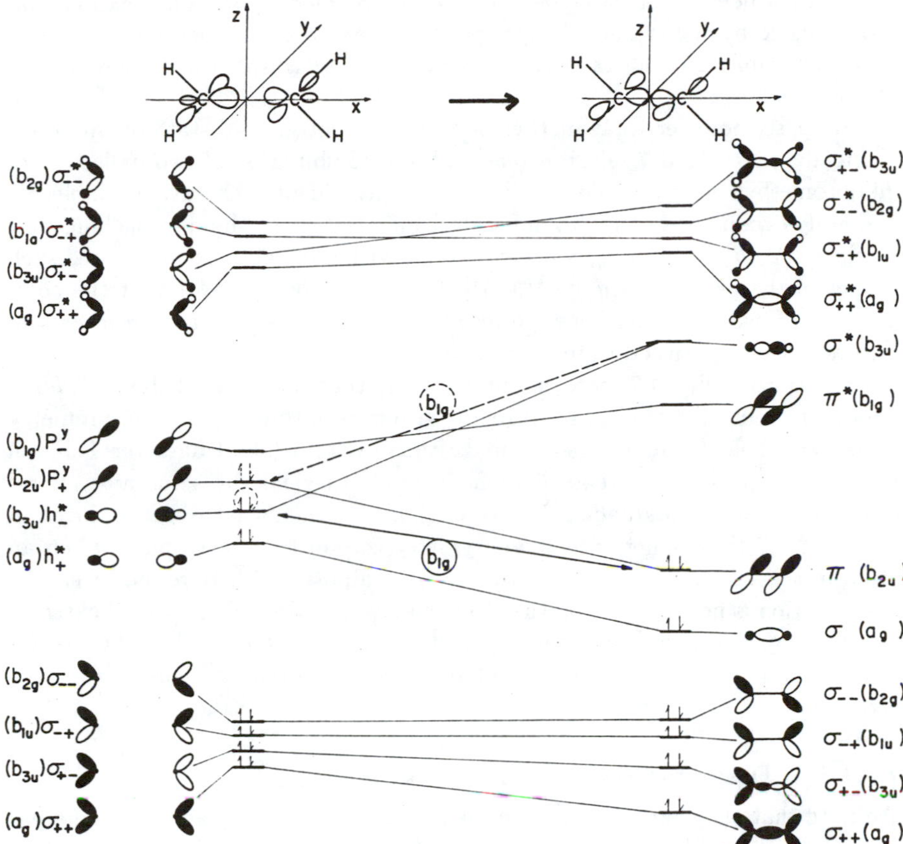

Figure 4.7. Correlation (and correspondence) diagram for dimerization of Methylene ($\mathbf{D}_{2h}$). (The meaning of the circled b_{1g} and the distinction between *correlation* and *correspondence* are explained in the text)

initially in its triplet ground state, in which – as discussed in Section 4.1.1 – one electron occupies an sp^2 hybrid and another, with parallel spin, is in the perpendicular p orbital. This is equivalent to saying that the system of interacting methylenes has two electrons with opposite spin in h^x_+, the more stable combination of sp^2 hybrids, and another electron-pair in the more stable combination of p_y orbitals, p^+_y.The two methylenes can combine to an overall spin-singlet, provided that the spins of the electrons in the second methylene are antiparallel to those in the first. The configuration adopted when two triplet methylenes of opposing net spin approach one another is therefore $[h^x_+(a_g)^2\, p^y_+(b_{2u})^2]$, which correlates directly with the ground-state configuration of ethylene. This finding is consistent with the results of calculations using different sophisticated computational techniques [10, 11], according to which there is no potential barrier at all to the dimerization of $^3CH_2$ along a least-motion pathway, i.e. one that retains $\mathbf{D}_{2h}$ symmetry throughout. Indirect experimental evidence for the ease with which triplet carbenes dimerize is provided by the observation that tetraphenylethylene is the principal product when diphenylmethylene – which also has a triplet ground state [12] – is generated by flash-photolysis of diphenyldiazomethane. [13]

Suppose, however, that methylene had a singlet ground-state or, more realistically, that Fig. 4.7 refers instead to the recombination of two molecules of difluoromethylene, which does [14]. The sp^2 hybrid of each carbene molecule is doubly-occupied, so the ground-state configuration of the approaching pair would be $[h_+^x(a_g)^2\, h_-^x(b_{3u})^2]$, which correlates with the highly excited closed shell singlet of the product $[\sigma(a_g)^2\, \sigma^*(b_{3u})^2]$. It is clear from Fig. 4.7 that the combination of two singlet carbenes is *forbidden* by *orbital symmetry conservation*, but more can be learned from it than that.

Scrutinizing Fig. 4.7 more carefully, we see that the cause of the *forbiddenness* is the refusal of $h_-^x(b_{3u})$ to correlate with $\pi(b_{2u})$. How should the symmetry be reduced below $\mathbf{D}_{2h}$ in order to make them match? The Character Table of $\mathbf{D}_{2h}$ can be consulted in two different ways; both yield the same answer, reproducing, *mutatis mutandis*, the two equivalent means that were employed in Section 2.2.7 for dealing with the desymmetrization that takes place when an atom in a quadrupolar field is perturbed by a dipolar field. Here, however, the perturbation is not imposed on the electrons by an artificially modified external field, but by a real change in the electrostatic field exerted on them by the nucleii as the nuclear frame is distorted away from its original, more symmetric, geometry.

4.3.1.1 I. Desymmetrization to a Subgroup

We note that the character of E and $\sigma(xy)$ is 1 in both B_{2u} and B_{3u}, whereas that of $C_2(z)$ and i is -1 in both irreps. The characters of all the other symmetry elements differ in the two representations: when it is 1 in the first it is -1 in the second, and vice versa. It follows that if only the four elements: E, $C_2(z)$, i, and $\sigma(xy)$ are retained – that is, if the symmetry along the reaction path is reduced from $\mathbf{D}_{2h}$ to $\mathbf{C}_{2h}^z$, in which the two recalcitrant orbitals have the same representation (b_u) – the electron configuration of the reactant pair of singlet carbenes will correlate with that of the product ethylene, and the previously *forbidden* dimerization will have become *allowed*.

4.3.1.2 II. Desymmetrization by a Perturbation

We recognize that a perturbation that permits a b_{2u} and a b_{3u} orbital to intercorrelate has the same irrep as their direct product. Just as in Equation 2.11.

$$\gamma(\text{distortion}) = b_{3u} \otimes b_{2u} = b_{1g} \tag{4.1}$$

Any distortion that has the representation b_{1g}, such as the one depicted in Fig. 4.8a, is symmetric with respect to E, $C_2(z)$, i, and $\sigma(xy)$ and antisymmetric with respect to $C_2(y)$, $C_2(x)$, $\sigma(zx)$ and $\sigma(yz)$. The first set of sym-ops comprise the subgroup $\mathbf{C}_{2h}^z$, which is the kernel of the representation: the symmetry

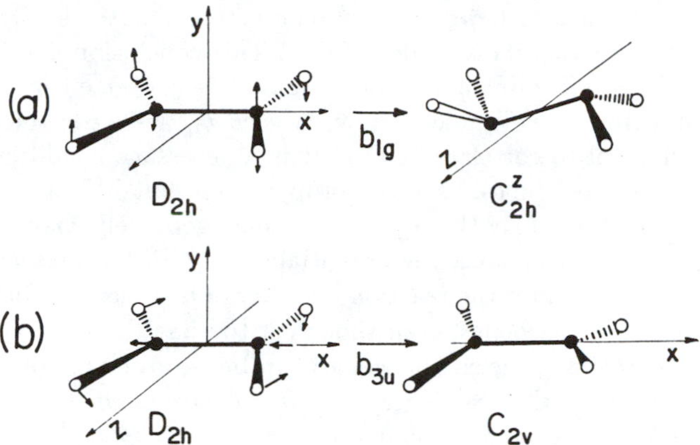

Figure 4.8a, b. Distortions of Ethylene away from $\mathbf{D}_{2h}$: (**a**) Along b_{1g} to $\mathbf{C}^z_{2h}$; (**b**) Along b_{3u} to $\mathbf{C}^x_{2v}$.

point group into which a molecule that originally has $\mathbf{D}_{2h}$ symmetry goes when distorted along a b_{1g} symmetry coordinate. It might be noted that the irrep of R_z is also b_{1g}, but a molecular rotation does not change the internuclear distances, leaving the molecule in its original symmetric geometry, and so has no effect on the potential energy of the electrons;[9] it is thus incapable of *allowing* the reaction.

4.3.1.3 WH-LHA and OCAMS: A Comparison

It was pointed out that the first of these two equivalent methods of analysis is related to the correlation procedures of Woodward and Hoffmann and of Longuet-Higgins and Abrahamson. In their discussion of the dimerization of ethylene, Hoffmann, Gleiter and Mallory [16] assert that "$\mathbf{C}_{2h}$ would have been sufficient for converting a forbidden reaction into an allowed one", but refrain from drawing a correlation diagram for the reaction in $\mathbf{C}_{2h}$, presumably because none of its sym-ops "bisects bonds made or broken in the reaction" [15, p. 31].[10] A complete application of the Longuet-Higgins–Abrahamson procedure would require repeatedly comparing the electron configurations of the carbene pairs, oriented according to the different subgroups of $\mathbf{D}_{2h}$, and eventually finding that correlation with the ground-state configuration of ethylene is achieved in $\mathbf{C}^z_{2h}$, but not in $\mathbf{C}^y_{2h}$, $\mathbf{C}^x_{2v}$, $\mathbf{D}_2$, *etc...*

The second aproach, that of OCAMS, leads directly to the conclusion that the reactants can circumvent the symmetry-imposed potential barrier present along the symmetric coplanar pathway, by approaching one another along a

[9] However, see footnote 21 of Chapter 3.

[10] A point, such as a center of inversion, can hardly be said to bisect a bond, except perhaps when it lies at its midpoint – as it does here. However, inversion can quite generally be regarded as an improper rotation of order 2: $i \equiv S_2$, and – as such – has a *virtual mirror plane* that can be placed so as to bisect any bond that we choose.

reaction coordinate that includes a b_{1g} symmetry coordinate as a component, and therefore only retains the sym-ops of $\mathbf{C}_{2h}^{z}$. This conclusion is incorporated in a *correspondence diagram*, into which Fig. 4.7 is converted by the simple expedient of drawing a two-headed arrow, labeled b_{1g}, between the two occupied orbitals that fail to correlate in $\mathbf{D}_{2h}$. It is a necessary condition that the *inducing* symmetry coordinate have the proper irrep, but it is not a sufficient one; in addition, motion along that coordinate must genuinely change the internuclear distances and thus affect the potential energy. In the present instance, the incorporation of R_z into the reaction coordinate is useless for inducing the dimerization of singlet carbenes, even though it too has the label b_{1g}, because during molecular rotation the x- and y-axes can be regarded as rotating about z as well.[11] As a result, the reacting system is not desymmetrized to $\mathbf{C}_{2h}^{z}$, the kernel of b_{1g}, but retains full $\mathbf{D}_{2h}$ symmetry.

Semi-empirical [16] and ab initio [17] calculations suggest that the dimerization of singlet carbenes does indeed bypass the totally symmetric barrier by desymmetrization to $\mathbf{C}_{2h}^{z}$; the latter reducing the activation energy nearly to zero. In both cases, the authors begin with the essentially perpendicular approach shown in Fig. 4.9a, in which $\sigma(xy)$ is the only symmetry element; it is an intuitively reasonable one, since it maximizes the interaction between the HOMO of one reactant molecule and the LUMO of the other. As they approach more closely, they take up the more symmetrical geometry of Fig. 4.9b, close to $\mathbf{C}_{2h}^{z}$, that maximizes two HOMO-LUMO interactions rather than only one. A corrective to this picture of $^1CH_2$ dimerization has been provided by more recent computations [18]: The singlet and triplet dimerization pathways are shown to occur in the same region of the potential energy surface. Since the methylene pairs have overall singlet multiplicity in both cases and correlation lines between states with the same space and spin symmetry cannot cross,[12] only $^3CH_2$ can dimerize directly to the ground-state, whereas the $^1CH_2$-pair correlates with

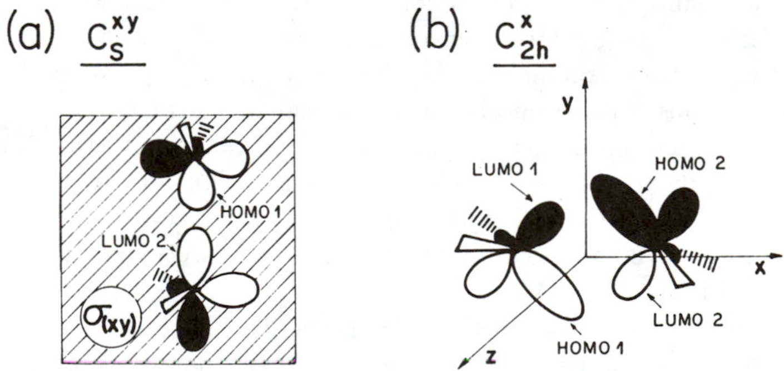

Figure 4.9a, b. Interaction of two singlet carbene molecules: (**a**) perpendicular coplanar approach ($\mathbf{C}_s^{xy}$); (**b**) $\mathbf{C}_{2h}^{z}$ approach.

11 We continue to limit ourselves to "pure" rotations, neglecting centrifugal distortion.
12 S. footnote 19 of Chapter 3.

a highly excited singlet of ethylene. However, the authors' parallel calculations on the pathway for dimerization of silylene [18], where this complication does not exist because $^1SiH_2$ is considerably more stable than $^3SiH_2$, confirm the qualitative features of the b_{1g} pathway admirably.

It may be somewhat puzzling that Fig. 4.7, which appears to disregard the unoccupied orbitals completely, leads to the same conclusion as Fig. 4.8, which explicitly invokes HOMO-LUMO interaction. The frontier orbital interaction is implicit in Fig. 4.7, even though the b_{1g} component of the reaction coordinate was chosen without direct reference to it. As in any correlation diagram, we must consider the HOMO and LUMO of the reacting system as a whole, rather than of the individual reactants. It is evidently the b_{1g} displacement which allows h^x_- to interact with p^y_+; after desymmetrization to $\mathbf{C}^z_{2h}$, both have the same irrep, b_u, so the non-crossing rule keeps them apart, obliging the former – which bears the frontier electrons – to correlate with π, and the latter to follow the dashed two-headed arrow to σ^*.

4.3.2 The Fragmentation of Ethylene

Consideration of the reverse reaction, fragmentation of ethylene (or better, silene) to two methylene (silylene) molecules in their singlet state, suggests that HOMO-LUMO interactions are not always the dominant factor. The energetically costly rupture of a double bond in a single step must nevertheless be formally *allowed* to retrace the b_{1g} pathway in reverse. π-bond rupture is facilitated by interaction of the HOMO, $\pi(b_{2u})$, with the *superjacent* $\sigma^*(b_{3u})$, while the LUMO, $\pi^*(b_{1g})$, assists σ-bond rupture by interacting with the subjacent MO, $\sigma(a_g)$.

4.3.2.1 Bader's Analysis of Molecular Fragmentation

The thermolysis of ethylene to two methylene fragments provides an opportunity for referring to Bader's [19] seminal contribution to our understanding of the relations between symmetry, geometry and energy. His ideas, which were subsequently developed by Pearson [20] and by Salem and Wright [21], are discussed in detail in Pearson's book [22], so they will only be touched upon briefly here.[13]

The approach is based on the premise that a thermally excited molecule will decompose preferentially along a pathway that incorporates the symmetry coordinate which mixes a low-lying excited state into the ground-state most effectively. In outline, his recipe is similar to our Equation 3.12, except that the configurations are not mixed by an external field, but by a vibrational distortion of the nuclear frame. The requirement can be stated[14]:

---

13 We will return to them briefly in Chapter 10, in connection with preferred modes of photofragmentation.

14 See footnote 18 in Chapter 3.

$$\Gamma_{\text{ground-state}} \otimes \Gamma_{\text{distortion}} \otimes \Gamma_{\text{excited state}} \ni \Gamma_I \tag{4.2}$$

When the symmetry point group is commutative, all of its irreps are non-degenerate and Equation 4.2 can be rewritten:

$$\Gamma_{\text{distortion}} = \Gamma_{\text{ground-state}} \otimes \Gamma_{\text{excited state}} \tag{4.3}$$

When, as will usually be the case, the ground-state is a closed-shell singlet, and therefore totally symmetric, the requirement reduces to:

$$\Gamma_{\text{distortion}} = \Gamma_{\text{excited state}} \tag{4.4}$$

Because the interaction between two states under the influence of a perturbation that mixes them varies inversely with the the energy difference between them, Bader assumes that the lowest excited singlet will be dominant in determining the symmetry species of the fragmentation pathway and, in consequence, the chemical identity of the fragments. This expectation is often borne out, but there are numerous exceptions, including our present example: In our axis convention, the excited singlet of ethylene has the configuration $[\sigma(a_g)^2\, \pi(b_{2u})\, \pi^*(b_{1g})]$ and its state label is $^1B_{3u}$[15]. Equation 4.4 suggests that the most effective perturbation should belong to the same irrep. A deformation like the one illustrated in Fig. 4.8b would hardly contribute to rupture of the CC-bond, though it might conceivably induce fragmentation of ethylene to vinylmethylene and hydrogen.[16] However, the next two excited singlets, with the configurations $[\sigma(a_g)\, \pi(b_{2u})^2\, \pi^*(b_{1g})]$ and $[\sigma(a_g)^2\, \pi(b_{2u})\, \sigma^*(b_{3u})]$, are both $^1B_{1g}$ states, and would be expected to mix under *configuration interaction*, stabilizing the lower of the two. The *allowedness* of the b_{1g} fragmentation pathway prescribed by Fig. 4.7, which reduces the symmetry of the molecule to $\mathbf{C}_{2h}^z$, then becomes fully consistent with Bader's ideas.

It should be kept in mind that whereas motion along a vibrational coordinate can mix an excited state into the ground-state, a translation or a "pure" rotation – neither of which affects the potential energy – cannot. As in the case of the dimerization of singlet carbenes, fragmentation of ethylene is not induced by $R_z(b_{1g})$, because the rotating ethylene molecule retains $\mathbf{D}_{2h}$ symmetry with regard to a coordinate system that has its origin at the center of mass and rotates with the molecule.

---

15 Different authors use different axis conventions, so this state may also be found labelled $^1B_{1u}$ or $^1B_{2u}$. In all of the figures in this book, the axis convention is specified unless it is evident from the context.

16 A more detailed analysis would show that it does not, unless an additional distortion in the zx plane is imposed. The symmetry requirements are essentially the same as those of the much discussed fragmentation of methane to CH_2 and H_2 [23]; small wonder that a $\pi \rightarrow \pi^*$ excitation is irrelevant to it.

4.4 Symmetry Coordinates and Normal Modes

The distribution of the $3N - 6$ vibrational symmetry coordinates of a non-linear polyatomic molecule among the irreducible representations of its symmetry point group can be determined by standard methods. [7] Ordinarily, not all of the symmetry species will be represented and several of them will include more than one coordinate. If the molecule belongs to a commutative symmetry point group, all of them will be assigned to one-dimensional symmetry species. If its group is non-commutative, and therefore has representations that are two- or three-dimensional, some of its vibrations may be degenerate; these are best discussed separately.

4.4.1 Non-Degenerate Vibrations

Let us take as our example the three vibrational coordinates of a non-linear triatomic molecule, illustrated in Fig. 4.3. Each of them describes an in-plane motion, so they are necessarily distributed between a_1 and b_1, the two irreps of $\mathbf{C}_{2v}$ that are symmetric to reflection in the molecular plane. The expression for the potential energy of the vibrating XYX molecule in the harmonic approximation is:

$$U = (k_1\xi_s^2 + k_2\eta^2 + k_3\xi_{as}^2)/2 + k_{12}\xi_s\eta \tag{4.5}$$

The potential energy (U) increases as the square of each of the the three symmetry coordinates, and also includes a *cross-term* between the two totally symmetric coordinates. The inclusion of this fourth quadratic term allows for the possibility that it may be easier – or more difficult – to simultaneously stretch the bonds and open the angle between them than to stretch them while closing the bond-angle. In order to get rid of the unwanted cross-term, the two coordinates are combined in such a way that the center of mass remains fixed; the arrows showing that Y moves to $-x$ as two X-atoms move to $+x$ (and vice versa) in the two a_1 vibrations is an attempt to depict this. When the symmetry coordinates are redefined so as to include the relative masses of the atoms, they become *normal coordinates*, and the potential energy of the non-linear XYX molecule becomes:

$$U = (\Lambda_1 Q_1^2 + \Lambda_2 Q_2^2 + \Lambda_3 Q_3^2)/2 \tag{4.6}$$

Q_1 and Q_2 both transform as a_1 but do not mix, provided that the nuclear displacements are small enough that the harmonic approximation to the potential energy is adequate. The harmonic *force constants* Λ_i $(i = 1,2,3)$, are all positive, because the potential energy increases as Q_i departs from its equilibrium value.[17] Also, barring an *accidental degeneracy*, they are all different, each

[17] $\Lambda_i = \partial^2 U/\partial Q_i^2$, which is positive with respect to all of the normal coordinates of a stable molecule at its geometry of minimum potential energy.

being proportional to the square of the corresponding vibrational frequency:

$$\Lambda_i = 4\pi\nu_i^2 \tag{4.7}$$

The conclusions just outlined can be summarized for the general case as follows:

1. The expression for the potential energy includes $3N - 6$ *square terms*, each proportional to the square of a symmetry coordinate, and – in addition – cross-terms between coordinates of the same symmetry species.

2. When proper account is taken of the relative masses of the moving nuclei, the symmetry coordinates are converted to $3N-6$ non-interacting normal coordinates, each of which contributes a single term of the form $\Lambda_i Q_i^2$, where $\Lambda_i = 4\pi\nu_i^2$. All of the Λ_is are positive and the frequencies (ν_i) real. Except for possible accidental degeneracies, which are sufficiently rare that they can usually be ignored, the numerical values of Λ_i – and hence of ν_i – will differ from one another.

4.4.2 Degenerate Vibrations

In order to illustrate the vibrational motions of a molecule belonging to a non-commutative symmetry point group, we return to the considerations of Section 2.3.2 and once more use as our example the square-planar complex, $NiF_4^=$. A non-linear penta-atomic molecule has nine independent vibrational coordinates, distributed among the symmetry species of $\mathbf{D}_{4h}$. These can be fully specified by standard methods [7], but the following simple qualitative considerations allow us to conclude that there are seven in-plane and two out-of-plane vibrations. Fig. 4.10 depicts several of the in-plane modes; the motion of the nickel atom to conserve the center of mass is implied.

The four NiF bond extensions can be combined in four ways. Two combinations belong to non-degenerate irreps: the totally symmetric stretching mode illustrated at the center of Figure 4.10 and the antisymmetric stretch labelled b_{1g}, in which the two pairs of *trans*-situated ligand atoms move out-of-phase with one another. In the *asymmetric* stretching mode illustrated, one pair moves together along x and the other – against one another – along y. Rotation by 90° about z or reflection in a diagonal plane, both perfectly legitimate sym-ops of $\mathbf{D}_{4h}$, transform this motion into an equivalent one in which the x and y axes are interchanged, so the coordinate illustrated is one of a doubly degenerate pair belonging to e_u.

The sum of the four internal angles is 360°, so there can be no totally symmetric in-plane bending mode, in which all four increase together. The one non-degenerate in-plane bending mode is the antisymmetric bend (b_{2g}), in which opposing pairs of angles open and close together. The remaining two in-plane bending modes, in which one angle closes at the expense of the one opposite, while the other two are unchanged, are interconvertible under the sym-ops of $\mathbf{D}_{4h}$, and so describe motion along another pair of e_u coordinates, differently

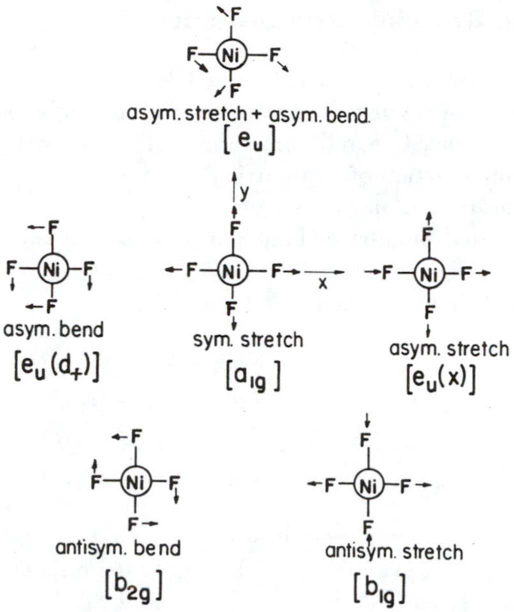

Figure 4.10. Several in-plane vibrational coordinates of $NiF_4^=$

aligned from the originally chosen pair of stretching coordinates and associated with different values of Λ_i and ν_i. Since symmetrically equivalent normal modes necessarily have the same Λ_i, any orthogonal pair of combinations is an equally good choice. The reader can confirm that the positive and negative combinations of $e_u(d_+)$ and its partner $e_u(d_-)$ are coordinates in which two adjacent angles increase and the other two decrease; in each case two *trans*-situated fluorine atoms move together at right angles to the bonds and the other two remain in place. The overall motion is parallel to x in one case and to y in the other.

The two out-plane-modes (not illustrated in the figure, are also easy to characterize; they are both non-degenerate: In one (a_{2u}), the F atoms move together parallel to z, and the central Ni atom moves in the opposite direction so as to conserve the center of mass. In the other (b_{2u}), one pair of *trans*-situated atoms moves up, the other pair moves down, and the central atom stays put.

Returning to the in-plane modes, we recognize that the asymmetric stretching and bending coordinates have the same two-dimensional irrep (e_u), and can therefore mix. Simultaneous excitation of both normal modes might produce the complex vibration illustrated at the top of the figure. However, if the degenerate bending coordinates had initially been replaced by the orthogonal pair of combinations aligned parallel to x and y,[18] each would mix only with the stretching mode that is similarly aligned. In that case the combined vibration would be described as a superposition of two motions: a stretching-bending combination aligned along x and a "pure" bending motion parallel to y.

[18] This is standard practice in vibrational spectroscopy.

4.4.3 Reducing Reducible Representations

A comparative vibrational analysis of the CH- and NiF-stretching modes in ethylene and $NiF_4^{=}$ respectively illustrates the distinction between the characters of the irreps of commutative and non-commutative symmetry point groups. It also allows the introduction of two particularly useful group theoretical terms: *direct sum* and *projection operator*.

The four localized bond-stretching coordinates, illustrated in Fig. 4.11 for ethylene ($\mathbf{D}_{2h}$) and the nickel tetrafluoride dianion ($\mathbf{D}_{4h}$), combine in both cases to the same four orthogonal linear combinations:

$$\xi_{++++} = (\delta r_1 + \delta r_2 + \delta r_1 + \delta r_2)/2 \tag{4.8}$$
$$\xi_{++--} = (\delta r_1 + \delta r_2 - \delta r_1 - \delta r_2)/2 \tag{4.9}$$
$$\xi_{+--+} = (\delta r_1 - \delta r_2 - \delta r_1 + \delta r_2)/2 \tag{4.10}$$
$$\xi_{+-+-} = (\delta r_1 - \delta r_2 + \delta r_1 - \delta r_2)/2 \tag{4.11}$$

In $\mathbf{D}_{2h}$, they are assigned very easily: $\xi_{++++}(a_g)$, $\xi_{++--}(b_{3u})$, $\xi_{+--+}(b_{2u})$, $\xi_{+-+-}(b_{1g})$. In $\mathbf{D}_{4h}$,[19] ξ_{++++} and ξ_{+-+-} belong to the respective one-dimensional representations a_{1g} and b_{1g}, whereas ξ_{++--} and ξ_{+--+} – that are interconverted by C_4 – are assigned to the two-dimensional irrep E_u.

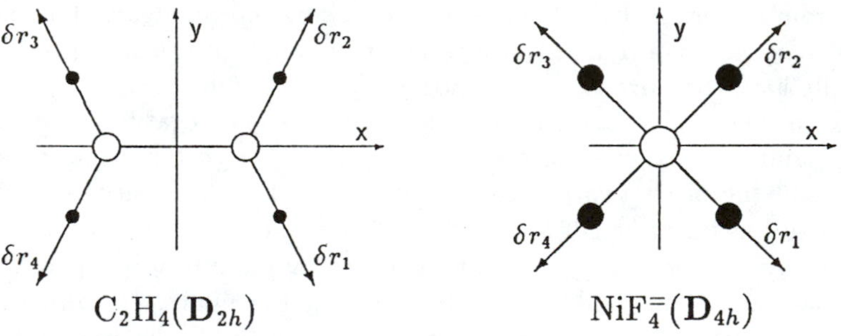

Figure 4.11. Stretching coordinates of $C_2H_4(\mathbf{D}_{2h})$ and $NiF_4^{=}(\mathbf{D}_{4h})$

Intuitive considerations of this sort are often sufficient, but familiarity with a few simple rules for determining beforehand the irreps of the combinations of coordinates (or orbitals) in a given symmetry point group and constructing them formally will be useful in the more complicated cases discussed in the following chapters. Their implementation will be illustrated first with the CH-stretching modes of ethylene and then with the NiF-stretching modes of $NiF_4^{=}$.

[19] The x and y axes are rotated about z by 45° relative to those in Fig. 4.10, for consistency with the axis convention appropriate to ethylene. B_{1g} and B_{2g} are interchanged as a result, as are B_{1u} and B_{2u}. Also the x and y components of E_u take different forms in the two axis coventions.

4.4.3.1 The CH-Stretching Coordinates of Ethylene

The CH-stretching coordinates of ethylene are treated in Table 4.1 as a four component vector; listed under each symmetry operation of $\mathbf{D}_{2h}$ is the result of applying that sym-op to the original vector – i.e. multiplying it by the matrix representing the sym-op. Each sym-op either leaves a given coordinate in place or converts it to another. As pointed out in Section 2.3.2.1, a component that remains in place under the sym-op contributes +1 to the character; one that is interchanged contributes nothing.[20] The resulting characters appear in the bottom row of the table. The four-dimensional representation $\Gamma(\delta r_i)$ is reducible to irreps of $\mathbf{D}_{2h}$, all of which are one-dimensional. It is done in two steps:

1) evaluation of the *direct sum* i.e. the distribution of the appropriate linear combinations among the irreducible representations, and

2) constructing the combination corresponding to each.

Table 4.1. Reduction of the representation of the CH-stretching coordinates of ethylene to the irreps of $\mathbf{D}_{2h}$

$\mathbf{D}_{2h}$	E	$C_2(z)$	$C_2(y)$	$C_2(x)$	i	$\sigma(xy)$	$\sigma(zx)$	$\sigma(yz)$
δr_1	δr_1	δr_3	δr_4	δr_2	δr_3	δr_1	δr_2	δr_4
δr_2	δr_2	δr_4	δr_3	δr_1	δr_4	δr_2	δr_1	δr_3
δr_3	δr_3	δr_1	δr_2	δr_4	δr_1	δr_3	δr_4	δr_2
δr_4	δr_4	δr_2	δr_1	δr_3	δr_2	δr_4	δr_3	δr_1
$\Gamma(\delta r_i)$	4	0	0	0	0	4	0	0

Direct Sum: $[a_g \oplus b_{2u} \oplus b_{3u} \oplus b_{1g}]$

The number of times a particular irrep appears in the reduced representation of the coordinates is given by its scalar product (Section 2.2.5) with the reducible representation. The formal procedure can be bypassed as follows: Since $\Gamma(\delta r_i)$ has 4 under E and $\sigma(zx)$ and 0 everywhere else, whereas $\mathbf{D}_{2h}$ has a total of 8 sym-ops, an irrep can appear no more than once in the direct sum. Moreover, the irreps that make it up are necessarily the four that are symmetrical to reflection in the molecular plane, *i.e* have 1 rather than −1 under $\sigma(xy)$. The direct sum is therefore: $[a_g \oplus b_{2u} \oplus b_{3u} \oplus b_{1g}]$, as shown in Table 4.1.

The four linear combinations are constructed as follows:
A representative coordinate, say δr_1 is chosen and operated upon in turn by the *projection operator* of each of the irreducible representations, Γ_j:

$$\xi(\gamma_j) = \mathcal{N} \sum_{i=1}^{g} \chi_j(R_i) R_i \, \delta r_1 \tag{4.12}$$

Reading Equation 4.12 from right to left: δr_1 is operated upon by each of the sym-ops of $\mathbf{D}_{2h}$ and the result is multiplied by the character of that operation

[20] Unlike a cartesian coordinate, δr_i cannot be converted to itself with change of sign, so it cannot contribute −1 to the character.

in Γ_j. The sum yields the unnormalized symmetry coordinate of irrep Γ_j, which is scaled by setting the normalization constant $\mathcal{N}$ equal to the reciprocal of the sum of the squares of the coefficients. As an example, let us construct the b_{2u} combination, using the first row of Table 4.1 and the $\mathbf{D}_{2h}$ Character Table.

$$\begin{aligned}\xi(b_{2u}) &= \mathcal{N}(\delta r_1 - \delta r_3 + \delta r_4 - \delta r_2 - \delta r_3 + \delta r_1 - \delta r_2 + \delta r_4)\\ &= (2\delta r_1 - 2\delta r_2 - 2\delta r_3 + 2\delta r_4)/\sqrt{16}\\ &= (\delta r_1 - \delta r_2 - \delta r_3 + \delta r_4)/2\end{aligned}$$

We have reproduced $\xi_{+--+}(b_{2u})$ (Equation 4.10); the other three combinations can be generated similarly. An attempt to construct a symmetry coordinate that transforms as one of the four excluded irreps (b_{2g}, b_{3g}, a_u, b_{1u}) would produce a vanishing "combination", all the coefficients of which are zero.

4.4.3.2 The NiF-Stretching Coordinates of $NiF_4^=$

Only the six symmetry operations of $\mathbf{D}_{4h}$ that contribute to the character of the reducible representation are included in Table 4.2. To the two present in $\mathbf{D}_{2h}$, E and $\sigma_h(\equiv \sigma(xy))$, are added the twofold rotations (C_2'') about perpendicular axes passing through two collinear NiF bonds, and reflection in the planes passing through these axes (σ_d). We note once more that sym-ops in the same class have the same character: it is 2 for both C_2'' rotations and – coincidentally – 2 for both σ_d reflections as well. When the scalar product of the reducible representation is taken with all the irreps of $\mathbf{D}_{4h}$ in turn, the direct sum is found to be $[a_{1g} \oplus b_{2g} \oplus e_u(2)]$; the single appearance of the doubly-degenerate e_u implies the presence of a pair of orthogonal symmetry coordinates of that irrep.

Table 4.2. Reduction of the representation of the NiF-stretching coordinates of $NiF_4^=$ to the irreps of $\mathbf{D}_{4h}$. (The notation is that of Table 2.4)

$\mathbf{D}_{4h}$	E	$C_2''(-)$	$C_2''(+)$	σ_h	$\sigma_d(-)$	$\sigma_d(+)$
δr_1	δr_1	δr_1	δr_3	δr_1	δr_1	δr_3
δr_2	δr_2	δr_4	δr_2	δr_2	δr_4	δr_2
δr_3	δr_3	δr_3	δr_1	δr_3	δr_3	δr_1
δr_4	δr_2	δr_2	δr_4	δr_4	δr_2	δr_4
$\Gamma(\delta r_i)$	4	2	2	4	2	2

Direct Sum: $[a_{1g} \oplus b_{2g} \oplus e_u(2)]$

The same conclusion can be reached more simply. The characters of the reducible representation add up to sixteen, the order of the group ($g = 16$). Therefore, the characters of the six contributing sym-ops have to be positive in all of the one-dimensional irreps included in the direct sum; the requirement is fulfilled only by a_{1g} and b_{1g}. The direct sum therefore has to include a two-dimensional representation as well; this can only be E_u, which has 2 rather than -2 as the character of σ_h.

Having gone thus far into the formalism, let us go a bit farther and show how to characterize all of the internal coordinates of a molecule. Each of N atoms can be displaced independently parallel to x, y and z, so $3N$ independent combinations can be formed from them. Taking the $3N$ cartesian coordinates as our basis vector, we find its reducible representation as follows:

Only atoms that remain in place contribute. Every such atom contributes 3 to E, since all three coordinates are retained. It contributes 1 to any reflection because the coordinate system can be rotated until the atom lies in a mirror plane, say xy, so that $\sigma(xy)$ retains the sign of two coordinates and reverses the sign of one. Inversion changes the sign of all three coordinates, so each atom at the inversion center provides -3. C_2 reverses the sign of two of the atom's coordinates and retains one, so it contributes -1. $C_n(z)$ retains the sign of z for any n and $S_n(z)$ reverses it, but when $n > 2$ both sym-ops mix x and y or – at best – interconvert them. It can be shown by elementary trigonometry that each unshifted atom contributes $1 + 2\cos\frac{2\pi}{n}$ and $-1 + 2\cos\frac{2\pi}{n}$ respectively to the character of C_n and S_n; since $\cos 90° = 0$, the respective contributions of C_4 and S_4 per stationary atom are 1 and -1.

In our example, $NiF_4^=$, $N = 5$, so there are 15 internal coordinates: 3 translations, 3 rotations and 9 vibrations. We note that the central Ni atom stays put under all 16 sym-ops; E and σ_h keep all four F atoms in place, whereas two are stationary and two are interchanged under the C_2'' rotations and σ_d reflections. Applying the rules set out above, we obtain the reducible representation:

$\mathbf{D}_{4h}$	E	$2C_4$	C_2	$2C_2'$	$2C_2''$	i	$2S_4$	σ_h	$2\sigma_v$	$2\sigma_d$
Γ_{coords}	15	1	-1	-1	-3	-3	1	5	1	3

Scalar multiplication with the irreps of $\mathbf{D}_{4h}$ reduces it to the direct sum:

$$a_{1g} \oplus a_{2g} \oplus b_{1g} \oplus b_{2g} \oplus 2 \times a_{2u} \oplus b_{1u} \oplus e_g(2) \oplus 3 \times e_u(2)$$

The translations transform as: $a_{2u} \oplus e_u(2)$ and the rotations as: $a_{2g} \oplus e_g(2)$, so the nine vibrational coordinates of $NiF_4^=$ are distributed:

$$a_{1g} \oplus b_{1g} \oplus b_{2g} \oplus a_{2u} \oplus b_{1u} \oplus 2 \times e_u(2)$$

Of these, the four NiF-stretching coordinates have already been accounted for as a_{1g}, b_{2g} and $e_u(2)$, and the five remaining vibrational coordinates were identified intuitively in Section 4.4.2. They are the antisymmetric (b_{1g}) and degenerate (e_u) pair of in-plane bending coordinates, and the out-of-plane bending (a_{2u}) and twisting (b_{1u}) coordinates.[21]

[21] Note that b_{1g} and b_{2g} have been interchanged in the new axis convention, as have b_{1u} and b_{2u}. (See preceding footnote.)

4.5 Motion Along the Reaction Coordinate

The mechanism of a given reaction, say the thermolysis of a stable molecule to yield two or more fragments, is almost universally discussed in terms of the motion of the constituent atoms on the potential energy surface along a reaction coordinate, from the reactant *via* a transition state to the products [24]. We have adopted this approach as a matter of course in our discussion of the fragmentation of ethylene (Section 4.3). The reaction coordinate is initially taken as some combination of the symmetry coordinates of the reactant, but – as the transition state is approached – it is better described as a combination of the symmetry coordinates of the latter. Let us consider the two regions separately, beginning with departure from the equilibrium geometry of the reactant.

It has been assumed implicitly that the nuclear displacements are sufficiently small for the harmonic approximation, expressed by Equation 4.6, to hold. Motion along a reaction coordinate involves large displacements, so a breakdown of the harmonic approximation is only to be expected. In particular, the two components of a doubly-degenerate vibration may cease to be equivalent for large displacements from equilibrium, as when a molecule embarks on a particular fragmentation pathway; the alignment in space of the components, which was touched upon lightly in the preceding section, then becomes important.

Consider a hypothetical reaction in which one NiF bond is ruptured and the three FNiF angles of the remaining fragment open to 120°. If it can be assumed that the five atoms remain in the xy plane, the reaction coordinate for the path of least motion can be constructed as a superposition of several in-plane modes: The (a_g) mode stretches all of the bonds equally, but a superposition of the x component of the e_u mode, in the opposite phase from that drawn in the figure, reverses the motion of all but the atom on the left, which becomes the "leaving group". The gradual increase of the FNiF angles from 90° to 120° is accomplished by incorporating into the reaction coordinate the asymmetric in-plane bending mode as well; but it too must be aligned along x rather than along a diagonal by superposing the coordinate depicted on that obtained by rotating it clockwise by 90° $(C_4^{(-)})$. Because only one suitably aligned component of modes was chosen from each of the e_u pairs, the reaction path retains the symmetry of the co-kernel, $\mathbf{C}_{2v}$, instead of being restricted to that of the kernel, $\mathbf{C}_s$. It should be kept in mind, therefore, that the subgroup to which a reacting system is taken by motion along a reaction path that includes degenerate symmetry coordinates depends critically on whether one or more components of each is incorporated in the reaction coordinate, and on their alignment in space.

The distinction between vibration and motion along a reaction path is perhaps best expressed as follows: The periodic displacements from equilibrium of a vibrating molecule are temporary; when the vibrational coordinate is not totally symmetric, the molecule goes into a subgroup of its original symmetry point group, but the full symmetry of the molecule is regenerated momentarily as it passes through the equilibrium geometry during the reverse phase of every vibration. In contrast, when a symmetrical molecule embarks on a reaction coordinate, the distortion, and any resulting reduction of symmetry, is permanent, though symmetry may well be increased at the transition state; in this sense it is analogous to substitutional desymmetrization, which was discussed in Section 2.3.2.

4.5.1 Distortional and Substitutional Desymmetrization Compared

A comparison of Fig. 4.10 with Fig. 2.12 allows us to pursue the analogy farther. The symmetric stretching vibration, like tetrasubstitution of another halogen for fluorine in Fig. 2.12, leaves the molecule in $\mathbf{D}_{4h}$. The antisymmetric stretching vibration desymmetrizes the molecule to $\mathbf{D}_{2h}$, the kernel of B_{1g}, twice every vibrational period, regenerating $\mathbf{D}_{4h}$ every time it passes through its equilibrium geometry. The combined motion along the asymmetric stretching modes, or any combination of differently aligned e_u modes, takes the molecule temporarily into $\mathbf{C}_s^{xy}$, the kernel of E_u; substitution of two fluorine atoms by two different halogens performs the same desymmetrization permanently. A substitutional desymmetrization analogous to the hypothetical reaction path described above, in which one $NiF^=$ bond is ruptured, is the conversion of $NiF_4^=$ to $NiF_3Cl^=$; both processes reduce the symmetry to $\mathbf{C}_{2v}$.

Both of the out-of-plane modes are non-degenerate. The kernel of the a_{2u} mode is $\mathbf{C}_{4v}$, the point group of a tetragonal pyramid, such as would be generated by the approach of a fifth ligand atom along z. Continued displacement along the b_{2u} coordinate takes square-planar $NiF_4^=$ into $\mathbf{D}_{2d}$, and eventually converts it into a tetrahedral complex; this interconversion will be discussed in detail in a Chapter 11.

4.5.2 At The Transition State

Up to now, we have been assuming that the molecular entity being considered is stable, in the sense that it is in a geometry of minimum potential energy, so that all of its $3N-6$ values of Λ_i are positive. This geometry need not be the most stable one on the potential energy surface; it can be at a *local minimum*. It follows that even a chemical species that would normally be classified as an *unstable intermediate*, as distinguished from a *transition state*, is stable in this restricted sense.

When the molecular entity involved is a transition state, one of its normal coordinates is the reaction coordinate, $Q^\ddagger$, motion along which is not opposed by a restoring force. The transition state is situated at a *saddle point* on the potential energy surface [25, Chap. 5]; [26, pp. 101–107], at which $\Lambda^\ddagger$, the *force constant* for motion along the reaction coordinate is negative and $\nu^\ddagger$ is consequently imaginary,[22] so a non-linear transition state can have no more than

[22] For mathematical convenience, $\Lambda^\ddagger$ is usually assumed to be zero in formal derivations of *Transition State Theory*, whereupon the reaction coordinate approximates a translation.

$3N - 7$ real frequencies. Nor can it have fewer than $3N - 7$, since a transition-state can have only one non-positive force constant. Murrell and Laidler [27] have shown that a point where there are two or more negative force constants does not lie along a *col* on the potential energy surface but on a *hill* - sometimes referred to as a *second-order saddle point*, descent from which along any arbitrary combination of the two normal coordinates lowers the energy. Such a point is necessarily bypassed by any reaction coordinate in its vicinity and so fails to qualify as a transition state. Specifically, when a highly symmetric transition state, one that belongs to a non-commutative group, is being postulated, the reaction coordinate cannot belong to a degenerate irrep. This follows from the fact that both components of a doubly-degenerate vibration - all three if it is triply-degenerate - have the same value of Λ_i, so there can be only $3N - 8$, or $3N - 9$, positive force constants.

4.6 References

[1] A.D. Walsh: J. Chem. Soc. *1953*, 2260, 2266, 2288, 2296, 2301, 2306, 2318, 2321, 2330.

[2] B.M. Gimarc: Accts. Chem. Research *7*, 389 (1974).

[3] R.J. Buenker and S.D. Peyerimhoff: Chem. Revs. *74*, 127 (1974). This comprehensive review summarizes the experimental and theoretical data on the molecules discussed.

[4] M.E. Casida, M.M.L. Chen, R.D. McGregor and H.F. Schaeffer III: Israel J. Chem. *19*, 127 (1980).

[5] For a summary of the physical properties of singlet and triplet methylene, see W.T. Borden and E.R. Davidson: Ann. Revs. Phys. Chem. *30*, 128ff. (1979).

[6] J. Katriel and R. Pauncz: Adv. Quantum Chem. *10*, 143 (1977).

[7] Two of several good introductory presentations are: a) G.M. Barrow: *Introduction to Molecular Spectroscopy.* McGraw-Hill, New York 1962; b) G.W. King, *Spectroscopy and Molecular Structure.* Holt, Rinehart and Winston, New York 1964. The definitive exposition is: c) E.B. Wilson, J.C. Decius and P.C. Cross: *Molecular Vibrations.* McGraw-Hill, New York 1955.

[8] For a thorough discussion of π-σ and through-bond *vs.* through-space interactions, see R. Gleiter: Angew. Chem. *86*, 770 (1974); Angew. Chem. Internat. Ed. (English) *13*, 6960 (1974).

[9] R.H. Hunt, R.A. Leacock, C.N. Peters and K.T. Hecht: J. Chem. Phys. *42*: 1931 (1965).

[10] H. Basch: J. Chem. Phys. *55*, 1700 (1971).

[11] O. Kikuchi: Jerusalem. Symp. Chem. Biol. Reactivity *6*, 189 (1974).

[12] E. Wasserman, L.C. Snyder and W.A. Yager: J. Chem. Phys. *41*, 1763 (1964).

[13] G.L. Closs and B.E. Rabinow: J. Amer. Chem. Soc. *98*, 8190 (1976).

[14] C.W. Mathews: Canad. J. Chem. *45*, 2355 (1967).

[15] R.B. Woodward and R. Hoffmann: *The Conservation of Orbital Symmetry.* Verlag Chemie, Weinheim and Academic Press, New York 1970.

[16] R. Hofffmann, R. Gleiter and F.B. Mallory: J. Amer. Chem Soc. *92*, 1460 (1970).

[17] P. Cremaschi and M. Simonetta: J. Chem. Soc. Faraday II *70*, 1801 (1974).

[18] K. Ohta, E.R. Davidson and K. Morokuma: J. Amer. Chem. Soc. *107*, 3466 (1985).

[19] R.F.W. Bader: Canad. J. Chem. *40*, 1164 (1962).

[20] R.G. Pearson: J. Amer. Chem. Soc. *91*, 1252, 4947 (1969).

[21] L. Salem and J.S. Wright: J. Amer. Chem. Soc. *91*, 5947 (1969).

[22] R.G. Pearson: *Symmetry Rules for Chemical Reactions.* Wiley, New York 1976.

[23] For a well documented discussion of this reaction, see: H.U. Lee and R. Janoschek: Chem. Physics *39*, 271 (1979).

[24] See any textbook of chemical kinetics, for example References [25, 26] below.

[25] A.A. Frost and R.G. Pearson: *Kinetics and Mechanism, 2nd ed.* Wiley, New York 1962.

[26] W.C. Gardiner: *Rates and Mechanisms of Chemical Reactions:* Benjamin, New York 1969.

[27] J.N. Murrell and K.J. Laidler: Trans. Faraday. Soc. *64*, 371 (1968).

Part III

The Classical Thermal Reactions

Chapter 5

Electrocyclic Reactions and Related Rearrangements

In their now classic monograph [1], Wooodward and Hoffmann concentrate on three basic types of "no mechanism" reaction: Electrocyclic reactions – notably polyene cyclizations, cycloadditions, and sigmatropic rearrangements. These three reaction types will be taken up in this and the next two chapters from the viewpoint of *Orbital Correspondence Analysis in Maximum Symmetry (OCAMS)* [2, 3, 4], the formalism of which follows naturally from that developed in Chapter 4. The similarities to the original WH-LHA approach [5, 6], and the points at which OCAMS departs from it, will be illustrated. In addition, a few related concepts, such as " allowedness" and "forbiddenness", *global vs. local symmetry*, and "concertedness" and "synchronicity", will be taken up where appropriate.

In these chapters we will concern ouselves for the most part with reactions that take place on the ground-state potential energy surface. The analysis of photochemical reactions and of those thermal proccesses that involve a change of electron-spin will be be taken up subsequently.

5.1 Rudimentary Analysis of Polyene Cyclization

Woodward and Hoffmann's analysis of this set of reactions [1, pp. 38–45] is too familiar to require more than a brief recapitulation: The first member of the series, s-*cis*-butadiene ↔ cyclobutene, will be discussed in some detail, and the *Rules* for its higher homologs will be shown to follow directly from the nodal properties of the polyenes.

The overall symmetry of both the reactant and product is $\mathbf{C}_{2v}$; each has a two-fold rotational axis and a mirror plane, both of which bisect the newly formed σ bond. However, in order for this bond to be formed, the methylene groups have to rotate out of the molecular plane, so only one of the two symmetry elements can be retained along the reaction pathway: the two-fold rotational axis during the *conrotation* or the mirror plane during the *disrotation*.

The two pathways are investigated individually, a separate orbital correlation diagram being set up for each [1, Fig. 19, p. 43]. The relevant MOs of the reactant and product are stacked in order of increasing energy and characterized according to whether they are symmetric or antisymmetric with respect to the symmetry element retained along the pathway. The orbitals occupied in the ground-state correlate along the *conrotatory* pathway, which is declared to be *allowed*, but not along the *forbidden disrotatory* pathway. Since a correlation diagram can be read in either direction, the same conclusion applies to the reverse reaction: ring-opening of cyclobutene.

OCAMS departs from the WH-LHA procedure by carrying out the analysis in the symmetry point group common to the reactant and the product, here $\mathbf{C}_{2v}$. The subgroup into which the reacting molecule has to be desymmetrized along the reaction path, or – equivalently – the irreducible representation of the non-totally symmetric symmetry coordinate(s) that has(have) to be incorporated into the reaction coordinate, is determined by a *Correspondence Diagram*. In its most rudimentary form, we take into account only the occupied orbitals on each side that are considered to be directly involved in the isomerization process.

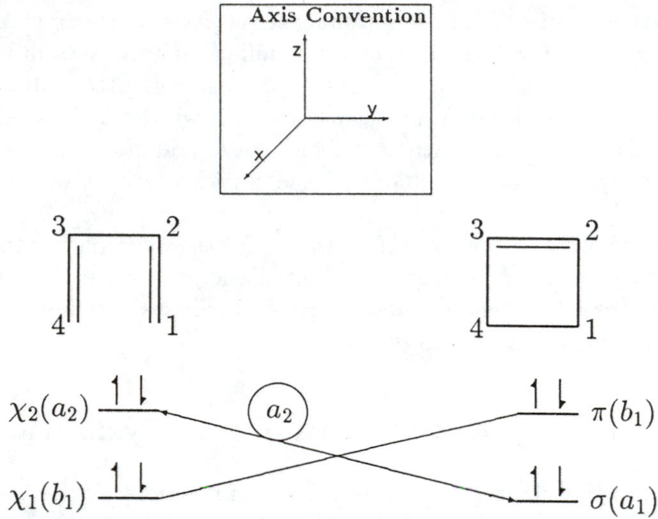

Figure 5.1. Rudimentary correspondence diagram for cyclization of Butadiene. (The orbitals are labeled as in Fig. 19 of ref. [1]; their irreps in $\mathbf{C}_{2v}^z$ are given in parenthesis)

5.1.1 Cyclization of Butadiene to Cyclobutene

At this level of analysis, it is sufficient to include in Fig. 5.1 the doubly-occupied MOs that appear in Fig. 19 of Reference [1]: χ_1 and χ_2 of butadiene and σ and π of cyclobutene. The latter molecule has $\mathbf{C}_{2v}$ symmetry, so the former is put into the same symmetry point group by adopting the s-*cis* conformation, on the reasonable assumption that the attainment of conformational equilibrium is a much faster process than cyclization. χ_1 and π have the same irrep (b_1), so they remain in correlation whether $\mathbf{C}_{2v}^z$ symmetry is retained or not. However, in order to deliver a pair of electrons from $\chi_2(a_2)$ to $\sigma(a_1)$ the reaction has to proceed along a pathway on which the distinction between a_1 and a_2 has disappeared. As in the case of dimerization of methylene (Section 4.3.1), we can consult the Character Table of $\mathbf{C}_{2v}^z$ from two equivalent points of view:

I. Only two sym-ops, E and $C_2(z)$, have the character 1 in A_2; therefore, the reaction will become *allowed* if the pathway is desymmetrized to $\mathbf{C}_2^z$, the kernel of A_2, in which A_1 and A_2 both map onto the same irrep (A). The conrotatory pathway, along which the two sym-ops that comprise $\mathbf{C}_2^z$ are the only ones retained, is therefore selected.

II. In analogy with Equation 4.1, the pathway must be displaced away from $\mathbf{C}_{2v}^z$ along a symmetry coordinate of the same irrep as the direct product of the irreps of the two orbitals between which the correspondence has to be *induced*[1]:

$$\gamma(\text{distortion}) = a_1 \otimes a_2 = a_2 \tag{5.1}$$

The reaction coordinate along the *allowed* pathway must therefore include, in addition to totally symmetric displacements, such as decreasing the C_1C_4 distance and shortening the C_2C_3 bond, at least one a_2 symmetry coordinate: Conrotation of the methylene groups evidently fulfills this requirement.

Any a_2 displacement desymmetrizes the pathway to $\mathbf{C}_2^z$, so the two ways of reading the Character Table are completely equivalent; their common conclusion is symbolized in Fig. 5.1, as in Fig. 4.7, by a two-headed arrow – here labeled a_2. It indicates that the reaction, which is *forbidden* in $\mathbf{C}_{2v}^z$, is made *allowed* by an *induced correspondence* between the non-correlating orbitals of the reactant (a_2) and product (a_1) that is produced when the pathway is desymmetrized to $\mathbf{C}_2^z$ by a conrotation (a_2).

5.1.2 *cis*-1,3,5-Hexatriene to 1,3-Cyclohexadiene

The cyclization product in this next member of the series also has $\mathbf{C}_{2v}$ symmetry, so here too the reactant is put into the same symmetry point group and the orbitals directly involved in the reaction are characterized accordingly by their irreps. Fig. 5.2 has one more orbital than Fig. 5.1 on each side: ϕ_1 and ϕ_2 respectively are symmetric and antisymmetric combinations of the two π orbitals of cyclohexadiene, whereas χ_1, χ_2 and χ_3 are the three π combinations of hexatriene, stacked in order of increasing energy.

The energy of π orbitals increases with the number of nodal surfaces perpendicular to the plane of the σ frame, here the yz plane. In $\mathbf{C}_{2v}^z$ with our axis convention, the irrep of the lowest-lying orbital – as well as all of those with an even number of nodal surfaces – is b_1, whereas those with an odd number of nodal surfaces all have the irrep a_2. The two π orbitals on the right of Fig. 5.2,

[1] This equation is identical in form and closely related in content to Pearson's *bond symmetry rule* [7], [8, p. 72].

ϕ_1 and ϕ_2 correlate directly with the two lowest on the left, χ_1 and χ_2, leaving two electrons in $\chi_3(b_1)$ that cannot be delivered to $\sigma(a_1)$ of the cyclization product unless correspondence between the two non-correlating MOs is induced by a disrotation (b_1).

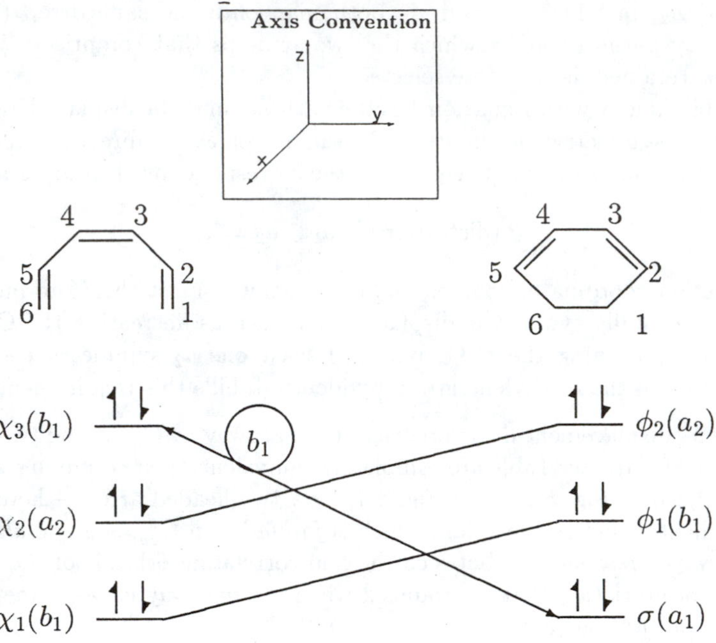

Figure 5.2. Rudimentary correspondence diagram for cyclization of *cis*-1,3,5-hexatriene

Extension to the cyclization of higher homologs is straightforward: The lower $N-1$ π orbitals of a polyene with N conjugated double bonds will correlate in $\mathbf{C}_{2v}$ with all of the π MOs of its cyclization product, leaving the Nth out of correlation with $\sigma(a_1)$. If N is even ($N = 2q$, where $q = 0, 1, 2...$), χ_N will be a_2 and a conrotation will be required in order to induce a correspondence with σ; if N is odd ($N = 2q + 1$), χ_N will be b_1 and a disrotation will be called for. Since all N orbitals are doubly-occupied, the number of electrons involved (k) is $2N$; the W.-H. Rules for thermal cyclization [1, pp. 38–45] follow directly.[2] Thus, for example, all-*cis*-octatetraene cyclyzes readily to cyclooctatriene *via* a conrotatory pathway. [9]

[2] A polyene with n carbon atoms has $k(= 2N)$ electrons in N doubly-occupied orbitals; so do a polyenyl anion with n-1 and a polyenyl cation with n+1 carbon atoms. The symmetry requirements for their cyclization are the same as for the polyene with the same number of electrons.

5.2 More Subtle Considerations

5.2.1 Correlation *vs.* Correspondence

The terms *correlation* and *correspondence*, that were so prominent in the preceding discussion, are not quite synonymous.[3] The distinction is illustrated by a comparison of Fig. 5.2 with Fig. 5.3, the correlation diagram actually calculated for the hexatriene-cyclohexadiene interconversion in the simple HMO approximation.

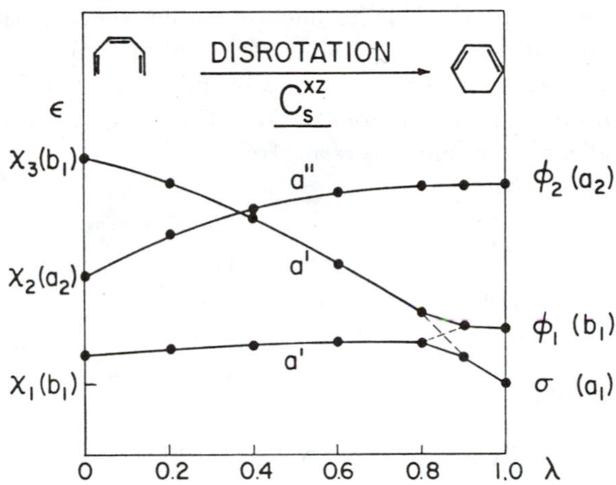

Figure 5.3. Computed (HMO) correlation diagram for cyclization of *cis*-1,3,5-hexatriene. (Adapted from ref. [10]) ϵ: energy; λ: extent of reaction.

The same correlation between the two a_2 orbitals appears in both diagrams but the b_1 orbitals behave differently: The uppermost on the left, χ_3, which is specified in Fig. 5.2 to correspond under a disrotation with σ, is seen in Fig. 5.3 to correlate with ϕ_1, leaving χ_1 to correlate with σ. Fig. 5.2 shows the orbital correlations, or *direct correspondences* in $\mathbf{C}_{2v}^{z}$; in obedience to the non-crossing rule, they are drawn between the two lowest b_1 orbitals on the left and the two with the same irrep on the right. As a result χ_3 has no partner on the right; like σ, it correlates in $\mathbf{C}_{2v}$ with one of the unoccupied antibonding orbitals of the other molecule that are not included in the figure. The diagram then tells us that in order to induce a correspondence between these two orbitals, and thus "allow" the reaction, the pathway has to be desymmetrized to $\mathbf{C}_s$ by a disrotation. Fig. 5.3 describes the correlations *after* the disrotation has been imposed; in $\mathbf{C}_s$, b_1 and a_1 map onto the same subgroup (a'). As predicted by Fig. 5.2, χ_3 starts off towards σ and χ_2 towards ϕ_1, but the crossing is avoided,

[3] The need for a term different from *correlation* was impressed upon the author by Professor Edgar Heilbronner [10], whereupon *correspondence* was adopted for the purpose. The term *corresponding orbitals* had been assigned another meaning, [11] but in a sufficiently different context that no confusion should arise.

the upper line being raised slightly by interaction between the two accidentally degenerate (a') orbitals and the lower being pushed down to the same extent. As a result, there is no net effect on the energy, which – at this lowest level of approximation – is simply the sum of the orbital energies, each multiplied by two, the number of electrons occupying it.

5.2.2 The Rôle of σ-Orbitals

The central rôle assigned by OCAMS to nuclear motion suggests that the alternative modes of desymmetrization capable of "allowing" polyene cyclization might profitably be described as symmetry coordinates with different irreps in C_{2v}^z: b_1 for disrotation and a_2 for conrotation. These are illustrated for the cyclization of butadiene in (**a**) and (**b**) respectively of Fig. 5.4.

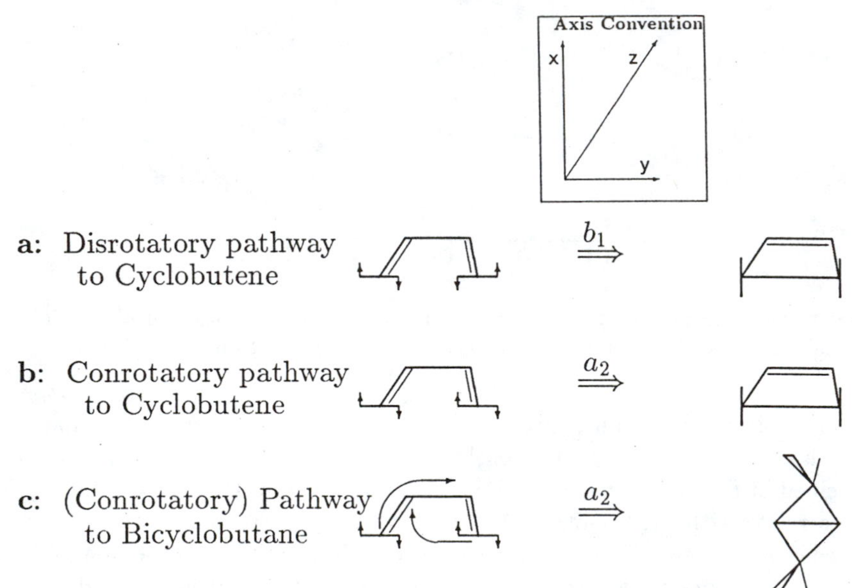

Figure 5.4a–c. b_1 and a_2 coordinates of s-*cis*-Butadiene (**a**) Disrotation; (**b**) Conrotation; (**c**) see text (Section 5.3.1)

As the methylene groups rotate, the atomic orbitals on carbon and hydrogen follow suit. The conventional practice is to ignore the CH bonds and say that the p orbitals on the terminal carbon atoms remain perpendicular to the methylene groups and rotate with them. It is closer to the spirit of OCAMS, and also more consistent with the way reaction-path computations are routinely carried out, to regard the coordinate axes as fixed in space and to label the atomic orbitals accordingly. [12] In s-*cis*-butadiene, the p_x orbitals of carbons 1 and 4 are used for CC bonding and their p_y orbitals for CH bonding; in cyclobutene, these functions are reversed. It follows that the preferred pathway will be one in which these bonding functions are transferred from one set of AOs to the other in a concerted manner at both ends of the molecule. From this point of view, it is clearly unjustified to ignore the CH-bonding orbitals on the terminal carbon atoms; the rudimentary correspondence diagram (Fig. 5.1) is therefore expanded in Fig. 5.5[4] to include them explicitly.

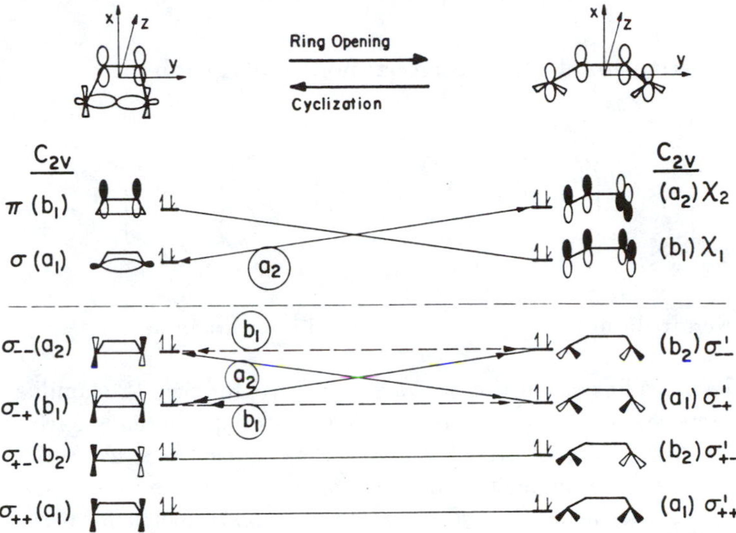

Figure 5.5. Correspondence diagram for cyclization of butadiene

The four CH bonds of cyclobutene interconvert under the four sym-ops of C_{2v} and their linear combinations span its four irreps.[5] In butadiene, all four bonds are in the molecular plane and come in two distinct pairs, an inner and an outer; as a result, only those irreps appear that are symmetric to reflection in $\sigma(yz)$, i.e. a_1 and b_2, each one twice.

Few organic chemists would contest the proposition that the CH-bonding electrons remain localized in the CH bonds as the methylene groups rotate.[6]

The CH-bonding orbitals would therfore be expected to correlate across the diagram. This can be accomplished either by a conrotation, which induces the correspondences: $\sigma_{-+}(b_1) \leftrightarrow \sigma'_{--}(b_2)$ and $\sigma_{--}(a_2) \leftrightarrow \sigma'_{-+}(a_1)$, or by a disrotation, which induces the alternative pair of correspondences: $\sigma_{-+}(b_1) \leftrightarrow \sigma'_{-+}(a_1)$ and $\sigma_{--}(a_2) \leftrightarrow \sigma'_{--}(b_2)$.

[4] The positions of cyclobutene and butadiene in the diagram have been reversed in order to emphasize the fact that the reactant and product have equal status in any symmetry analysis.

[5] Formally, they belong to the *regular representation* of the symmetry point group.

[6] The basic quantum mechanical principle that electrons are indistinguishable is not contravened: It is not claimed that a specific pair of electrons is localized in the bond, but merely that two electrons with opposite spin continue to occupy the region between the bonded nuclei as they change their position in space.

The CC-bonding orbitals behave as in Fig. 5.1: They can be induced to correlate if the reaction path is desymmetrized by displacing the nuclei along an a_2 symmetry coordinate, i.e. a conrotation. Taking both parts of Fig. 5.5 together, we see that while CH bonding is released at both termini by p_x and taken over by p_y along both pathways, only the conrotation releases the appropriate p orbitals of C_1 and C_4 for concerted CC bonding.

5.2.3 Substitutional Desymmetrization: Norcaradiene

Norcaradiene Bisnorcaradiene

In the preceding subsection we have merely reproduced the familiar prediction once more, but the viewpoint adopted above may perhaps have provided additional insight into its origin. When it is recognized that a reaction that is *forbidden* by orbital symmetry conservation to take place in a group of high symmetry can be made *allowed* by an suitable desymmetrizing perturbation, it becomes appropriate to ask whether substitutional desymmetrization, such as that discussed in Section 2.3.2, may not be as effective for that purpose as distortional desymmetrization.

The experimental enthalpy of activation for disrotatory thermal isomerization of *cis*-1,3,5-hexatriene to 1,3-cyclohexadiene in the gas phase at 100°C is 29.2 kcal/mol [13]. The reaction is exothermic by 14.5 kcal/mol [14, p. 127], so $\Delta H^{\ddagger}$ of the reverse reaction is 43.7 kcal/mol, but – in spite of its high activation energy – it is characterized as *allowed* by all of the common orbital symmetry criteria. In norcaradiene ([4.1.0]hepta-2,4-diene), the cyclopropane ring bridging C_1 and C_6 of cyclohexadiene has built the disrotation into the molecule, desymmetrizing it – and its monocyclic isomer, cycloheptatriene – to $\mathbf{C}_s$, in which the b_1 and a_1 orbitals correlate directly (Fig. 5.3). The rate of isomerization is so much faster that it had to be measured at low temperature (ca. 100° K) in a hydrocarbon glass; [15] $\Delta H^{\ddagger}$ is only 6.3 kcal/mol!

5.2.4 Local *vs.* Global Symmetry:

5.2.4.1 Bisnorcaradiene

The isomerization of bisnorcaradiene ([4.4.1]propellatetraene) to 1,6-methano-[10]annulene[7] under similar conditions is also very fast: $\Delta H^{\ddagger}_{100} = 5.1$ kcal/mol

[7] The two H atoms explicitly shown in the illustration of [10]annulene at the top left of Fig. 5.6 are replaced by a methylene bridge.

[16] The conventional procedure would be to regard the molecule as a derivative of norcaradiene, in which two H atoms have been replaced by a butadiene moiety with a "frozen" pair of conjugated double bonds. The reaction would then be analysed in $\mathbf{C}_s$ the *local symmetry* of norcaradiene. Such an assumption implies that these two π bonds remain isolated from the other three in the product isomer, but methano[10]annulene is a nearly planar molecule with $4N + 2$ mobile electrons that has been established by spectroscopic methods to be aromatic. [17] A computational study [18] supports this conclusion, indicating that the *global symmetry* of the annulene, like that of bisnorcaradiene – from which it is separated by a very low barrier – is $\mathbf{C}_{2v}$.

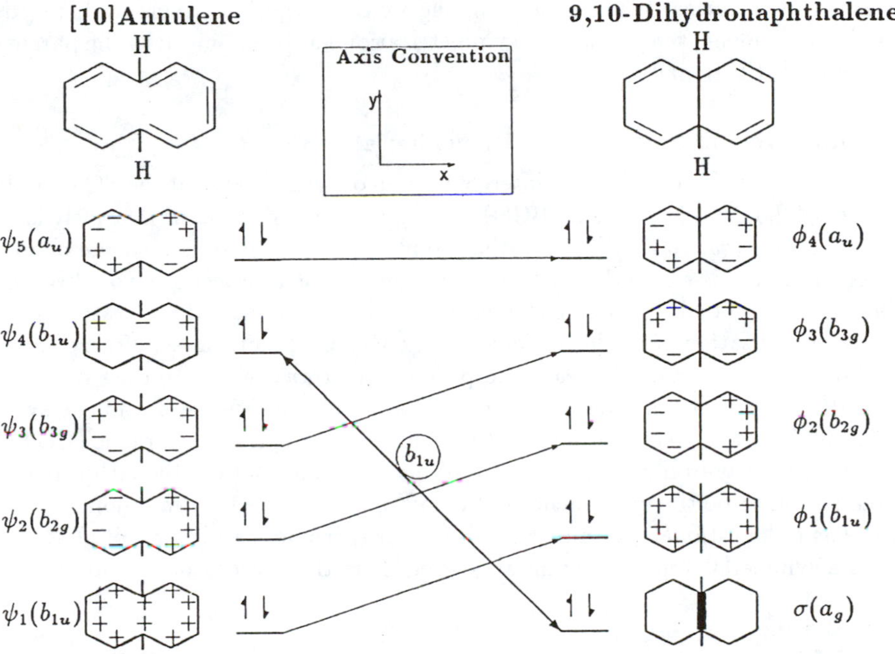

Figure 5.6. Correspondence diagram ($\mathbf{D}_{2h}$) for interconversion of 9,10-dihydronaphthalene and [10]annulene. (Viewed along the positive z axis. The positive lobes of the p_z orbitals are denoted by + and their negative lobes by −)

It will be shown in Section 5.3 that the conceptual division of a conjugated system into two non-interacting subunits can lead to ambiguity. Accordingly, the isomerization is more properly analysed in its global $\mathbf{C}_{2v}$ symmetry, in which all of the occupied orbitals would be expected to correlate across the diagram, since it would be most surprising if such a facile reaction turned out to be *forbidden*. In Fig. 5.6 we go farther: regarding the reaction as being a substitutionally desymmetrized variant of the interconversion between 9,10-dihydronaphthalene and [10]annulene, we take the liberty of setting up the reactant and product in $\mathbf{D}_{2h}$, even though dihydronaphthalene would be much too strained if forced into planarity.[8] It can be formed from [10]annulene with either a *cis* or a *trans* junction, going into $\mathbf{C}_{2v}^{z}$ or $\mathbf{C}_{2h}^{x}$ repectively; Fig. 5.6 determines which of these two modes of desymmetrization is preferred.

The one orbital mismatch in Fig. 5.6 is between $\sigma(a_g)$ on the right and one of the two $\pi(b_{1u})$ orbitals on the left; they can be induced to correspond under a b_{1u} perturbation, the kernel of which is $\mathbf{C}_{2v}^{z}$. Note once more that the induced correspondence connects the higher of these to $\sigma(a_g)$ whereas the lower corresponds directly with a similar $\pi(b_{1u})$ orbital; the two lines necessarily cross. Along the isomerization pathway, both irreps map onto a_1 of $\mathbf{C}_{2v}^{z}$, so the crossing is avoided and the partners are exchanged in the eventual correlation diagram. In the parent system illustrated in Fig. 5.6, the desymmetrization can only be effected by a nuclear displacement; in the isomerization of bisnorcaradiene to methanoannulene, the requisite desymmetrization has been built into the parent system substitutionally.

5.2.4.2 Cyclooctatetraene ↔ Bicyclooctatriene

It has long been known that the interconversion of cyclooctatetraene (COT) and bycyclo[4.2.0]octa-2,4,7-triene (BCO) is rapid in both directions at relatively low temperatures (ca. 100°C) [19, 20]; the product has a *cis* junction, so the disrotatory pathway is evidently preferred. The conventional procedure would be to "freeze" one π bond and regard the reaction as being an *allowed* disrotatory six-electron cyclization of the hexatriene moiety. One could, of course, "freeze" one pair of adjacent π bonds instead, and produce the *trans*-joined product by a no less *allowed* conrotatory four-electron cyclization of the residual butadiene system. In view of the fact that adjacent double bonds of cyclooctatetraene (COT) are twisted away from each other in its stable *tub* conformation, the latter mode would seem to be the more facile, so the failure to observe the *trans*-joined isomer has to be rationalized in terms of its lesser thermodynamic stability, rather than a symmetry-imposed potential barrier along the conrotatory pathway.

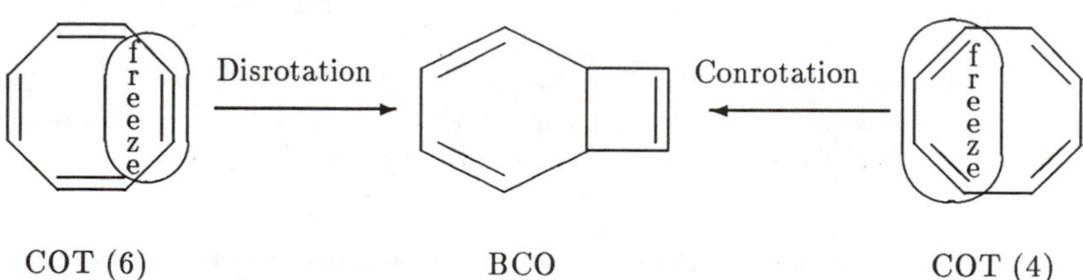

That this is indeed the case can be ascertained from the pair of *WH-LHA* correlation diagrams (Fig. 5.7), in which COT is desymmetrized separately from the global symmetry of the *tub* conformation ($\mathbf{D}_{2d}$) to $\mathbf{C}_s$ and $\mathbf{C}_2$, the respective subgroups corresponding to *cis*- and *trans*-BCO. The result suggests that the 0.01% of BCO that is in thermal equilibrium with COT at 100°C [19] contains

[8] This procedure will be justified in Chapter 7.

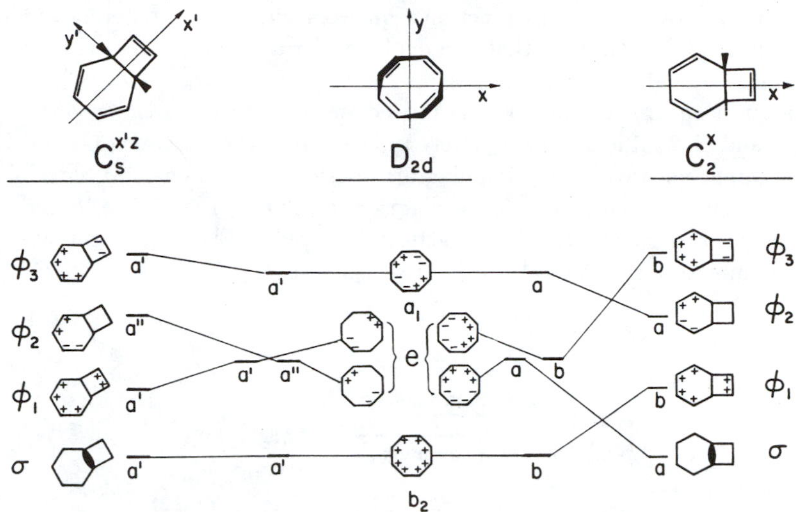

Figure 5.7. WH-LHA correlation diagrams for isomerization of cyclooctatetraene to bicyclooctene. (Viewed along positive z axis. The positive lobes of the p_z orbitals are denoted by + and their negative lobes by −)

a minute amount of *trans*-BCO, that interconverts rapidly with the more stable *cis* isomer *via* reversible rearrangement to COT.

It is clear from the two examples just given that all of the mechanistic information obtainable from an analysis in local symmetry can be deduced from correlation or correspondence diagrams formulated in the global symmetry of the reactant and product. As we will see in the following sections, the converse is not true: the arbitrary separation of a molecular system into moieties with different local symmetries can be misleading.

5.3 "Allowedness" and "Forbiddenness"

5.3.1 Rearrangement of s-*cis*-Butadiene to Bicyclobutane

Let us return to Fig. 5.4 and compare two closely related symmetry coordinates illustrated in it: (**b**) is the familiar conrotation that has been confirmed to be essential for *allowing* the cyclization of s-*cis*-butadiene to cyclobutene, whereas (**c**) leads directly to bicyclobutane. The irrep of both coordinates in C_{2v}^z is a_2; i.e. both are conrotatory. They differ by virtue of the concerted out-of-plane motion of C_1 and C_4, which brings them within bonding distance of C_3 and C_2 respectively, rather than of each other.

Woodward and Hoffmann consider this reaction to be a [2+2]-cycloaddition [1, p. 76], which – if it occurs in one step – the *Rules* require to be $[_\sigma2_s +_\sigma 2_a]$, and cite confirmatory stereochemical evidence [21]. They note, however, that the experimentally observed activation energy (41 kcal/mol) is rather high, and outline several possible stepwise reaction sequences with different stereochemical consequences, [22] suggesting that the evidence for a single process reaction is perhaps less compelling than it might be.

The question arises whether it is indeed justified to treat the π bond between C_1 and C_2 as if it were completely isolated from that between C_3 and C_4. [1, p. 34] Might not two correlation diagrams between butadiene and bicyclobutane, or an OCAMS correspondence diagram similar to Fig. 5.5, afford additional insight into the reaction? The latter, displayed in Fig. 5.8, is constructed in the common global symmetry of the two isomers, $\mathbf{C}_{2v}$.

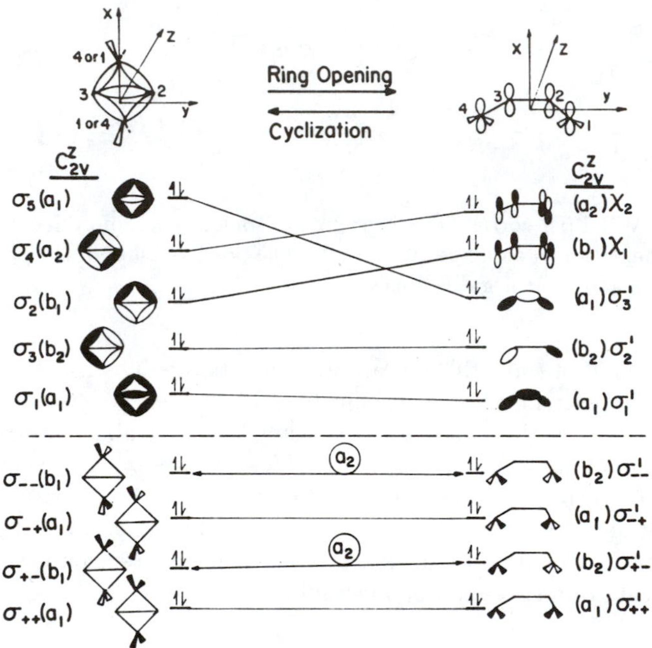

Figure 5.8. Correspondence diagram for interconversion of butadiene and bicyclobutane

In bicyclobutane, the CH bonds to C_1 and C_4 all lie in the xz plane and are not interconvertible. Accordingly, two of the four CH-bonding combinations are totally symmetric (a_1) and the remaining two are labelled b_1. In contrast, the CC bonds that form the four-membered ring are interconverted by the symops of $\mathbf{C}_{2v}$, and their combinations span its four irreps. For completeness, we add the fifth CC-bonding orbital. In zeroth order, it is localized in the central bond between C_2 and C_3, but it interacts – favorably and unfavorably – with the totally symmetric σ_{CC} combination to produce two a_1 orbitals, σ_1 and σ_5. Since the analysis of the reaction is being formulated as an 18-electron problem, butadiene too has to be represented by nine doubly-occupied orbitals: To the six in Fig. 5.5 we add three σ_{CC} orbitals to house the three electron pairs in the

σ bonds that are not broken during the ring opening process but change their orientation in space.

As in Fig. 5.5, the two totally symmetric CH-bonding orbitals correlate across the diagram, but the other two, both of them b_1 on one side and b_2 on the other, can only be induced to correspond by displacement along an a_2 coordinate, *viz.* a conrotation. The five CC-bonding orbitals correlate across the diagram without requiring any non-totally symmetric displacement at all. Evidently, the reaction has to be characterized formally as *allowed* in the subgroup $\mathbf{C}_2$, the kernel of a_2. This conclusion not only differs from that of the *Rules*, but stands in apparent contradiction to experiment: Bicyclobutene is thermodynamically much less stable than butadiene; if its conversion to the latter can take place with conservation of orbital symmetry, why is it so difficult?

The answer requires a reexamination of the all too familiar concepts: *allowed* and *forbidden*. In the case under consideration, it is instructive to compare Figs. 5.5 and 5.8, with specific regard to the atomic orbitals affected by motion of the H atoms bonded to C_1 and C_4. Fig. 5.5 shows disrotation and conrotation to be equally capable of transferring the electrons initially localized in the CH bonds of the reactant to CC-bonding orbitals of the product, and *vice versa*. The conrotation is chosen because the AOs released from CH bonding as the methylene groups rotate are taken over smoothly for CC bonding at both ends; this is not the case along the disrotatory pathway, which is therefore rejected. In contrast, Fig. 5.8 selects the conrotation as the only symmetry coordinate that allows the CH-bonding electrons to stay in the CH bonds as they rotate, whereas the CC-bonding orbitals correlate across the diagram without any reduction of symmetry below $\mathbf{C}_{2v}$. Since the CC-bonding orbitals correlate in the parent symmetry point group, they cannot fail to correlate in any of its subgroups, including $\mathbf{C}_2$, the only non-trivial one[9] that is consistent with the conrotatory motion dictated by the methylene groups. It follows that the reaction is formally *allowed* when analysed by means of the C_2 axis alone, the sym-op retained along the reaction path.

The passive sort of "allowedness" just described is qualitatively different from that illustrated in Fig. 5.5, where the AOs released from CH bonding are taken up smoothly for CC bonding. It is clear from Fig. 5.8 that neither the conjugated π system of butadiene nor the bicyclic σ system of bicyclobutane has anything to gain from the conrotatory motion that describes their interconversion, since the carbon p orbitals released from CH bonding for CC bonding and *vice versa* do not provide a driving force for the rearrangement. Like butadiene and cyclobutene, that are connected by a convenient conrotatory reaction path, bicyclobutane lies in a local minimum on the potential energy surface. Though considerably higher than the other two, it has no incentive to emerge from its potential well, as this would require beginning to break two σ bonds before any energetic advantage is gained from concerted π bond formation.

[9] Every group has the trivial subgroup $\mathbf{C}_1$, which only contains the identity element.

5.3.2 The Bond-Bisection Requirement: Benzvalene

The question posed in the preceding paragraphs as to the need for a reevaluation of the concept of "allowedness", can be dismissed as a "non-problem" as long as it is taken as axiomatic that, for an orbital symmetry analysis to be of any use, "the symmetry elements [retained along the pathway][10] must bisect bonds made or broken in the process". In contrast to the *allowed* conrotatory cyclization of butadiene to cyclobutene, in which the C_2 axis bisects a newly formed σ bond, the only bond bisected by the axis in its conversion to bicyclobutane is the one between C_2 and C_3, which is essentially single in both the reactant and the product.

It has already been pointed out that the "bond-bisection requirement" is imprecise.[11] Moreover, even if this restriction is accepted provisionally, the problem reappears as soon as we get to the higher homologs of bicyclobutane: benzvalene, in which the bicyclobutane moiety is fused onto ethylene, and naphthvalene, in which it is fused to a benzene ring. Both of these molecules are much less stable thermodynamically than their respective aromatic valence isomers, benzene and naphthalene, but – like cyclobutadiene – are remarkably resistant to thermal isomerization. Naphthvalene will be discussed at some length in a subsequent chapter, in connection with its unusual photochemical properties. We will here restrict our attention to the reluctance of benzvalene to undergo thermal isomerization to benzene.

Soon after benzvalene was synthesized and its remarkable kinetic stability observed, [23] van der Hart, Mulder and Oosterhoff [24] analyzed the reaction using a valence bond method. They declared the reaction to be *symmetry allowed* or, at any rate, more so than the genuinely *forbidden* thermal isomerization to benzene of its other two familiar valence isomers, Dewarbenzene and prismane. A few years later, the present author [25] showed that the molecular orbitals of benzvalene correlate with those of benzene, not only in $\mathbf{C}_2$, the group of highest symmetry that can be retained along the pathway, but also in $\mathbf{C}_{2v}$, the symmetry point group of benzvalene itself. It was pointed out that the "allowedness" of a reaction path having the correct symmetry to "allow" a reaction that is *forbidden* in a group of higher symmetry is qualitatively different from the more passive "allowedness" of a reaction in which the electron configurations of the reactant and product correlate in a symmetry point group higher than that prescribed by the geometric requirements of the reaction.

The rôle of the CH-bonding orbitals in specifying these geometric requirements had not yet been recognized, so they were not included in the correspondence diagram. They are included in Fig. 5.9, which is completely analogous to Fig. 5.8 and leads to identical conclusions. The orbitals of the bicyclobutane

---

[10] Woodward and Hoffmann's original statement, that refers to "the symmetry elements chosen for the analysis", has been broadened slightly. In OCAMS, the symmetry elements are not "chosen", but comprise all of those common to the reactant and product, some of which are not retained along the reaction path. In the classic WH-LHA correlation procedure [1, p. 31], the elements are necessarily "chosen" from among those "retained".

[11] See Section 1.4 and also footnotes 12 of Chapter 1 and 10 of Chapter 4.

moiety are labeled as in Fig. 5.8, taking due account of the new orientation and axis convention,[12] according to which the symmetry of the molecule is $\mathbf{C}_{2v}^{x}$. The π orbital completes the CC-bonding subsystem, to which we add the two CH-bonding combinations involving C_1 and and C_6.

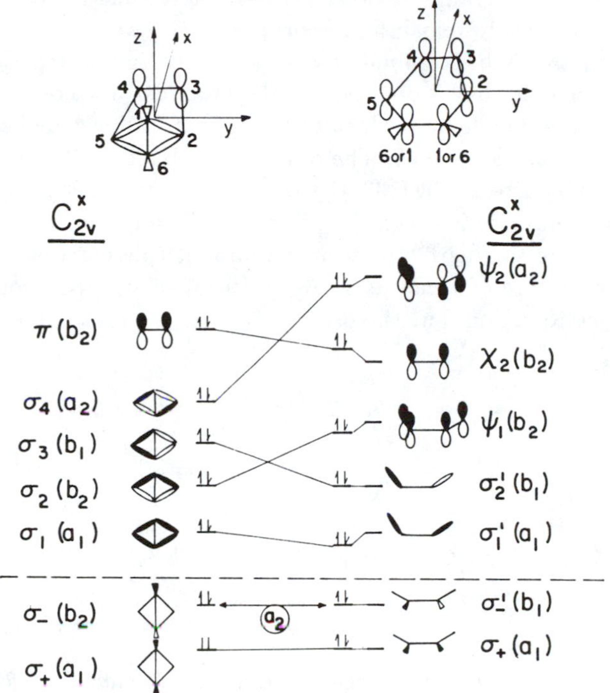

Figure 5.9. Correspondence diagram for isomerization of benzvalene to benzene

The symmetry of benzene is reduced from $\mathbf{D}_{6h}$ to $\mathbf{C}_{2v}^{x}$ by shortening the bond between C_3 and C_4 and lengthening equally those between C_2 and C_3 and between C_4 and C_5. The former becomes more like a double bond and the latter two more like single bonds, as illustrated schematically by the form given to the π orbitals. This description is grossly exaggerated, but it should be kept in mind that even the slightest displacement of the nuclei that desymmetrizes benzene from $\mathbf{D}_{6h}$ to $\mathbf{C}_{2v}^{x}$ splits the degeneracy of the two upper occupied orbitals of benzene (e_{1g} in Fig. 1.2) to a_2 and b_2. The two CH-bonding combinations are added; so are two σ_{CC} orbitals, to accommodate the electron pairs in the bonds that change their orientation in space but are not broken.[13]

[12] The molecules in the diagram have been reoriented for clarity. The axis convention has been changed from that of the previously published diagram [25, Fig. 4] in order to bring it into line with established practice: the z axis is perpendicular to the benzene ring, x axis bisects two opposing bonds of benzene and the y axis passes through an opposing pair of carbon atoms.

[13] The σ bond between C_1 and C_6 is omitted; its inclusion would merely have added an additional a_1 orbital on each side of the diagram.

Correspondence of the CH-bonding orbitals is induced by a conrotation (a_2), but the CC-bonding systems on both sides of the diagram correlate in the higher symmetry of the parent group. As in the case of bicyclobutane, the MOs of benzvalene necessarily correlate in $\mathbf{C}_2$ with those of benzene, but the desymmetrization to the subgroup provides no driving force for the rearrangement, so benzvalene is content to remain in its local potential well.

Does the C_2 axis bisect bonds made or broken in the reaction? It would be hasty to conclude that it does not on the grounds that bond C_1C_6 is single and bond C_3C_4 is double on both sides of Fig. 5.9. In benzene both bonds are aromatic, with bond order 1.67. The order of the bond C_1C_6 increases by 0.67 and that of C_3C_4 decreases by 0.33; the absolute values of the changes in the two bisected bonds thus add up to one bonding unit. It might be noted at this point that the *Rules* are singularly uninformative about this reaction: The choice of one Kekulé structure of benzene in Fig. 5.10 makes the reaction a *forbidden* $[_{\sigma}2_a +_{\sigma} 2_a]$ cycloreversion, but the other makes it $[_{\sigma}4_a +_{\sigma} 2_a]$, and consequently *allowed*.

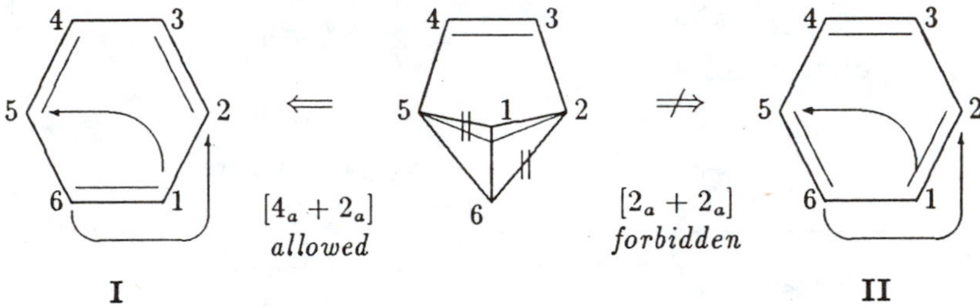

Figure 5.10. The benzvalene-benzene interconversion: *allowed* or *forbidden*? (For clarity the arrows are drawn to depict the reverse reaction: $[_{\pi}2 +_{\pi} 2]$ cycloaddition, to which the same *Rules* apply)

Nguyen Trong Anh [26] argues that the rôle of the double bond between C_3 and C_4 is negligible, because the transition state is an early one; as a result, the reaction is very similar to the bicyclobutane-cyclobutene isomerization. This is no doubt true, but it is so because the carbon π orbitals released from CH bonding by the conrotation do not facilitate electron release from the bicyclic σ system of the reactant to the conjugated π system of the product. Abandoning the global symmetry of the molecule in favor of an "artificial symmetry axis" that bisects the bonds actually broken – on the prior assumption that the double bond of benzvalene is virtually unshifted in the transition state – is equivalent to the selection of Kekulé structure *II* and neglect of *I* in Fig. 5.10. While it certainly provides an *a posteriori* rationalization of the resistance of benzvalene to isomerization, it has very little predictive value.

Empirically, there is no doubt that the energetic demands of simultaneous rupture of both σ bonds is too demanding. Computations with extensive configuration interaction [27] confirm the results of an earlier SCF calculation [28] according to which the two bonds are lengthened unequally in the transition state for thermal isomerization. However, only one transition state is found along the pathway, so the bond-breaking process has to be characterized as *concerted* but not *synchronous*.[14]

5.3.3 Genuinely *Forbidden* Valence Isomerizations

In contrast to the thermal isomerization of benzvalene to benzene, those of the other two well known valence isomers, Dewarbenzene and prismane are genuinely *forbidden*.[15] Not only do the occupied orbitals fail to correlate along

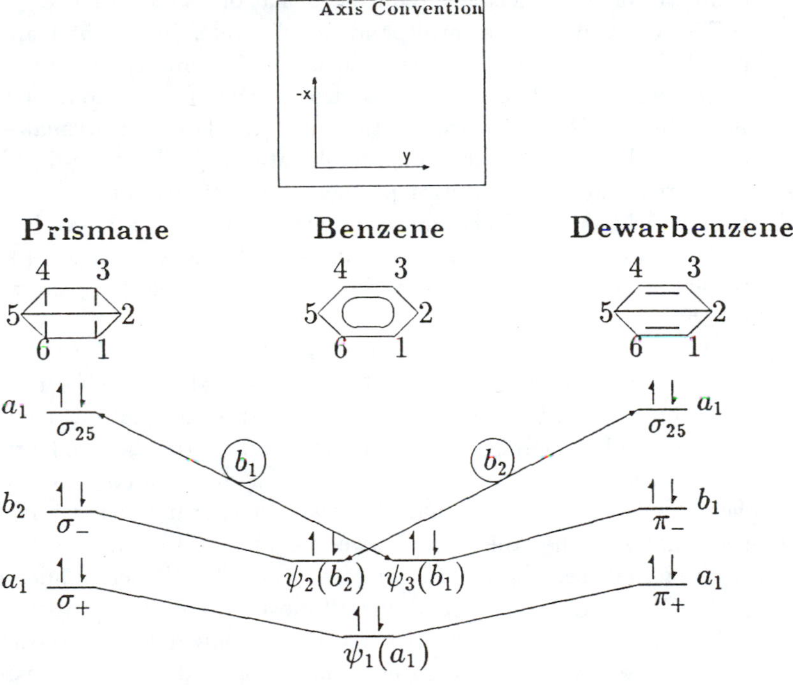

Figure 5.11. Correspondence diagrams for thermal isomerization of prismane and dewarbenzene to benzene ($\mathbf{C}_{2v}^{z}$). (The molecules are viewed along the positive z-axis. See text for identification and labeling of the orbitals)

---

[14] *Concerted*: Two or more *primitive changes* are said to be *concerted* (or to constitute a concerted process) if they occur within the same elementary reaction. Such changes will normally (though perhaps not inevitably) be "energetically coupled".

Synchronous: A *concerted* process in which the primitive changes concerned (generally bond rupture and bond formation) have progressed to the same (or comparable) extent at the transition state is said to be *synchronous* [29]

[15] The relatively facile spin-non-conservative isomerization of Dewarbenzene [30] will be taken up in a subsequent chapter.

the pathway of lowest common symmetry, but the symmetry coordinates which have the proper irrep to induce a correspondence between the non-correlating orbitals in each case lead the system away from the postulated product rather than towards it.

The MOs in the correspondence diagrams displayed in Fig. 5.11 are labeled according to their irreps in $\mathbf{C}_{2v}$, the group of highest common symmetry. Dewarbenzene genuinely belongs to this symmetry point group, which is a subgroup of both $\mathbf{D}_{6h}$ and $\mathbf{D}_{3h}$. Benzene is desymmetrized from $\mathbf{D}_{6h}$ to $\mathbf{C}^z_{2v}$ by raising C_2 and C_5 ever so slightly out of plane, i.e. displacing these atoms along a properly oriented e_{2u} symmetry coordinate. The displacement that desymmetrizes prismane from $\mathbf{D}_{3h}$ to $\mathbf{C}_{2v}$ by lengthening any one of its three long bonds relative to the other two, transforms as e'. Having fixed C_2 and C_5 in the yz plane and chosen to lengthen the bond connecting them, the z component of e' is chosen and the symmetry of the molecule is reduced to that of its co-kernel, $\mathbf{C}^z_{2v}$.[16]

Unlike Figs. 5.1 and 5.2, the correspondence diagrams in Fig. 5.11 are not "rudimentary", because they incorporate all of the information necessary to deduce the "allowedness" of the thermal isomerizations. The six combinations of the σ_{CC}-bonding orbitals of benzene retain their $\mathbf{C}_{2v}$ labels in prismane and Dewarbenzene as do its six CH-bonding combinations.[17] The analysis of the reaction is thus reduced to a six-electron problem; the occupied orbitals involved are: the three Hückel-MOs of benzene – labelled as in Fig. 1.2, the long σ-bonding orbital (σ_{25}) in its two valence isomers, and the symmetric (+) and antisymmetric (−) combinations of σ_{13} and σ_{46} in prismane or of π_{16} and π_{34} in Dewarbenzene.

Fig. 5.11 tells us first what has been known for a long time: both valence isomerizations are *forbidden* on the ground-state potential surface if symmetry is not reduced below $\mathbf{C}_{2v}$. It adds, however, that both isomerizations could be induced to occur under appropriate reductions of symmetry: by displacement along a b_1 symmetry coordinate in one case and a b_2 symmetry coordinate in the other. The former implies that the reaction coordinate must include a component that stretches the bond between C_1 and C_6 of prismane while compressing that between C_3 and C_4, or vice versa. The latter implies that atoms C_3, C_4, C_1 and C_6 of Dewarbenzene all move in one direction while C_2 and C_5 – which would have to move apart as the molecule flattens towards the geometry of benzene – are displaced together in the opposite direction instead.

The consequences of the requirement that a non-totally symmetric displacement must be incorporated into the reaction coordinate can be interpreted in two complementary ways:

a) Superposition of a non-totally symmetric displacement onto the "least motion" ($\mathbf{C}_{2v}$) reaction coordinate produces a vibrationally excited transition state, and thus a gratuitously high energy – and free energy – of activation.

---

[16] Caution! In order to retain the same axis convention for all three isomers, it is necessary to specify y, rather than the commonly chosen z, to be the threefold rotational axis of prismane.
[17] It is easily, if unnecessarily, confirmed that the two sets transform as $[2\times a_1 \oplus a_2 \oplus 2\times b_1 \oplus b_2]$ and $[2\times a_1 \oplus a_2 \oplus b_1 \oplus 2\times b_2]$ respectively.

b) The displacement moves the atoms away from the geometry of the putative product so as to circumvent the symmetry-imposed barrier. In the case of a rigid reactant molecule, the same displacement in the reversed phase might get it back on the pathway at an additional cost in vibrational energy. Alternatively, as will be seen in the following chapter in connection with less constrained systems, the motion dictated by the induced correspondence may move the reactant(s) in a direction leading to a different product, and thus open an alternative reaction pathway for investigation.

5.3.4 Quantifying "Allowedness": Cubane ↔ Cyclooctatetraene

The fact that the benzvalene-benzene interconversion is *allowed* in global symmetry led Mulder [31] to conclude that the *Woodward-Hoffmann Rules*, that were designed to deduce stereochemistry, should be distinguished from the use of orbital correlation diagrams, and that correlation of the occupied orbitals of the reactant and product is no guarantee that the reaction will not have a high activation energy.

The reliability of stereochemical criteria of "allowedness" will be looked into more carefully in the following chapters. Let us here consider the validity of the "equations":

$$\text{facile} = \textit{allowed} \quad ; \quad \text{difficult} = \textit{forbidden},$$

which have been approved by the authoritative statement [1, p. 33]: "In fact, whether a reaction is symmetry-allowed or forbidden is determined by the height of the electronic hill that reactant or product must climb in reaching the transition state."

There is a logical difficulty here: If the fact that certain reactions proceed more readily than others is *explained* by the fact that the former are "symmetry-allowed" and the latter are not, it is not very enlightening to *define* "symmetry-allowed" reactions as those that proceed readily and "symmetry-forbidden" as those that do not. Needless to say, the distinction between *symmetry-allowed* and *symmetry-forbidden* reactions is useful only if it can be shown to have energetic consequences, but the criteria that distinguish between them surely have to be based on symmetry rather than on energy.

There is a practical difficulty as well. Even in the case of those reactions cited in this chapter that are commonly considered to be *allowed*, the activation energies are spread over a wide range, and the reaction conditions required for their kinetic investigation range from isolation in a glass matrix at −100°C to gas kinetics at well over 200°C. For example, the *allowed* disrotatory decyclization of 1,3-cyclohexadiene to *cis*-1,3,5-hexatriene has an activation enthalpy well in excess of 40 kcal/mol (Section 5.2.3), but it is so much more facile than decyclization along the alternative *forbidden* conrotatory pathway that the energetic consequences of orbital symmetry conservation are confirmed.

A particularly instructive case in point is the reluctance of cyclooctatetraene (COT) to isomerize to cubane in one step [1, pp. 32–33]. Cubane is substantially less stable than COT, so it is more reasonable to check the "allowedness" of the reverse process, which has the lower activation energy. It can be argued that if $\Delta H^\ddagger$ for the concerted rupture of one σ bond and formation of one π bond can be as high as 40 kcal/mol, the simultaneous conversion of four π bonds to four σ bonds might well cost 160 kcal/mol and still be formally *allowed*! It turns out that the activation energy for the thermal isomerization of cubane to COT is surprisingly low: $\Delta H^\ddagger$ at 250°C is only 42. kcal/mol [32, 33]. The authors prefer to interpret the reaction as proceeding *via* initial rupture of a single bond followed by a series of steps leading to vibrationally excited COT; an alternative interpretation that has been recently proposed invokes two successive *forbidden* cycloreversions [34]. However, the observed increase with increasing gas pressure of the ratio of COT to its fragmentation products suggests instead that vibrationally excited COT is the primary product, formed from cubane in a single step.

Since the energetic criterion of "allowedness" is logically questionable and its application to the Cubane-COT interconversion is so uncertain in practice, it had best be abandoned in favor of a distinction that is based firmly on symmetry properties; the relative utility of local symmetry *vs.* global symmetry for this purpose arises once more.

5.3.4.1 Analysis in Global Symmetry

Each of the two molecules has a total of 40 valence electrons, lodged in 20 doubly-occupied MOs. The 8 σ_{CC} and 8 σ_{CH} bonds that make up the σ frame of COT have identical symmetry properties in cubane, so it is enough to consider the MOs comprising the four bonds that are interconverted during the isomerization.

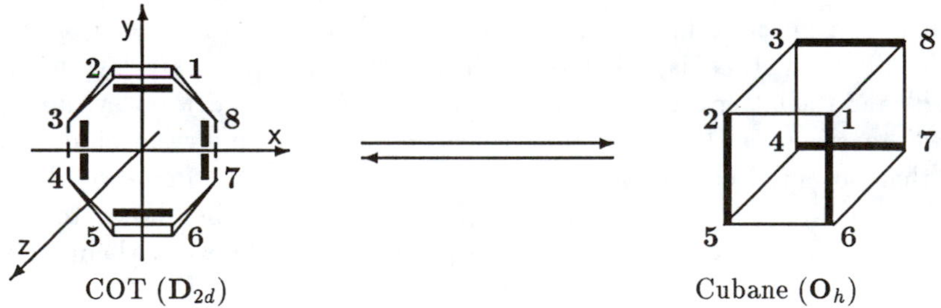

COT ($\mathbf{D}_{2d}$) Cubane ($\mathbf{O}_h$)

We place COT in the xy plane and allow it to relax to its stable *tub* conformation. Atoms 1, 2, 5 and 6 move to positive z and 3, 4, 7 and 8 to negative z, behind the plane of the diagram, but the four π bonds can still be thought of essentially as being linear combinations of their p_z AOs. The four linear combinations of bond orbitals (LCBOs), labeled by their irreps in $\mathbf{D}_{2d}$, are[18]:

$$\psi_1(b_2) = \pi_{12} + \pi_{34} + \pi_{56} + \pi_{78} \tag{5.2}$$

$$\psi_2(e_{yz}) = \pi_{12} - \pi_{56} \tag{5.3}$$

[18] The reader who is reluctant to take Equations 5.2–5.9 on faith is invited to apply the procedures outlined in Section 4.4.3.

$$\psi_3(e_{xz}) = \pi_{34} - \pi_{78} \tag{5.4}$$
$$\psi_4(a_1) = \pi_{12} - \pi_{34} + \pi_{56} - \pi_{78} \tag{5.5}$$

Cubane has cubic ($\mathbf{O}_h$) symmetry, but a slight elongation of the four σ bonds emphasized in the digram, desymmetrizes the molecule to $\mathbf{D}_{2d}$. The four LCBOs are characterized as follows:

$$\phi_1(a_1) = \sigma_{16} + \sigma_{25} + \sigma_{38} + \sigma_{47} \tag{5.6}$$
$$\phi_2(e_y) = \sigma_{47} - \sigma_{38} \tag{5.7}$$
$$\phi_3(e_x) = \sigma_{25} - \sigma_{16} \tag{5.8}$$
$$\phi_4(b_2) = \sigma_{16} + \sigma_{25} - \sigma_{38} - \sigma_{47} \tag{5.9}$$

The electron-configurations of the two molecules correlate directly in $\mathbf{D}_{2d}$, so the reaction must be characterized as *allowed*.

5.3.4.2 Analysis in Local Symmetry

Woodward and Hoffmann [1, p. 33] state that although the occupied orbitals correlate along the pathway of least motion, the reaction is *forbidden* because two of the occupied π orbitals of COT correlate at the MO level of approximation with two unoccupied σ orbitals of cubane that have the same symmetry labels, and vice versa. This apparent non-correlation, which is not evident when the reaction in analyzed in global symmetry, induced the authors to invoke local symmetry instead and describe the reaction as comprising two *forbidden* [2+2]-cycloadditions.

The source of the offending bonding-antibonding correlations is revealed when the relevant bond-orbitals in Equations 5.1–5.8 are rewritten as combinations of their contributing AOs. For COT:

$$\psi_1(b_2) = p_z^1 + p_z^2 + p_z^3 + p_z^4 + p_z^5 + p_z^6 + p_z^7 + p_z^8 \tag{5.10}$$
$$\psi_2(e_{yz}) = p_z^1 + p_z^2 - p_z^5 - p_z^6 \tag{5.11}$$
$$\psi_3(e_{xz}) = p_z^3 + p_z^4 - p_z^7 - p_z^8 \tag{5.12}$$
$$\psi_4(a_1) = p_z^1 + p_z^2 - p_z^3 - p_z^4 + p_z^5 + p_z^6 - p_z^7 - p_z^8 \tag{5.13}$$

For cubane[19]:

$$\phi_1(a_1) = p_y^6 - p_y^1 + p_y^5 - p_y^2 + p_x^3 - p_x^8 + p_x^4 - p_x^7 \tag{5.14}$$
$$\phi_2(e_y) = p_x^4 - p_x^7 - p_x^3 + p_x^8 \tag{5.15}$$
$$\phi_3(e_x) = p_y^5 - p_y^2 - p_y^6 + p_y^1 \tag{5.16}$$
$$\phi_4(b_2) = p_y^6 - p_y^1 + p_y^5 - p_y^2 - p_x^3 + p_x^8 - p_x^4 + p_x^7 \tag{5.17}$$

Let us mimic the pathway that retains $\mathbf{D}_{2d}$ symmetry: Rotate C_1 and C_2 of COT about the x axis in one sense while rotating C_5 and C_6 in the other,

[19] Note that in order for two p AOs to form a σ bond they have to point towards each other, so they must have opposite signs.

both pairs moving towards positive z. Simultaneously rotate C_3 and C_4 about y in one sense and C_7 and C_8 in the other, both pairs moving towards negative z. On the conventional assumption that atomic orbitals rotate with the atoms, when the geometry of cubane has been attained, each of the eight p_z orbitals will have been converted to either $\pm p_x$ or $\pm p_y$ as listed in Table 5.1.

Table 5.1. Transformation of p-AOs of COT to those of cubane (see text)

COT :	p_z^1	p_z^2	p_z^3	p_z^4	p_z^5	p_z^6	p_z^7	p_z^8
cubane :	$-p_y^1$	$-p_y^2$	$-p_x^3$	$-p_x^4$	p_y^5	p_y^6	p_x^7	p_x^8

Substituting the transformed AOs into Equations 5.10–5.13 and rearranging them in the same order as in Equations 5.14–5.17, we obtain:

$$\psi_1 \longrightarrow p_y^6 - p_y^1 + p_y^5 - p_y^2 - p_x^3 + p_x^8 - p_x^4 + p_x^7 \quad (5.18)$$
$$\psi_2 \longrightarrow -p_y^5 - p_y^2 - p_y^6 - p_y^1 \quad (5.19)$$
$$\psi_3 \longrightarrow p_x^3 + p_x^8 + p_x^4 + p_x^7 \quad (5.20)$$
$$\psi_4 \longrightarrow p_y^6 - p_y^1 + p_y^5 - p_y^2 + p_x^3 - p_x^8 + p_x^4 - p_x^7 \quad (5.21)$$

Comparison of Equations 5.18 with 5.17 and of Equation 5.21 with 5.14 shows that ψ_1 and ψ_4 go smoothly to ϕ_4 and ϕ_1 of cubane respectively. The p orbitals in Equations 5.19 and 5.20, however, all point in the same direction; they evidently do not correspond to the bonding σ orbitals of Equations 5.16 and 5.15 respectively, but rather to a degenerate pair of σ^* orbitals that are unoccupied in cubane. The "forbiddenness" of the COT-cubane interconversion would seem to have been confirmed.

Plausible as the foregoing argument may appear, it is seen to be flawed when the conclusion that the two π MOs of COT with irrep e correlate with σ^* orbitals of cubane is examined critically. Adopting the viewpoint of Section 5.2.2, in which the nucleii are seen to be moving into AOs that retain their orientation in space, we realize that the p_x, p_y and p_z orbitals exchange their bonding functions during the reaction: The p_z orbitals, which in COT are restricted mainly to π bonding,[20] are increasingly incorporated in the eight σ_{CC}- and eight σ_{CH}-bonding orbitals that were not considered explicitly in the analysis. When the irreps of all forty orbitals are considered explicitly, they are found to contain no less than ten pairs of MOs that are of symmetry species E – five of them occupied and five unoccupied. In the higher symmetry of cubane, these are distributed among the four triply degenerate irreps of $\mathbf{O}_h$, but they can be expected to mix thoroughly on desymmetrization to $\mathbf{D}_{2d}$, reducing the energy of activation to well below that required to rupture four isolated, if strained, σ bonds or, for that matter, to force two simultaneous genuinely *forbidden* $[{}_\pi 2_s + {}_\pi 2_s]$ cycloreversions.

[20] They are entirely so when COT becomes planar ($\mathbf{D}_{4h}$) during tub-tub inversion.

5.3.5 The Bottom Line So Far

The considerations developed in this chapter lead to the following conclusions:

1. An orbital symmetry analysis can be carried out reliably and unambiguously only in the global symmetry of the reacting system, i.e. a symmetry point group common to the reactant and product.

2. Occupied orbitals can be omitted from the correlation diagram provided that their irreps are obviously in one-to-one corrspondence. Otherwise they should be included, even if they do not appear to be directly involved in the reaction.

3. When the electronic configurations of the reactant and product correlate in the point group of their highest common symmetry, their interconversion is evidently not *forbidden* by orbital symmetry conservation. It can therefore hardly be regarded as other than *allowed*, even when a substantial investment in activation energy is required in order to nudge a thermodynamically unstable molecule out of its comfortable local minimum on the potential energy surface.

4. Whenever the electronic configurations of the reactant and product fail to correlate in the highest symmetry point group common to them, their interconversion is *forbidden* by orbital symmetry conservation along any pathway that retains that symmetry point group. The irrep of the displacement(s) required to *allow* the reaction, by desymmetrizing the pathway into its kernel (or co-kernel) subgroup, is determined with a correspondence diagram. However, a formally *allowed* pathway is practically feasible only if incorporation of symmetry coordinates of the *allowing* irrep(s) into the reaction coordinate is geometrically and energetically consistent with conversion of the reactant into the postulated product.

As defined above, the term *forbidden reaction* poses no problem. An *allowed reaction*, however, is defined simply as one that is not *forbidden* in the symmetry point group in which the reaction is being analyzed; this is, of course, at variance with the rather vague way in which the term *allowed* is ordinarily understood. In order to avoid confusion, the latter term will be used sparingly in the subsequent chapters.

The points noted above have so far only been shown to apply to reactions that proceed on the closed shell ground-state surface. Even within this restricted, if extensive, class of reactions, several questions arise, that will be addressed – among others – in the following three chapters:

1. How reliable is the orbital approximation, and is it possible to identify in advance those cases in which it is apt to break down?

2. Most organic molecules are not highly symmetrical. Does the proposed restriction of orbital symmetry analysis to the global symmetry of the reactant and product exclude their reactions from consideration, or can the approach be extended so as to take them into account?

3. How dependent is the result of a symmetry analysis on the relative geometry in which the reactant and product are set up in the diagram? This

question, which did not arise in connection with the isomerizations used as examples in this chapter, attains considerable importance in bimolecular reactions, where the initial relative orientation of the reacting molecules is unrestricted.

5.4 References

[1] R.B. Woodward and R. Hoffmann: *The Conservation of Orbital Symmetry.* Verlag Chemie, Weinheim and Academic Press, New York 1970.
[2] E.A. Halevi: Helvet. Chim. Acta. *58*, 2136 (1975).
[3] J. Katriel and E.A. Halevi: Theoret. Chim. Acta *40*, 1 (1975).
[4] E.A. Halevi: Angew. Chem. *88*, 664 (1976); Angew. Chem. Internat. Ed. (English) *15*, 593 (1976).
[5] R.B. Woodward and R. Hoffmann: J. Amer. Chem. Soc., *87*, 2046 (1965).
[6] H.C Longuet-Higgins and E.W. Abrahamson: J. Amer. Chem. Soc. *87*, 2045 (1965).
[7] R.G. Pearson: Theoret. Chim. Acta. *16*, 107 (1970).
[8] R.G. Pearson: *Symmetry Rules for Chemical Reactions.* Wiley, New York 1976.
[9] R. Huisgen, A. Dahmen and H. Huber: J. Amer. Chem. Soc. *89*, 7130 (1967).
[10] E. Heilbronner: *Personal communication.* March 22, 1974.
[11] A.T. Amos and G.G. Hall: Proc. Roy. Soc. (London) *A-220*, 483 (1961).
[12] E.A. Halevi, J. Katriel, R. Pauncz, F.A. Matsen and T.L. Welsher: J. Amer. Chem. Soc. *100*, 359 (1978).
[13] K.E. Lewis and H. Steiner: J. Chem. Soc. *1964*,3080.
[14] J.J. Gajewski: *Hydrocarbon Thermal Isomerizations.* Academic Press, New York 1981.
[15] M.B. Rubin: J. Amer. Chem. Soc. *103*, 7779 (1981).
[16] H. Frauenrath, M. Kapon, M.B. Rubin and H.D. Scharf: Israel J. Chem. *29*, 307 (1989).
[17] E. Vogel and H.D. Roth: Angew. Chem. *76*, 145 (1964); Angew. Chem. Int. Ed. (English) *3*, 228 (1964).
[18] L. Farnell and L. Radom: J. Amer. Chem. Soc. *104*, 7650 (1982).
[19] R. Huisgen and F. Mietzsch: Angew. Chem. *76*, 36 (1964); Angew. Chem. Int. Ed. (English) *3*, 83 (1964).
[20] E. Vogel, H. Kiefer and H.D. Roth: Angew. Chem. *76*, 432 (1964); Chem. Int. Ed. (English) *3*, 442 (1964).
[21] G.L.Closs and P.E. Pfeffer: J. Amer. Chem. Soc. *90*, 2452 (1968).
[22] See [1, pp. 76–78] and references cited therein.
[23] T.J. Katz, E.J. Wang and N. Acton: J. Amer. Chem. Soc. *93*, 3782 (1971).
[24] W.J. van der Hart, J.J.C. Mulder and L.J. Oosterhoff: J. Amer. Chem. Soc. *94*, 5728 (1972).
[25] E.A. Halevi: Nouveau J. Chim. *1*, 229 (1977).
[26] Nguyen Trong Anh: The Use of Aromaticity Rules, Frontier Orbitals and Correlation Diagrams: Some Difficulties and Unsolved Problems. In: R. Daudel (ed.) *Quantum Theory and Chemical Reactions.* Reidel, Dordrecht 1980, pp. 177–189.
[27] S. Oikawa, M. Tsuda, Y. Okamura and T. Urabe: J. Amer. Chem. Soc. *106*, 6751 (1984).

[28] M.J.S. Dewar and S. Kirschner: J. Amer. Chem. Soc. *97*, 2932 (1975).

[29] V. Gold (Compiler and Editor): *Glossary of Terms Used in Physical Organic Chemistry*. Pure and Appl. Chem. *55*, 1281 ff. (1983).

[30] R. Lechtken, R. Breslow, A.H. Schmidt and N.J. Turro: J. Amer. Chem. Soc. *95*, 3025 (1973).

[31] J.J.C. Mulder: J. Amer. Chem. Soc. *99*, 5177 (1977).

[32] H.-D. Martin, T. Urbanek, P. Pföhler and R. Walsh: J .Chem. Soc. Chem. Commun. **1985**, 964.

[33] K. Hassenrück, H.-D. Martin and R. Walsh: Chem. Revs. *89*, 1125 (1989).

[34] W. von E. Doering, W.R. Roth, R .Brueckmann, L. Figge, H.-W. Lennarz, W.D. Fessner and H. Prinzbach: Chem. Ber. *121*, 1 (1988).

Chapter 6

Cycloadditions and Cycloreversions: I. [2+2]-Cycloaddition

It was pointed out in the preceding chapter that correlation diagrams can be read from right to left as easily as from left to right. This is certainly true from the formal viewpoint of orbital symmetry, but does not preclude the need to examine the geometric and energetic consequences of the nuclear motions involved. Thus, for example, the necessity for including a conrotatory (a_2) displacement in the reaction coordinate for the cyclization of s-*cis*-butadiene (Fig. 5.1) implies that a coordinate with the same irrep has to be incorporated into the reaction coordinate for ring opening of cyclobutene. However the energetic requirements of the nuclear motions involved differ greatly. Internal rotation about the central bond of butadiene – with concomitant conrotation of its terminal methylene groups – is quite facile, whereas the reverse of the same motion in cyclobutene is opposed by a substantial restoring force.

Considerations of this kind, that were not emphasized in connection with the unimolecular reactions dealt with in the preceding chapter, attain crucial importance when the geometric requirements of cycloadditions and cycloreversions are compared. Like the isomerizations previously discussed, cycloreversions are unimolecular; a non-totally symmetric vibrational motion that may be called for by the correspondence diagram will ordinarily be opposed by a restoring force. Cycloadditions, at least the prototypical ones, are bimolecular: the two reactants can approach each other in a variety of ways, their reorientation in space costing no energy at all. It then becomes reasonable to ask how the conclusions which may be reached by the orbital symmetry analysis depend on the initial geometry assumed for the approach of the reactants towards one another.

6.1 Addition of Singlet Carbenes to Ethylene

The simplest cycloaddition, if the dimerization of two methylenes to form ethylene (Section 4.3) can be excluded, is the (1+2)-addition[1] to ethylene of a singlet carbene CX_2, where X is a halogen or another electronegative substituent. The reactions of methylene itself, and other carbenes that have triplet ground-states, will be deferred until the symmetry properties of electron spin are taken up. An incidental practical advantage of setting up the correlation digrams in Fig. 6.1

[1] In this notation [1] the numerals refer to the number of reacting atomic centers in each reactant; in the more common notation that specifies the number of electrons in the two reacting systems [2, p. 70], the reaction is a legitimate member of the family of [2+2]-cycloadditions.

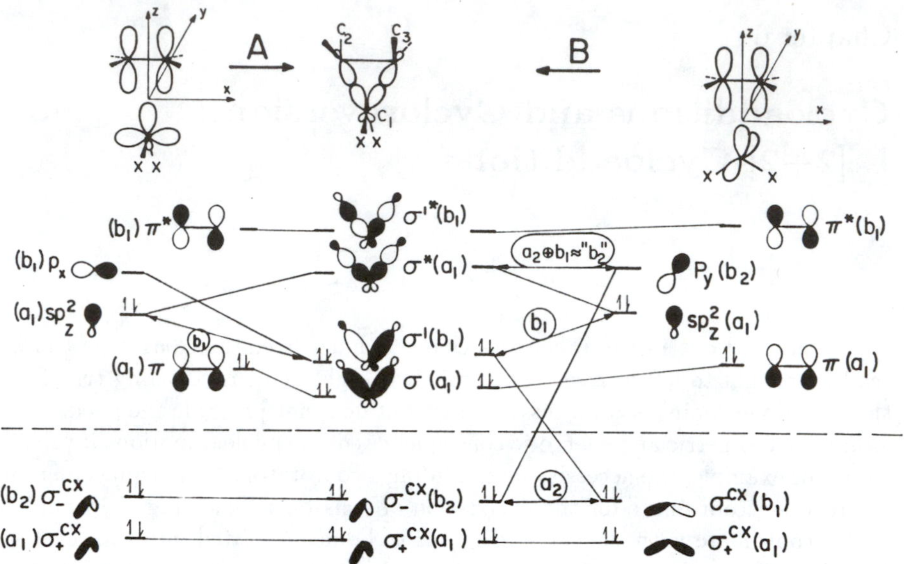

Figure 6.1. Correspondence diagrams ($\mathbf{C}_{2v}^z$) for addition to Ethylene of properly (**A**) and and improperly (**B**) oriented CX_2.

with CX_2, rather than CH_2, is that its cycloaddition product with ethylene has $\mathbf{C}_{2v}$ and not $\mathbf{D}_{3h}$ symmetry.

The reactant molecules on both sides of Fig. 6.1 are set up in $\mathbf{C}_{2v}$, with the C_2 axis specified to be z. The direct approach is depicted on the left (**A**); the carbene is in the yz plane, as is the CX_2 group in the product cyclopropane. On the right (**B**) the carbene is – unreasonably from the point of view of the product – in the zx plane. If the method of analysis is reliable, both diagrams should yield the same mechanistic conclusions.

6.1.1 The Direct Approach

In the more straightforward orientation (**A**) the two CX bonds of the carbene combine to an a_1 and a b_2 bonding orbital, just as they do in the cycloaddition product, so they could just as well have been omitted from the correspondence diagram – as they were in the originally published version [3, Fig. 1]. The totally symmetric hybrid orbital (sp_z^2) is sufficiently lower in energy than p_x to be occupied by a lone pair. It correlates with a vacant antibonding orbital (σ^*), so the reaction is formally *forbidden* in $\mathbf{C}_{2v}$, but a rotation of the carbene in the xz plane has the proper b_1 symmetry to *allow* it. This motion, which costs no energy when the reactants are far enough apart, brings the HOMO of ethylene (π) into overlap with the LUMO of the carbene. [4], [1, Fig. 2] The latter is no longer simply p_x, but – though still a "pure" p orbital – has acquired some p_z character. Similarly, the HOMO of the carbene, which – though still an sp

hybrid – now includes p_x as well as p_z, is brought into overlap with the LUMO of ethylene (π^*). These two favorable interactions are possible because all of the four orbitals involved have the same irrep (A') in $\mathbf{C}_s^{zx}$, the kernel of B_1.

As a result, the reaction is extremely rapid: Computations at various levels of sophistication confirm the geometry of the pathway and the very low activation barrier. [5, p. 188] The experimental activation enthalpies are very low, and can even become negative, [1] a somewhat disturbing circumstance that has been rationalized in terms of a pre-equilibrium between the reactants and a loose complex. [6] An extended computational study by Houk *et al.* [7], however, does not confirm the intermediacy of a complex along the reaction path for cycloaddition of CCl_2 and CF_2 to ethylene. As their most sophisticated computations show no potential barrier to the reaction,[2] they ascribe the finite rate constants, which necessarily imply that $\Delta G^\ddagger$ is positive, to an overriding negative entropy of activation. Be the detailed description of the reaction profile as it may, it is clear that the reaction, *forbidden* in $\mathbf{C}_{2v}^z$, has become virtually unactivated under the influence of a mutual reorientation of the reactants that desymmetrizes the reaction path to $\mathbf{C}_s^{xz}$ at no energetic cost.

6.1.2 Correcting a Geometrically Unreasonable Approach

No rational organic chemist would expect the reaction path for (1+2)-addition of singlet carbenes to ethylene to retain the reactant orientation shown on the right hand side of Fig. 6.1. Nonetheless, there are several reasons for paying some attention to correspondence diagram (**B**): First, most reactions are less simple than the one being considered, and it will often be more difficult to judge whether the reactants are properly oriented or not. Second, the CX_2 molecule may approach its reaction partner along any one of many different trajectories, very few of which can serve as "reasonable" reaction pathways. It is therefore of interest to see whether the correspondence diagram can "correct" the mutual orientation of the reactants and guide them onto an energetically favorable reaction path. Finally, the diagram can be used to show how to deal with *composite* motions, i.e. those that have to be treated as a superposition of displacements belonging to two or more irreps of the symmetry point group adopted in the construction of the diagram.

The only difference between the right- and left-hand sides of Fig. 6.1 that results from rotating the CX_2 molecule into the zx plane is an interchange of the labels of σ_-^{CX} and the vacant p orbital, which is changed from p_x to p_y.

---

[2] There is a serious conceptual difficulty here: Transition state theory requires a barrier on the potential energy surface; otherwise the transition structure cannot be defined. However, the observation of a negative activation enthalpy does not necessarily imply the absence of a potential barrier. Bell's [8, p. 71] qualitative argument runs briefly as follows: At 0°K, $\Delta G_0 = \Delta H_0$ for any reaction, both being equal to the potential energy difference corrected for vibrational zero-point energy. At any finite temperature, ΔH is increased by $\int_0^T C_p dT$ and ΔS by $\int_0^T (C_p/T) dT$. Since $\Delta G_T = \Delta H_T - T\Delta S_T$, the two T-dependent terms oppose one another, so $\Delta G^\ddagger$ is plausibly a better approximation to the potential barrier than $\Delta H^\ddagger$.

As a result, it is the latter that correlates with the antisymmetric CX-bonding combination of the product, in violation of the principle stated in Section 5.2.2, according to which the electrons in a σ bond that is merely reoriented in space during the reaction remain localized within the bond. In order to comply with this requirement, σ_-^{CX} has to retain its identity, and this can be accomplished simply by rotating the CF_2 molecule into the yz plane, as it is on the left of the figure. Rotation of a single reactant has the same irrep as rotation of the entire system, a_2, and thus formally induces the required correspondence: $b_1 \otimes a_2 = b_2$. This relative rotation, however, is insufficient; the lone-pair in sp_z^2, which is unaffected by the rotation, has to be induced to correlate with σ' by a displacement in the zxplane, in order to put the reacting system onto a feasible reaction coordinate, precisely the same one that was prescribed by (**A**).

6.1.2.1 Composite Motions

The motion just described is a composite one, belonging to a reducible representation of $\mathbf{C}_{2v}$: $a_2 \oplus b_1$. The diagram does not tell us the sequence in which the two components of the motion occur or whether they are concerted; it merely spells out the geometric requirements of the pathway. If the initial geometry of approach is as shown on the right side of Fig. 6.1, the correspondence diagram is consistent with any of the following sequences of "corrective" motions:

1) The approaching carbene rotates about the symmetry axis, taking the system into $\mathbf{C}_2$ – the kernel of a_2, and then moves parallel to x along a coordinate that has the irrep b_2 in $\mathbf{C}_{2v}$ but maps onto b in the lower $\mathbf{C}_2$ symmetry.

2) It slides sideways in the xz-plane, reducing the symmetry of the reaction path to $\mathbf{C}_s^{zx}$ and then rotates about its own axis (a'' in $\mathbf{C}_s^{zx}$) in order to adopt the proper orientation of the CX bonds.

3) Most reasonably, it moves along a composite trajectory that incorporates both symmetry coordinates.

Note that pathway (**B**) retains no symmetry elements at all, but the correspondence diagram has determined its geometric features quite well. The fact that diagram (**A**) is more informative, confirming that reflection symmetry in the zx plane may be retained, can be taken as evidence that chemical intuition is not an altogether valueless commodity.

The correspondence diagrams in the preceding chapter did not include unoccupied orbitals. Nor was it necessary to refer to them in the present connection; the features of the favored pathway emerge clearly from the symmetry properties of the occupied orbitals alone, despite the obvious energetic importance of HOMO-LUMO interactions. The unoccupied orbitals – often referred to as *virtual orbitals* – are included in Fig. 6.1 not only because p_y is too close in energy to sp_z^2 to be omitted, but because its induced correlation with σ^* illustrates an important additional point about composite motions. If this correlation is considered in isolation from the rest of the reacting system, we recognize that it calls for a b_2 displacement, i.e. motion in the yz plane that would bring p_y into overlap with π of ethylene and eventually contribute to stabilization of the product's $\sigma(a_1)$ orbital. The composite motion chosen by the occupied orbitals does not include b_2, but the effect of a b_2 displacement can be simulated *in second order* by a composite motion, composed of a_2 and b_1. Its nature and geometric aspects are best brought out by means of an analogy with a superposition of two electrostatic fields:

The dominant term of a dipolar field aligned along x, such as that illustrated in Fig. 2.9, transforms as x; in $\mathbf{C}_{2v}^z$ it has the irrep b_1. If the field is non-linear, as it may well become at high field strengths, it will include a second-order term that transforms as $x^2(a_1)$ and higher order terms ($x^3(b_1)$, $x^4(a_1)$, *etc...*) as well. A similar field along y will have a second order term in $y^2(a_1)$ in addition to its dominant term, $y(b_2)$. If the two perpendicular fields are superimposed, a new quadratic term appears that is proportional to the product of the two field strengths and transforms as $xy(a_2)$. In general, a mixed second-order term has the same irrep as the product of its components. A term of this kind will be distinguished from a first order term with the same irrep by putting its symmetry label in parentheses; $\approx$ is used rather than $=$ in order to make it clear that the irrep of a composite displacement is equivalent to the product of the irreps of its components only in this very restricted sense.

In (**B**) of Fig. 6.1 the $p_y \longleftrightarrow \sigma^*$ correspondence is induced by the second-order perturbation: $[a_2 \oplus b_1] \approx$ "b_2". It has no energetic consequences in the present example, because the orbitals involved are unoccupied. Moreover, successive or simultaneous displacement along coordinates belonging to any two non-totally symmetric irreps of $\mathbf{C}_{2v}^z$ desymmetrizes the system all the way down to $\mathbf{C}_1$, in which – subject to the non-crossing rule – any pair of orbitals can be brought into correlation. The energetic consequences of composite displacements will be explored in subsequent examples, in which they are called upon to induce correspondence between occupied orbitals and where the initial symmetry is sufficiently high that they lead to only partial desymmetrization.

6.2 Concerted $[_\pi 2+_\pi 2]$-Cycloaddition

The application of OCAMS to the paradigmatic cyclodimerization of ethylene was dealt with in detail in the primary publication on the method [9].

6.2.1 The $[_\pi 2_s+_\pi 2_s]$ Approach

Fig. 6.2 is set up in the $[_\pi 2_s +_\pi 2_s]$-orientation; it reproduces the left hand side of Fig. 1.11, but twists both ethylenes back into the zx plane and assuumes – for the purpose of analysis – that the product cyclobutane is planar. The latter is desymmetrized from $\mathbf{D}_{4h}$ to $\mathbf{D}_{2h}$ by a slight elongation of bonds C_1-C_3 and C_2-C_4 and the orbitals on both sides of the diagram are labeled accordingly. As expected, the reaction is *forbidden* by the familiar HOMO-LUMO crossing, but the "forbiddenness" can be removed by displacement along a b_{2g} symmetry coordinate.

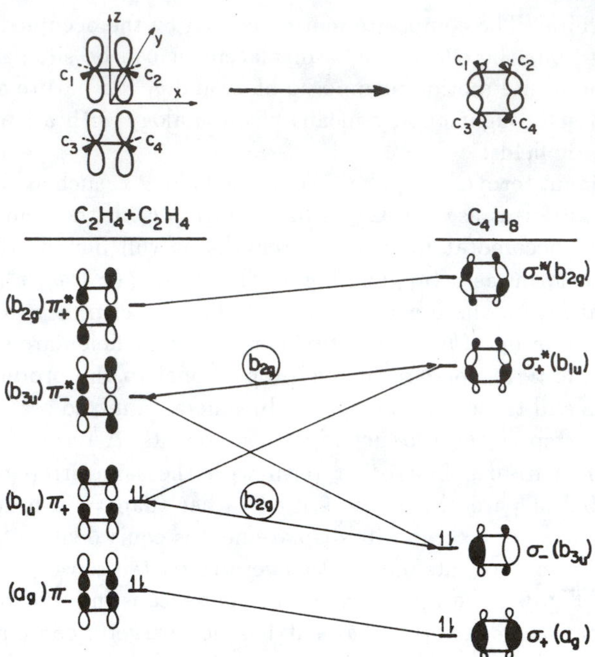

Figure 6.2. Correspondence diagram ($\mathbf{D}_{2h}$) for [$_\pi 2_s + _\pi 2_s$]-cyclodimerization of ethylene

The principal component of the reaction coordinate is the approach of the two ethylene molecules towards one another with retention of the full symmetry assumed in the construction of the correspondence diagram; as Fig. 6.2 was set up in $\mathbf{D}_{2h}$, this "least motion" approach has the irrep a_g. The diagram then tells us that the reaction coordinate for concerted conversion of the two π bonds into the two σ bonds of cyclobutane also has to include a b_{2g} component. Several symmetry coordinates, and the subgroups of $\mathbf{D}_{2h}$ to which they desymmetrize the reaction path, are shown in Fig. 6.3. If the correspondence diagram had called for an a_u displacement, the relatively facile formation of cyclobutane in its stable puckered $\mathbf{D}_2$ conformation would have been expected. If a b_{3u} component were required to induce the neccesary correspondence, the favored pathway would generate a *cisoid* biradical, which would immediately collapse to cyclobutane.[3] The nominally stepwise reaction would then be kinetically indistinguishable from one in which the formation of both bonds is synchronous.

The b_{2g} displacement, which desymmetrizes the reaction path to $\mathbf{C}_{2h}^{y}$, has the proper symmetry to *allow* direct closure to cyclobutane, but has the unfortunate geometric property of moving one diagonally situated pair of atoms,

[3] Dewar and Kirschner [10] have shown that the *cis*-stable biradical is *homomeric* with cyclobutane (in the sense of having the same electronic configuration), and that there is no potential barrier between them.

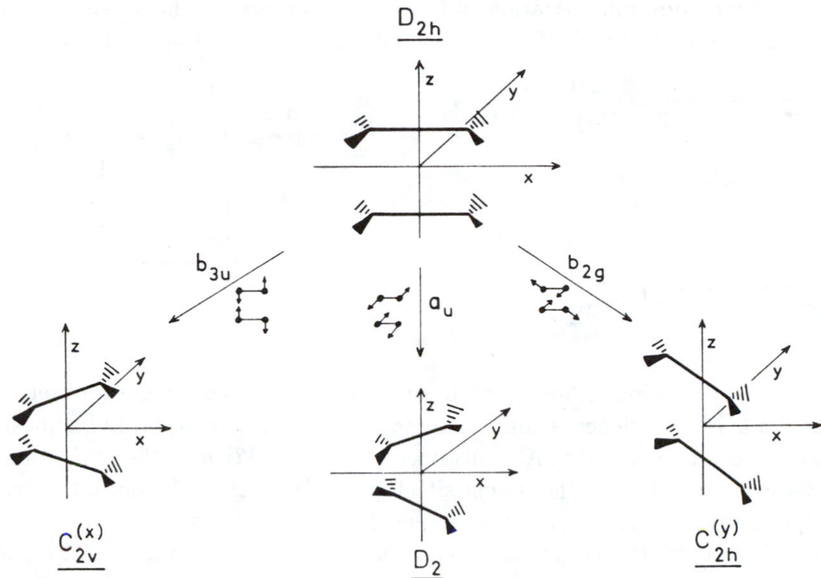

Figure 6.3. $\mathbf{D}_{2h}$ symmetry coordinates of two approaching ethylene molecules

say C_1 and C_4, farther apart while bringing the second pair into bonding range. Comparison with the addition of CX_2 to ethylene is instructive. Fig. 6.1 and Fig. 6.2 both require the incorporation of a similar displacement into the reaction coordinate: a relative in-plane reorientation of the two components. In the former case, the second bond to the carbenic C atom can close immediately after formation of the first, or simultaneously with it. In the latter, once C_2 and C_3 have been joined, C_1 and C_4 are too far apart to bond; the newly formed *transoid* tetramethylene biradical has to undergo internal rotation about its central bond and cross a potential barrier[4] before it can collapse to cyclobutane.

6.2.1.1 Substitutional Desymmetrization: Dimerization of Silaethylene

Before taking up the biradical pathway and its extension to a zwitterionic mechanism, let us reconsider the possibility of reaction in a single step. The only way of reducing the symmetry of the pathway from $\mathbf{D}_{2h}$ to $\mathbf{C}_{2h}^{y}$ in the parent system is to displace the nuclei along a b_{2g} symmetry coordinate, such as the one shown in Fig. 6.3. We recall, however, that the effect of such a nuclear displacement is to perturb the electrostatic field acting on the electrons. It follows that a

[4] The existence of a *trans* biradical, presumably homomeric (see preceding footnote) with the reactants [10], has been confirmed computationally, whereas a distinct *cis*-stable one has not [11]. A second biradical in a *gauche* conformation was reported in the same computational study; it was identified as a loosely coupled complex of two ethylene molecules, but is separated from them by a higher barrier than the *trans* biradical.

substitutional desymmetrization to $\mathbf{C}_{2h}^{y}$ of the electrostatic field experienced by the electrons as the reactants approach one another should have a similar effect.

$H_2C{=}SiR_2 + R_2Si{=}CH_2 \Longrightarrow$ 1,3-disilacyclobutane (H_2C–SiR_2–CH_2–SiR_2 ring, head-to-tail)

This expectation is borne out beautifully by the experimental observation that dialkylsilaethylenes readily undergo thermal dimerization, but almost exclusively head-to-tail. [16] Ab initio computations [17] find the activation energy for dimerization of the parent silaethylene (R = H) to be quite low (ca. 20 kcal/mol) even when the reactant molecules are constrained to approach each other along a strictly rectangular approach. There is also strong kinetic evidence [18] that the head-to-tail cycloaddition of dimethylsilaethylene (R $=$ CH_3) is virtually unactivated. These two findings are not in conflict: while the strict rectangular pathway already has $\mathbf{C}_{2h}^{y}$ symmetry, further energetic advantage can be gained by in-plane readjustment of the CSiC angles without further desymmetrization.

A word of caution: The mere fact that symmetry has been reduced to $\mathbf{C}_{2h}^{y}$ does not guarantee concerted cycloaddition. If it did, isobutene would also be expected to undergo facile concerted cycloaddition with strict head-to-tail stereospecificity.[5] The presence of methyl substituents on one of the carbon atoms does not introduce enough of an electronegativity difference between it and its neighbor to break the essential symmetry of the π system and overcome the "forbiddenness" of $[_\pi 2_s +_\pi 2_s]$-cycloaddition; its replacement by the more electropositive silicon atom, producing a strongly polarized C=Si bond, evidently does.

6.2.2 $[_\pi 2_s +_\pi 2_a]$-Cycloaddition

A few words have to be said here about the concerted $[_\pi 2_s +_\pi 2_a]$ pathway [2, p. 69ff.] in the approach depicted on the right-hand side of Fig. 1.11. It was pointed out in Section 1.4.2 that this pathway is formally *allowed* only if one of the reacting ethylene molecules is singled out beforehand to react suprafacially and the other to react antarafacially. The various ways in which two free ethylene molecules, initially oriented in $\mathbf{D}_{2h}$, can approach one another were broken down to their component symmetry coordinates (see Section 6.1.2.1) in the primary publication on OCAMS. [9] It was shown that this particular

[5] Pushing the argument to an absurd extreme, it might be applied to 1,1-dideuterioethylene as well!

mode of approach cannot be made "equivalent" to b_{2g} in higher order unless the pathway is desymmetrized all the way down to $\mathbf{C}_1$.

Computations [12] show that concerted $[_\pi 2_s +_\pi 2_a]$-cycloaddition of ethylene is disfavored relative to biradical formation even when the recommended orientation is adopted. The qualitative argument has been made [13] that the HOMO-LUMO interactions along this pathway are less favorable than in the orientation leading to a *transoid* biradical. All in all, it is hardly surprising that $[_\pi 2_s +_\pi 2_a]$ cycloaddition of alkenes has only been observed in exceptional circumstances – if at all. For example, Kraft and Koltzenburg [14] observed that bicyclo[4.2.2]-deca-*trans*-3,*cis*-7,9-triene dimerizes to a product with one *cis* and one *trans* junction:

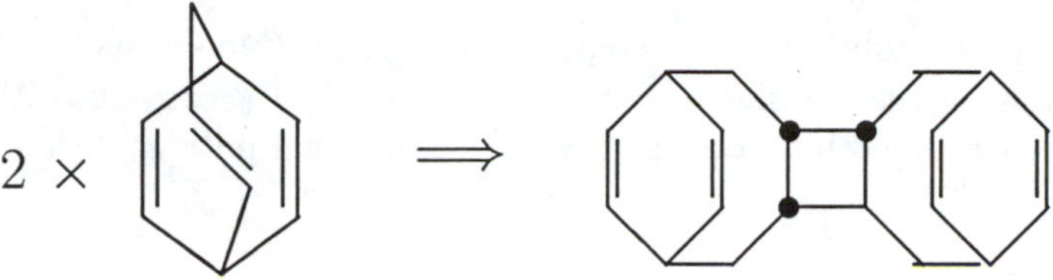

The unusual stereochemistry of the product was taken to be compelling evidence for concerted $[_\pi 2_s +_\pi 2_a]$ cycloaddition [2, p. 75 ff.], until the reaction was shown [15] to proceed by a stepwise mechanism.

6.3 Cycloaddition via a Tetramethylene Intermediate

6.3.1 The Biradical Mechanism

The stepwise nature of olefin cycloaddition is well established. Bartlett and his coworkers [19, 20, 21] demonstrated conclusively that [2+2]-cycloaddition of a tetrahaloethylene to a conjugated diene competes effectively with the concerted [4+2] process and occurs through the intermediacy of a tetramethylene biradical that is sufficiently stable for ring formation to occur with at least partial loss of configuration. They then showed [22] that a *cisoid* tetramethylene biradical, formed by extrusion of N_2 from a cyclic azo-compound, collapses stereoselectively to a four-membered ring. To cite Bartlett [23]: "By forming the singlet biradical in the *cis* conformation, we have made it behave stereochemically like the 1,4-dipolar ions which occur in ionic cycloaddition[6]: ring closure occurs from the *cis* conformation more rapidly than rotation can take place in the biradical." The experimental evidence thus indicates that a *trans*-stable tetramethylene biradical is the first intermediate; this then rotates about its newly formed bond and crosses a barrier to a *cis*-stable biradical that closes rapidly to the product. It is fully consistent with the conclusions that were drawn from Fig. 6.2, and

[6] See Section 6.3.2 below.

is singled out by Houk [13] as the preferred mode of approach on the basis of favorable HOMO-LUMO interactions.

The symmetry analysis is, however, still incomplete: The correspondence diagram in Fig. 6.2 was drawn for concerted closure of both bonds; this pathway was shown to be formally *allowed* under a b_{2g} perturbation, a prediction – perhaps more properly a retrodiction – borne out by the facile head-to-tail dimerization of silaethylene. Where there is no effective substitutional desymmetrization, the "allowing" perturbation is necessarily a b_{1g} displacement that foils the concerted process. To be sure, the nuclei move in the right direction to form a *transoid* biradical, but it has yet to be confirmed that the stepwise process is consistent with the requirements of orbital symmetry conservation. This is done in Fig. 6.4 with the aid of a correspondence diagram specifically set up in $\mathbf{C}_{2h}^y$ to analyse the first step, formation of a *trans*-stable bradical.

The MOs on the left side of Fig. 6.2 are relabelled according to their $\mathbf{C}_{2h}^y$ irreps. The better p-orbital overlap across the diagonal puts π_+^* below π_-^*, but

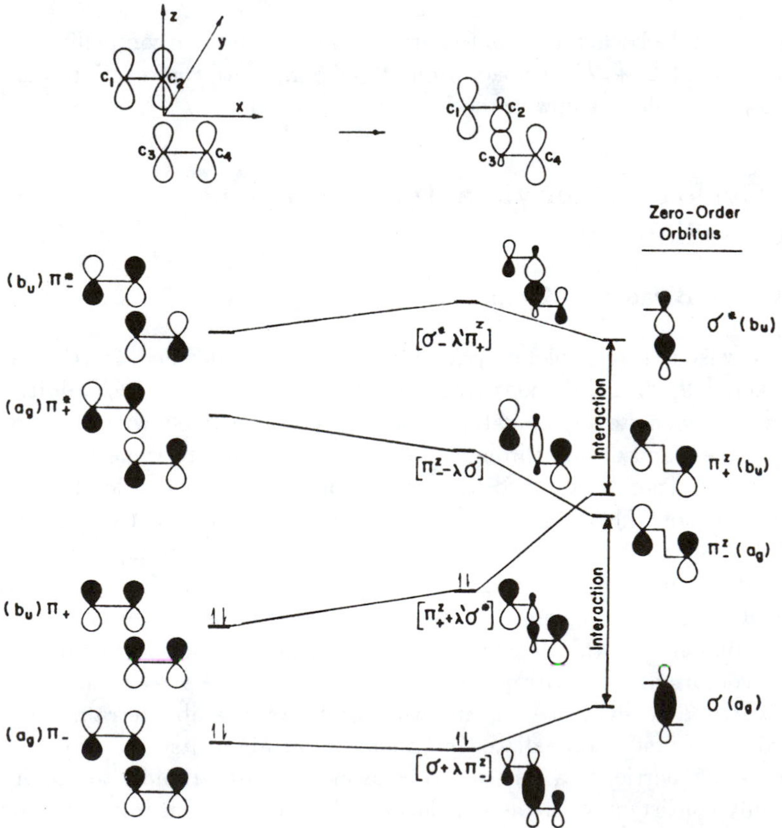

Figure 6.4. Correspondence diagram ($\mathbf{C}_{2h}^y$) for stepwise dimerization of ethylene (Taken from ref. [9])

the energetic reordering of these two virtual orbitals is irrelevant to the correspondence diagram. In analogy with the construction of Fig. 4.1, the MOs of the biradical are characterized in two stages:

1) The four zero-order orbitals chosen are a bonding-antibonding pair of σ orbitals localized in the new C_2-C_3 bond lined up along z, and the two combinations of p_z orbitals on C_1 and C_4. Of the latter two, *through-space* interaction is assumed to stabilize π^z_+ slightly relative to π^z_-.

2) The orbitals of like symmetry interact, the lower being stabilized at the expense of the upper. The resulting *through-bond* interaction [24] reverses their order, so that the electronic configuration of the *trans*-stable radical becomes $[a_g^2 b_u^2]$ like that of the reactant pair of ethylenes.

Ring closure must be preceded by rotation about the C_2-C_3 bond, which can occur with retention of $\mathbf{C}_2^y$ symmetry. In this subgroup of $\mathbf{C}_{2h}^y$, $a_g \rightarrow a$ and $b_u \rightarrow b$, so the *trans*-stable biradical still has to cross a symmetry-imposed barrier between it and its *cis*-stable conformer before it can collapse to cyclobutane. It is generally recognized that *configuration interaction (CI)* is essential for the quantitative or even semiquantitative understanding of biradicals [5, p. 130], but the molecular orbital approximation nevertheless retains its qualitative validity: CI reduces the barrier for interconversion of biradical conformers substantially; Segal [11] computes it to be 3.6 kcal/mol for the parent system. However, as Bartlett [20] points out, a biradical only has to live 10^{-10} seconds ($\Delta G^\ddagger \approx 4$ kcal/mol at 300°) to undergo ten *cis* $\rightleftharpoons$ *trans* interconversions and cyclize with substantial loss of stereoselectivity.

6.3.1.1 Stereochemistry of Biradical Cycloaddition

Conventional wisdom has it that whenever stepwise cycloadditions are stereoselective, the thermodynamically more stable stereoisomer will be the one formed preferentially. For example, an olefin that has subsituents with different steric requirements would be expected whenever possible to avoid dimerization to an isomer of the product in which bulky substituents eclipse one another. In those cases when the preferred product turns out to be the more sterically strained one, the mechanistic argument – not always explicit – usually runs as follows:

(a) The reaction occurs under kinetic rather than thermodynamic control; it follows that

(b) the reaction must be taking place in a single step; therefore

(c) the relative orientation of the reactants as they approach one another is governed by orbital interactions;

and, since the *allowed* mode of concerted cycloaddition is $[_\pi 2_s + _\pi 2_a]$,

(d) the approach must be that illustrated in Fig. 1.7 [2, p. 69].

Brief reflection on the stepwise process shows that (b) does not necessarily follow from (a), and since (c) may be true even when (b) is not, it does not imply (d)! The observed stereochemistry is fully consistent with a stepwise biradical mechanism; in the absence of specific electronic effects, only two plausible assumptions need be made:

1) The first biradical is formed in the *trans* conformation, as required by Fig. 6.4.

2) Both bond-forming steps occur preferentially from the conformation of the reactant(s) that leads to the least congested transition states.

The preferred stereochemical course for dimerization of an olefin with two large substituents and two small substituents in the E (or *trans*) configuration is traced in Fig. 6.5.

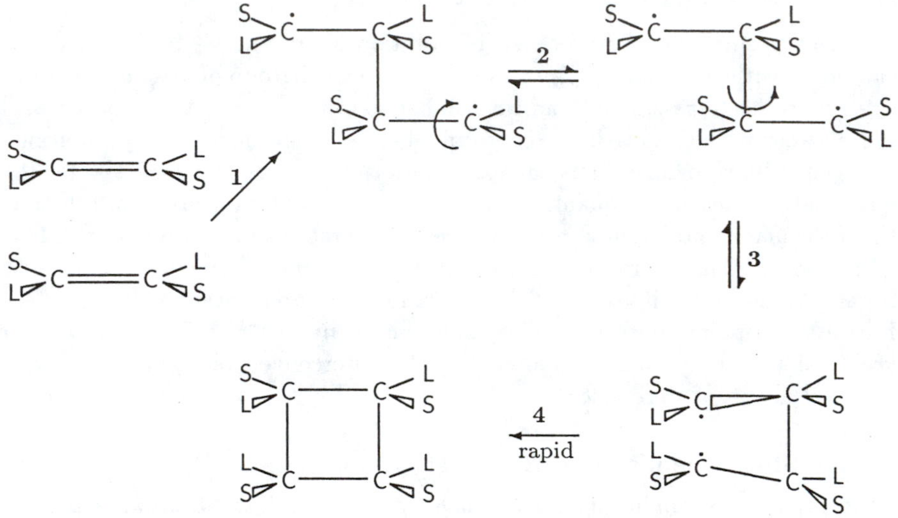

Figure 6.5 Stereochemistry of biradical $[_\pi 2 +_\pi 2]$-cycloaddition

In order to minimize the steric repulsions in *Step 1*, the two bulky substituents must be antiperiplanar with respect to the bond being formed; the alternative initial orientation of the ethylene molecules can therefore be rejected. For the second bond to be formed with minimum strain in the transition state, the bulky substituents have to be staggered with respect to it; if rotation about the central bond, as in *Step 3*, were to take place without further conformational change, they would be in the eclipsed conformation. Evidently, this step must be preceded by rotation about one of the terminal bonds (*Step 2*). Rotation towards the *cis* conformation (*Step 3*) brings the *trans*-stable biradical up against the barrier between it and the *cis*-stable state when it is still in a *gauche* conformation. Crossing the barrier and collapse to the cyclobutane are assumed to occur so rapidly that these two *primitive changes* are telescoped in the same *elementary reaction*[7], (*Step 4*).

In this example of a simple dimerization, the observed stereochemistry of the product is dominated by purely steric effects. When there are radical-stabilizing

[7] For a formal definition of these terms, see Reference [25].

sites on one or both of the reactant molecules, the rationalization of product stereochemistry requires that a third postulate, the validity of which was established by Bartlett many years ago [19, 20, 21], be added to the preceding two:

3) The first bond will be formed between the atoms that are least suited for bearing an unpaired electron.

For example [19], in the addition of $CCl_2{=}CF_2$ to 1,3-butadiene, the first bond formed is between the F-substituted C atom of the olefin and C_1 of the diene, because Cl stabilizes an adjacent radical center more than F, and an unpaired electron is delocalized more effectively on C_2 than on C_1.

Careful application of the three simple postulates listed above can yield insight into the mechanism and stereochemistry of biradical reactions as complex as the thermal dimerization of *cis, trans*-1,5-cyclooctadiene [26] or the isomerization of allyl-substituted cyclopropanes *via* internal [2 + 2]-cycloaddition [27]. An attempt to do so here would take us too far afield, in view of the ease with which biradical intermediates interconvert. Instead let us move on to the considerably more stereoselective cycloaddition of reactant pairs with complementary polarity, that proceeds stepwise along a zwitterionic pathway. [23]

6.3.2 The Zwitterionic Mechanism

The most familiar example of zwitterionic $[_\pi 2 +_\pi 2]$-cycloaddition is that of tetracyanoethylene (TCNE) to activated ethylenes, such as vinyl ethers. Huisgen [28] has carefully reviewed the experimental evidence, which is due in large measure to him and his collaborators. It can be summarized briefly as follows:

1. As a rule, cycloaddition is only partially stereoselective, though examples are cited in which stereoselectivity, which decreases with solvent polarity, is essentially complete. However, as Bartlett has pointed out [23]: "In general, configuration loss appears to be a sufficient but not a necessary criterion for stepwise cycloaddition, at least when the intermediate is a dipolar ion".
2. A kinetic analysis of the isomer ratios in the product and the unconsumed enol ether shows zwitterion formation to be reversible. In highly polar solvents, the cyclobutane opens in a slow back reaction to the same zwitterion that is an intermediate in its formation.
3. The large substituent effect and strong dependence of the rate on solvent polarity support substantial charge separation in the transition state. The consistently large negative entropy and volume of activation are also in accord with this interpretation.
4. The zwitterions can be trapped by alcohols and their stereochemistry deduced from the structure of the products.

The symmetry of the reaction path can be derived directly from Figs. 6.2 and 6.4 when the effect of electron-withdrawing substituents on one of the reaction partners is taken into account. If it can be assumed as a first approximation that the alkoxy substituent in the vinyl ether does not disrupt the essential symmetry of the ethylenic π system, the prototypical zwitterionic [2+2]-cycloaddition can be chosen to be that of TCNE and ethylene. The least-motion coplanar pathway is clearly *forbidden*:

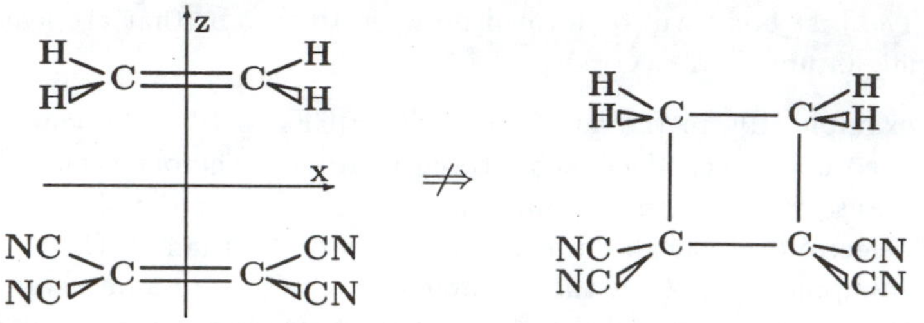

Tetrasubstitution of one of the ethylene molecules in the symmetric coplanar arrangement desymmetrizes the pathway of least motion – or highest symmetry – from $\mathbf{D}_{2h}$ to $\mathbf{C}_{2v}^{z}$, in which the highest occupied orbital of the reactant pair in Fig. 6.6 correlates no better with that of the tetrasubstituted cyclobutane than do the analogous MOs in Fig. 6.2:

$$B_{1u}(\mathbf{D}_{2h}) \Rightarrow A_1(\mathbf{C}_{2v}^{z}) \text{ and } B_{3u}(\mathbf{D}_{2h}) \Rightarrow B_1(\mathbf{C}_{2v}^{z})$$

So – at the orbital level of approximation – an in-plane glide (b_1 in $\mathbf{C}_{2v}^{z}$) is still needed in order to circumvent the symmetry-imposed barrier to cycloaddition.

Epiotis has argued that $[{}_{\pi}2_s + {}_{\pi}2_s]$-cycloaddition is here *allowed*, because "the excited intermediate ... of a nonionic reaction becomes the ground intermediate

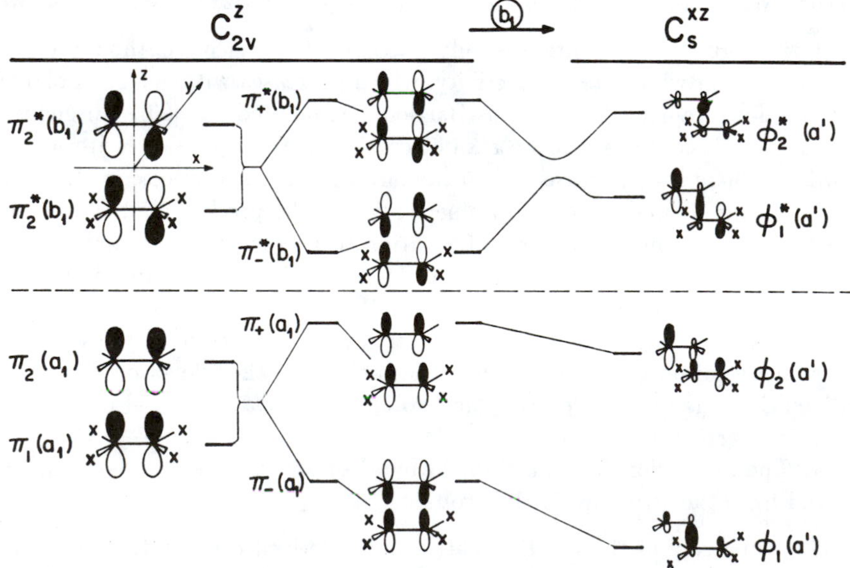

Figure 6.6. Correspondence diagram for zwitterion formation

of an ionic reaction". [29, p. 66] His argument can be recast as follows in the formalism adopted in this book: Substitution by electron-attracting substituents stabilizes both the π and π^* orbitals of ethylene. As illustrated schematically on the left side of Fig. 6.6, substitutional desymmetrization to $\mathbf{C}_{2v}^{z}$ maps both HOMOs onto a_1 and both LUMOs onto b_1; the ensuing interaction depresses the lower of each pair and raises the upper. As the reactants come closer to one another the energy of π_-^* approaches that of π_+, and their order is inverted. More realistically, configuration interaction between them becomes increasingly important, until – near the crossing – the orbital approximation loses its validity and the "forbiddenness" of concerted $[_\pi 2_s + _\pi 2_s]$-cycloaddition is lifted.

The experimental evidence cited above indicates that this does not occur. As suggested at the right of Fig. 6.6, the in-plane glide (b_1) begins well before the intended HOMO-LUMO crossing, which is *avoided* because all four MOs have the same irrep (a') in $\mathbf{C}_s^{xz}$. The HOMO and LUMO of the extended *transoid* zwitterion are qualitatively similar to those in the biradical illustrated in Fig. 6.4, but are more widely separated in energy: As a result, the orbital approximation – and the symmetry analysis based upon it – is no less reliable than for the biradical mechanism. The zwitterionic mechanism can be accomodated by Fig. 6.5, with self-evident modifications arising from the polar nature of the tetramethylene intermediate.[8]

As a result of the asymmetry of charge, the extended zwitterion is stable only in polar solvents. In solvents of low or moderate polarity, the oppositely charged termini attract one another and pull the developing tetramethylene zwitterion into a *gauche* conformation: i.e. *Steps 1–3* coalesce to a single elementary reaction.[9] The larger HOMO-LUMO separation, and specific solvation at the oppositely charged termini, raise the activation energy of *Step 4* and permit the zwitterion to be trapped before it can collapse to the cyclobutane. Huisgen, who – with his collaborators – trapped the intermediate and established its *gauche* conformation [31], states that "there is no experimental evidence for this contortion"[10]; [28, p. 206] nor is there any evidence against it. In order to confirm or refute it, experiments have to be performed in which both reactant molecules bear substituents with different steric requirements.

6.4 Ketene Cycloadditions

The last bastion of $[_\pi 2_s + _\pi 2_a]$-cycloaddition is the reaction between ketenes and activated olefins to form cyclobutanones. The experimental evidence for this

[8] A model MNDO calculation on the zwitterion from cycloaddition of tetrafluoroethylene to ethylene constrained to the geometry of Fig. 6.6 [30] yields a dipole moment of *ca.*3 Debye units, with the negative charge concentrated on the substituent atoms.

[9] The cited experimental evidence indicates that this step – and even *Step 4* – are reversible, but the strong dependence of the rate on solvent polarity is consistent with its being the rate-determining step under the conditions of the kinetic experiments. [33]

[10] Adapting the "old axiom" of Sherlock Holmes [32], we might say :"When all other mechanisms fail, the reaction path that remains, however contorted, is the true one".

mechanism has been documented and summarized concisely by March [34, pp. 761–762]:

1) The reactions are highly stereoselective, the thermodynamically less stable isomer being formed preferentially.

2) The rate is rather insensitive to solvent polarity or to polar substituent effects.[11]

The stereochemical evidence, [36, 38, 39] the much weaker solvent dependence for ketene cycloaddition than for reactions that are known to proceed *via* polar transition states, [35] and the predominance of steric over electronic substituent effects in the cycloaddition of ketenes to vinyl ethers [37, 42, 43] convinced Huisgen that the reaction occurs mainly, if perhaps not uniquely, as a concerted $[_{\pi}2_s + _{\pi}2_a]$-cycloaddition.

Meanwhile, however, experimental support for cycloaddition *via* a stepwise zwitterionic mechanism was being accumulated by Moore and his coworkers, [44] culminating in a study in which the zwitterionic intermediate was generated independently from a cyclic precursor and shown to produce a cyclobutanone with the same stereochemistry as the cycloaddition product of *tert*-butyl cyanoketone (TBCK) and *trans*-trimethylsiloxypropene. [45]

6.4.1 Diversion: Secondary Isotope Effects

Several attempts were made to use secondary isotope effects as an additional mechanistic criterion, with apparently conflicting results – as well as contradictory conclusions drawn from similar results. The criterion had been used successfully by Dolbier and Dai [46] to establish the stepwise nature of the cycloaddition of acrylonitrile to 1,1-dideuterioallene. They argued that if bond formation to a terminal atom occurs in the rate-limiting step, the hybridization change from sp^2 to sp^3 and the consequent higher frequency of the out-of-plane bending mode in the transition state [47], [48, p. 145] should produce an inverse isotope effect ($k_H < k_D$); as a result, deuterium would be preferentially incorporated in the ring and protium in the exocyclic methylene group.

The authors found just such an effect in the presumably concerted $[_{\pi}4 + _{\pi}2]$-cycloaddition of allene to hexachlorocyclopentadiene, but not in its $[_{\pi}2 + _{\pi}2]$-cycloaddition to acrylonitrile. There, athough no isotope effect was observed on the rate of reaction, there was a substantial direct isotope effect ($k_H > k_D$) on product formation: protium was preferentially incorporated in the ring. The authors therefore concluded that the reaction takes place in two steps.

According to Fig. 6.7, the rate-limiting step is bonding C_2 of the olefin to the central C atom of allene to form a biradical;[12] this is followed by competitive closure of C_1 to either C'_1 or C'_3. On the face of it, the direction of the isotope ef-

[11] The latter is something of an overstatement, in view of the fact that the rates of addition of diphenylketene (DPK) to various olefins span a range of seven powers of ten [35]!

[12] The initial orientation chosen is the favorable one for formation of the most stable biradical according to Postulate 3 of Section 6.3.1.1.

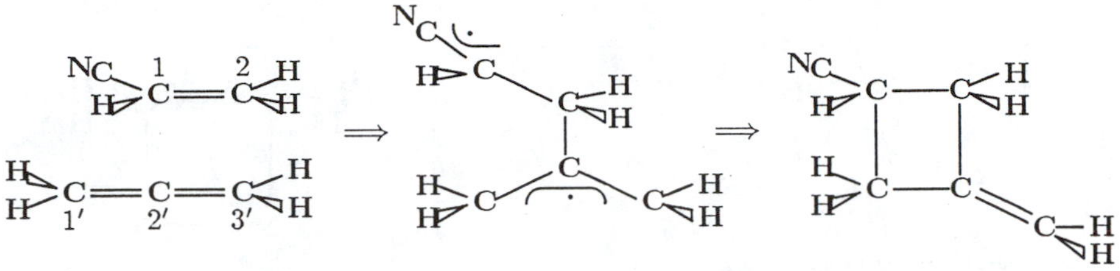

Figure 6.7. Allene-acrylonitrile cycloaddition

fect is unexpected: One would anticipate the product-determining step, closure of the four membered ring, to be subject to an inverse isotope effect ($k_H < k_D$). Following Crawford and Cameron [49], the authors ascribe the observed direct effect ($k_H > k_D$) to slower rotation of the deuterated methylene group before ring-closure. A fuller explanation would have to take into account the fact that the vibrational frequencies, particularly that for the out-of-plane bending mode, increase at *both* termini of the allene moiety as the biradical closes to form a methylenecylobutane. The observed isotope effect can be rationalized qualitatively if it is assumed that formation of the exocyclic double bond is nearly complete at the transition state, while ring-closure has just begun. Like rupture of the two bonds in the isomerization of benzvalene discussed in Section 5.3.2, the two primitive changes that comprise the elementary reaction can be said to be *concerted* but not *synchronous* [25]. As a result, the frequency of the out-of-plane bending mode at the unbonded terminus is greater than at the end where bonding has just begun.

A similar set of experiments was carried out by Baldwin and Kapecki [50] on the cycloaddition of diphenylketene (DPK) to styrene.

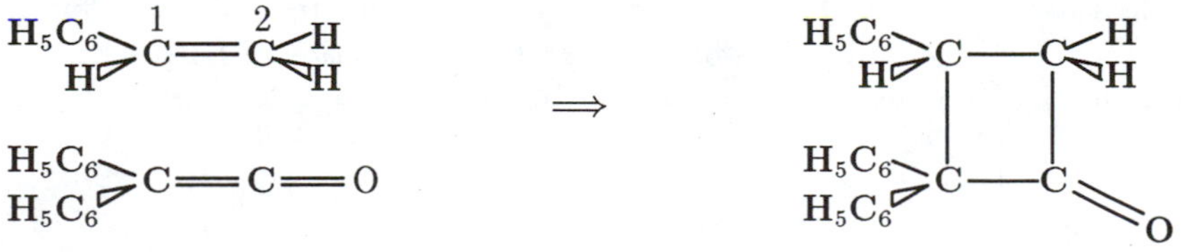

Deuteration at C_2, which bonds to the carbonyl C-atom of the ketene, accelerates the reaction ($k_H/k_D \approx 0.91$ per D-atom) and deuteration at C_1 retards it ($k_H/k_D \approx 1.23$), suggesting a stepwise mechanism *via* a biradical or zwitterion, but the negligible solvent effect and the known stereoselectivity of ketene cycloadditions was taken as presumptive evidence that the reaction nevertheless occurs in a single step: Closure of the two bonds was adjudged to be *concerted* but *non-synchronous*.

At about the same time, Koerner von Gustorf and his associates [51, 52] carried out a parallel investigation of the formation of diazetidines by [2 + 2]-cycloaddition of diazodicarboxylates to vinyl ethers.

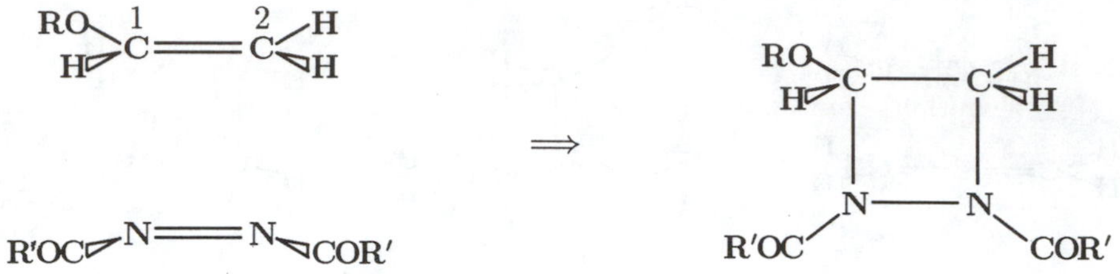

The isotope effects were similar: The rate of cycloaddition to dimethyl azodicarboxylate of ethyl vinyl ether, is accelerated substantially by deuteration at C_2 ($k_H/k_D \approx 0.83$ per D-atom), whereas deuteration at C_1 retards it ($k_H/k_D \approx 1.12$). Although the solvent effect is miniscule, the authors were able to trap a zwitterionic intermediate, in which the bond to C_2 has been formed but the ring has not yet closed. The absence of a large solvent effect, like retention of stereochemical configuration, was thus shown to be neither a necessary nor a sufficient condition for concerted cycloaddition.

This confusing situation elicited the following trenchant comment from the referee of a paper reporting the effect of methyl deuteration on the cycloaddition rate of DPK to α-methylstyrene [53, footnote 7]: “If Koerner von Gustorf's non-concerted mechanism for one [2+2]-cycloaddition is accepted, and if his $k_H/k_D = 1.12$ and Baldwin's $k_H/k_D = 1.23$ are readily interpreted as indicating a two-step reaction, then Huisgen's concerted mechanism for the ketene-styrene mechanism can't be accepted, as it is. Either both azo-olefin and ketene-olefin additions are concerted, or they are not, or either K. von Gustorf or Baldwin has made a large error in measuring k_H/k_D,[13] or perhaps $k_H/k_D > 1$ means *concerted* in one situation and *non-concerted* in another.”

The confusion was compounded when Isaacs and Hatcher [54] reported an inverse isotope effect ($k_H/k_D \approx 0.8$) at C_1 of styrene in its cycloaddition to dimethylketene and cited it as further evidence for a concerted reaction. Finally, Holder *et al.* [55] measured the isotope effect for the reaction of DPK with 5,5-dimethylcyclopentadiene.

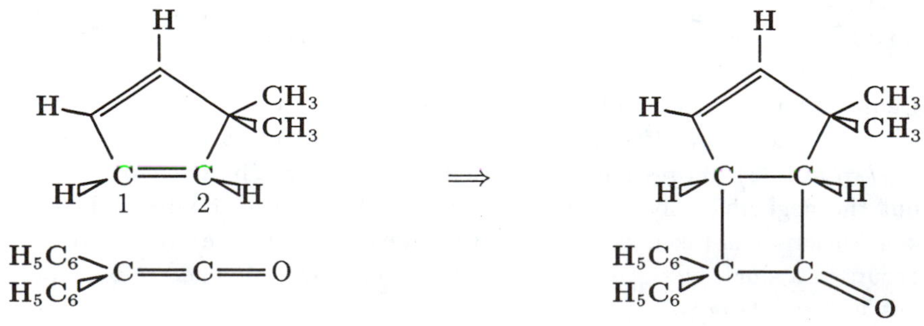

[13] Since the experimental results in the two systems are in qualititative agreement and only the interpretations differ, both investigators would have had to be guilty of a similar error!

Deuteration at the end of the conjugated π system of the ketenophile (Atom 2 in the illustration), which adds to the carbonyl C atom, produces a large inverse effect ($k_H/k_D = 0.84$) whereas deuteration at C_1 had no effect on the rate at all, so the authors came down firmly for a stepwise – biradical or zwitterionic – mechanism.

6.4.2 Reconciling the Evidence

The generally contrathermodynamic stereochemistry, as well as the disparate substituent, solvent and isotope effects, are consistent with the zwitterionic mechanism illustrated in Fig. 6.8 [56], in which the orbital symmetry analysis plays a small but essential role. It is assumed that substitutional desymmetrization is insufficient to destroy the essential symmetry of the π orbitals, which sets the reactants on a path in which the four interacting C atoms are initially coplanar, but are induced by orbital symmetry conservation to bond along the diagonal. The preferred direction of approach and the stereochemical consequences then follow directly.

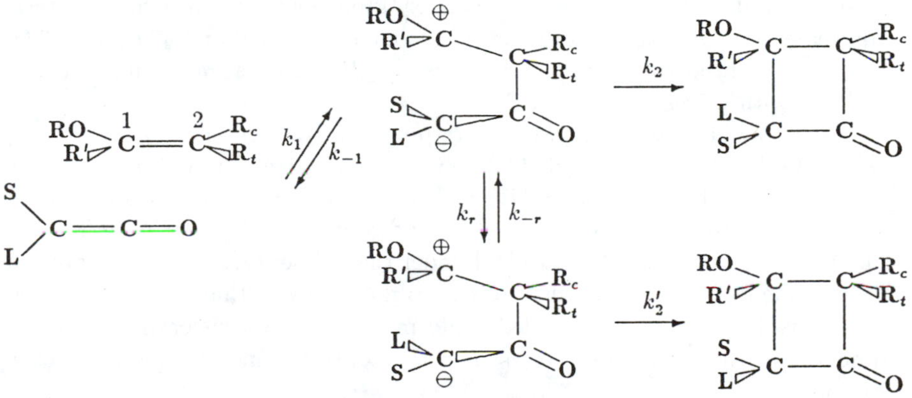

Figure 6.8 . Steric course of zwitterionic ketene cycloaddition: ($E_s^{R'} > E_s^{OR}$ and $E_s^{R_c} > E_s^{R_t}$)

6.4.2.1 Product Stereochemistry

The initial stabilizing interaction is between the HOMO of the ethylene and the LUMO of the ketene, the localized π^*_{CO} orbital that lies in the same plane as the substituents on its terminal C atom. [58, Fig. 3, p. 360] Of the four such mutual orientations, only the one depicted at the left of Fig. 6.8 fulfils two requirements:

1) The electropositively substituted carbon atom of the alkene (C_1) and the oxygen atom of the ketene are diagonally disposed across the rectangle[14]

[14] More precisely, "the trapezoid".

formed by the two interacting double bonds, so that they can acquire positive and negative charge respectively while the zwitterion is being formed in the required *trans* orientation.

2) The ketene's less bulky substituent (**S**), as expressed by a less negative value of Taft's steric parameter (E_s)[15] [59], [60, Chap. 4], faces its reaction partner in order to minimize steric repulsion with the substituents on C_1 as bonding to C_2 proceeds.

As the ketene glides in plane relative to its reaction partner, it executes an additional motion that can be described as a superposition of two displacements: (i) an out-of-plane translation to the side of the less bulky of the two substituents on C_1; and (ii) rotation about its own CC bond to further reduce repulsion between these two substituents and, at the same time, to allow progressive localization of negative charge on the substituted carbon atom of the ketene moiety, and thus to stabilize the zwitterion in the proper conformation for its eventual closure.

There is an additional steric effect that affects the rate of formation of the intermediate. As the bond to C_2 closes, the carbonyl oxygen atom is brought against the substituent *cis* to the smaller of the two substituents on C_1; in our example it is the one *trans* to the alkoxy group. As a result, when its steric requirements are greater than that of the second substituent on C_2 ($E_s^{R_c} > E_s^{R_t}$), the *cis*-substituted vinyl ether reacts more rapidly than the *trans*, an effect that increases with increasing bulk of R_t.

The case illustrated in Fig. 6.8 corresponds to Huisgen and Mayr's [42, 43] 1-alkenyl ethyl ethers (**R′** = H; **OR** = C_2H_5) and to Al-Husaini and Moore's [45] *cis*-trimethylsiloxypropenes (**R′** = H; **OR** = $OSi(CH_3)_3$). The substituted C atom of the ketene moves towards **R′** and repulsion with **S** (CN in TBCK) causes **L** (*tert*-butyl in TBCK) to fold under the bulkier **OR**. Then, if repulsion between the latter two bulky substituents is not so severe that internal rotation – presumably *via* a more extended conformation of the zwitterion – becomes competitive with ring closure ($k_r \approx k_2$), the contrathermodynamic product is obtained.

When both **OR** and **L** are so large that k_2 is reduced below k_r, the reaction is shunted to the alternate pathway, which leads to the thermodynamically stable product.[16] This occurs in the reaction of TBCK with *trans*-trimethylsiloxypropene (**L**=*tert*-butyl and **OR**=$OSi(CH_3)_3$); in the less sterically hindered reaction of TBCK with the *cis*-isomer, the bulk of the silyloxy group has to be increased (**OR**=$OSi(CH_3)_2{}^tC_4H_7$) in order to produce the thermodynamically stable cyclobutanone [45].

---

[15] E_s is increasingly negative for bulky substituents, so $E_s^1 > E_s^2$ expresses the fact that the second substituent is bulkier than the first.

[16] In the extended conformation, rotation about bond C_1-C_2 can also occur. Then, when k_2 is rate-limiting and the disparity between $E_s^{R_c}$ and $E_s^{R_t}$ is very large, as between H and *tert*-butyl, the *cis*:*trans* rate ratio may also decrease [42, Table 1].

6.4.2.2 Substituent, Solvent and Isotope Effects

As long as excessive steric compression of **OR** and **L** does not reduce k_2 below k_r, the relative rates of formation, dissociation and collapse of the zwitterionic intermediate are irrelevant to the stereochemistry of the product cyclobutanone. In contrast, the substituent, solvent and isotope effects cannot be understood unless the timing of the various steps is taken into account.

An elementary *steady state* analysis of the stepwise mechanism illustrated in Fig. 6.8, on the assumption that $k_r << k_2$, leads to the following expression for the rate of formation of the contrathermodynamic cyclobutanone:

$$k_{\text{exp}} = k_1 k_2 / (k_{-1} + k_2) \tag{6.1}$$

This reduces to two familiar limiting forms when k_2 and k_{-1} are very unequal:
Case 1 ($k_2 >> k_{-1}$): $k_{\text{exp}} \approx k_1$; zwitterion formation is rate-limiting.
Case 2 ($k_2 << k_{-1}$): $k_{\text{exp}} \approx (k_1/k_{-1})k_2$; ring-closure occurs from a low concentration of zwitterion maintained in a pre-equilibrium with the reactants.

Stabilization of the zwitterion by polar substituents decreases both k_{-1} and k_2, but is not expected to have a dominant effect on their ratio. Bulky substituents on the ketene and on C_1 of the olefin should therefore favor *Case 2* by reducing k_2, provided that it is not reduced so drastically that $k_2 \approx k_r$, in which case rotation and closure to the thermodynamically more stable isomer will become competitive. A kinetic probe into the timing of the successive steps that should avoid the latter possibility would therefore be a series of reactions in which **R′**=H and the steric requirements of one or both ketene substituents are varied. As the bulk of these substituents (**L** and **S** in Fig. 6.8) is increased, the reaction should shift from *Case 1* towards *Case 2*, the change manifesting itself in a gradual reduction in the sensitivity to polar substituent effects and to solvent polarity. Pending such an investigation, the published evidence is supportive of the proposed mechanism:

The substantial decrease in the absolute value of the polar reactivity constant from $\rho = -1.4$ for cycloaddition of substituted styrenes with dimethylketene (DMK) [39] to $\rho = -0.73$ for their cycloaddition to the much more sterically hindered (DPK) [40] ($E_s^{\text{CH}_3} - E_s^{\text{C}_6\text{H}_5} = 2.55$ [60, Table 4.1]) is consistent with a transition from *Case 1* to *Case 2*. [41]

As noted above, Baldwin and Kapecki [40] observed that increased solvent polarity has a negligble effect on the cycloaddition of DPK to styrene, where k_2 should be subject to massive steric retardation; this too is consistent with *Case 2*. DPK was also the ketene used in the classic study of Huisgen et al. [35], but their ketenophiles, butyl vinyl ether and dihydropyran, offer less hindrance to ring closure. The solvent dependence is relatively mild but not negligible, and is appreciably less for the more rigid cyclic substrate, strongly suggesting that the reaction belongs in the intermediate region where $k_2 \approx k_{-1}$, and Equation 6.1 cannot be taken to either limit. Cycloaddition to less hindered ketenes like DMK, which was assigned to *Case 1* on the basis of its greater sensitivity to polar substituent effects, would be expected to show a stronger dependence on solvent polarity as well.

It has been recognized that the different isotope effects measured in various laboratories can be accomodated by assuming a stepwise mechanism like that shown in Fig. 6.7; in an isotopic study of a related reaction, Bayne [61, ref. 6] refers to this possibility in detail. The difficulty, however, is as follows: If it is maintained that an inverse isotope effect can arise only on bond formation, as a result of the hybridization change from sp^2 towards sp^3, an inverse effect would be observed only in *Case 2*, and even then would be unlikely to lead to nearly identical isotope effects at both positions, because the hybridization change at the transition state of the second step is complete at C_2 and only partial at C_1. [61] More serious still is the inconsistency with the solvent and substituent effects: In the cycloaddition of DPK to styrene, which the solvent and substituent effects assign firmly to *Case 2*, the isotope effect at C_2 is appropriately inverse, but – disconcertingly – it is direct at C_1 [50]. In contrast, the inverse effect at C_1 shows up in the cycloaddition of DMK to styrene, [54] that has to be relegated to *Case 1*, into which k_2 does not enter at all! Finally, in the reaction of DPK with 5,5-dimethylcyclopentadiene – a substrate with similar steric requirements to Huisgen's dihydropyran [35], that was assigned above to the intermediate region – the effect at C_2 remains inverse but that at C_1 vanishes [55].[17]

The consistently inverse isotope effect at C_2 poses no problem; it is a *secondary isotope effect of the first kind* that results primarily from the increased frequency of the out-of-plane CH-bending vibrations that accompany the hybridization change from sp^2 to sp^3 during the bonding process.[18] The varying isotope effect at C_1, however, has to be interpreted as composite: An inverse effect on k_1, moderated and eventually reversed as the reaction moves from *Case 1* towards *Case 2* and k_2 becomes increasingly important in the expression for k_{exp}.

The two compensating effects must both be *of the second kind* [48, p. 111, p. 180ff.]: 1) As the zwitterion is formed, positive charge is localized on C_1, raising the frequency of the CH-stretching and bending modes,[19][48, p. 159-160ff.] and thus leading to an inverse isotope effect that is fortuitously similar in magnitude to that on C_2. 2) Because of the severe steric congestion[20], bond formation at C_1 will have progressed to only a slight extent at the transition state of *Step 2*. The positive charge will, however, have been transferred to its bonding partner at the transition state for ring closure, which therefore resembles an extended singlet biradical. As noted in connection with biradical cycloaddition, the methylene H atoms at a radical center are "looser" than in the corresponding olefin; the

[17] A similar timing of the two steps is probably responsible for Katz and Dessau's [62] early observation that the carbonyl C atom of DPK bonds preferentially to the deuterated C atom of cyclohexene-d_1.

[18] The experimental results are too sparse to allow a distinction between the effect of partial bonding in the transition state of *Step 1* and that of full bonding but increased HCH angle in the transition state of *Step 2*.

[19] This is the rationale for the "inductive effect of deuterium" [48, p. 137].

[20] The congestion does not directly involve the H atom, so an inverse "steric isotope effect" [63], [48, pp. 143–144] would not be expected.

observed direct isotope effect ($k_2^H > k_2^D$), though perhaps not predictable in advance, is by no means unreasonable.

Although the interpretation offered above for the isotope effect is consistent with the solvent and substituent effects, as well as with secondary isotope effects on other reactions, it will probably be resisted as "counterintuitive" unless its predictive power is established experimentally. A finding that the isotope effect at C_1 is inverse in the cycloaddition of DPK to relatively unhindered alkyl vinyl ethers and reversed as the steric requirements of the substituents on C_1 increase will go a long way towards its confirmation. Such an investigation was undertaken by the late Professor E.A. Koerner von Gustorf but discontinued on his untimely death in September 1975.[21]

6.5 Apologia

The discussion in this chapter has ranged well outside the main theme of the book. In addition to the writer's early involvement with secondary isotope effects, which can serve as partial extenuation, the mechanism of [2+2]-cycloaddition has sufficient intrinsic interest to justify the digression. Orbital symmetry conservation plays but a small part in its mechanistic analysis, but it is a crucial one. Fig. 6.2 applies strictly only to the cyclodimerization of ethylene, or to an olefin symmetrically tetrasubstituted by substituents that do not add to the "essential" number of electrons involved in the reaction. Nevertheless, the principal conclusion drawn from it, that the initial plane-rectangular interaction of the two π systems leads to formation of a bond between diagonally situated atoms, is remarkably robust. It can be applied to a variety of reactions with different electronic and steric requirements, provided that the specifics of each reacting system are kept firmly in mind. The wealth of diverse, superficially contradictory, experimental results cannot be fit into a consistent logical framework without it.

6.6 References

[1] R.A. Moss: Accts. Chem. Research *13*, 58 (1980); ibid. *22*, 15 (1989).
[2] R.B. Woodward and R. Hoffmann: *The Conservation of Orbital Symmetry.* Verlag Chemie, Weinheim and Academic Press, New York 1970.
[3] J. Katriel and E.A. Halevi: Theoret. Chim. Acta. *40*, 1 (1975).
[4] R. Hoffmann: J. Amer. Chem. Soc. *90*, 1475 (1968).
[5] T.A. Albright, J.K. Burdett and M.-H. Whangbo: *Orbital Interactions in Chemistry.* Wiley, New York 1985.

[21] In a letter to the author dated September 2, 1975, Professor Koerner von Gustorf reported preliminary results indicating substantial, but unequal, inverse deuterium isotope effects at both C_1 and C_2 of *cis-* and *trans-*propenylpropyl ether on the rate of its cycloaddition to DPK.

[6] I.R. Gould, N.J. Turro, C. Doubleday, Jr., N.P. Hacker, G.F. Lehr, R.A. Moss, D.P. Cox, W. Guo, R.C. Munjal, L.A. Perez and M. Fedorynski: Tetrahedron *41*, 1587 (1985).
[7] K.N. Houk, N.G. Rondan and J. Mareda: Tetrahedron *41*, 1555 (1985).
[8] R.P. Bell: *The Proton in Chemistry.* Cornell University Press, Ithaca 1959.
[9] E.A. Halevi: Helvet. Chim. Acta. *58*, 2136 (1975).
[10] M.J.S. Dewar and S.Kirschner: J. Amer. Chem. Soc. *96*, 9246 (1974).
[11] G.A. Segal: J. Amer. Chem. Soc. *96*, 7892 (1974).
[12] T. Okada, K. Yamaguchi and T. Fueno: Tetrahedron *30*, 2293 (1974).
[13] K.N. Houk: Acc. Chem. Res. *8*, 361 (1975).
[14] K. Kraft and G. Koltzenburg: Tetrahedron Letters *1967* , 4357, 4723.
[15] A. Padwa, W. Koehn, J. Massarachia, C.L. Osborn and D.J. Trecker: J. Amer. Chem. Soc. *93*, 3633 (1971).
[16] L.E. Gusel'nikov,N.S. Nametkin and V.M. Volovin: Acc. Chem. Res. *8*, 18 (1975) and papers cited therein.
[17] R. Ahlrichs and R. Heinzmann: J. Amer. Chem. Soc. *99*, 7452 (1977).
[18] P. Potzinger, B. Reimann and R.S. Roy: Ber. Bunsenges. Phys. Chem. *85*, 1119 (1981).
[19] P.D. Bartlett, L.K. Montgomery and B. Seidel: J. Amer. Chem. Soc. *86*, 616 (1964).
[20] L.K. Montgomery, K. Schueller and P.D. Bartlett: J. Amer. Chem. Soc. *86*, 622 (1964).
[21] P.D. Bartlett and L.K. Montgomery: J. Amer. Chem. Soc. *86*, 628 (1964).
[22] P.D. Bartlett and N.A. Porter: J. Amer. Chem. Soc. *90*, 5317 (1968).
[23] P.D. Bartlett: Quart. Revs. Chem. Soc. *24*, 473 (1970).
[24] For a thorough discussion of π–σ and *through-bond–through-space* interactions, see R. Gleiter: Angew. Chem. *86*, 770 (1974); Angew. Chem. Internat. Ed. (English) *13*, 696 (1974). See also Reference [5, p. 195ff.].
[25] V. Gold (Compiler and Editor): *Glossary of Terms Used in Physical Organic Chemistry*. Pure and Appl. Chem. *55*, 1281ff. (1983).
[26] J. Leitich: Internat. J. Chem. Kinet. *11*, 1249 (1980).
[27] A. Padwa and T.J. Blacklock: J. Amer. Chem. Soc. *101*, 3390 (1979); ibid. *102*, 2797 (1980).
[28] R. Huisgen: Acc. Chem. Res. *10*, 117, 199 (1977) and references cited therein.
[29] N.D. Epiotis: *Theory of Organic Reactions*. Springer, Berlin Heidelberg 1978.
[30] E.A. Halevi: *Unpublished calculations*.
[31] I. Karle, J. Flippen, R. Huisgen and R. Schug: J. Amer. Chem. Soc. *97*, 5285 (1975).
[32] A. Conan Doyle: *The Adventure of the Bruce-Partington Plans*.
[33] G. Steiner and R. Huisgen: J. Amer. Chem. Soc. *95*, 5056 (1973).
[34] J. March: *Advanced Organic Chemistry, Third edition*. Wiley, New York 1985.
[35] R. Huisgen, L.A. Feiler and P.Otto: Chem. Ber. *102*, 3444 (1969).
[36] R. Huisgen, L.A. Feiler and G. Binsch: Angew. Chem. *76*, 892 (1964); Angew. Chem. Internat. Ed. (English) *3*, 753 (1964).
[37] R. Huisgen, L.A. Feiler and G. Binsch: Chem.Ber. *102*, 3460 (1969).
[38] R. Montaigne and L. Ghosez: Angew. Chem. *80*, 194 (1968); Angew. Chem. Internat. Ed. (English) 7, 221 (1968).
[39] N.S. Isaacs and P. Stanbury: J. Chem. Soc. Perkin II, *1973*, 166.

[40] J.E. Baldwin and J.A. Kapecki: J. Amer. Chem. Soc. *92*, 4868 (1970).
[41] See e.g. J.E. Leffler and E. Grunwald: *Rates and Equilibria of Organic Reactions*. Wiley, New York 1963.
[42] R. Huisgen and H. Mayr: Tetrahedron Letters *34*, 2965 (1975).
[43] R. Huisgen and H. Mayr: Tetrahedron Letters *34*, 2969 (1975).
[44] H.W. Moore: Accts. Chem. Research *12*,125 (1979).
[45] A.H. Al-Husaini and H.W. Moore: J. Org. Chem. *50*, 2595 (1985).
[46] W. R. Dolbier, Jr. and S.-H. Dai: J. Amer. Chem. Soc. *90*, 5028 (1968); S.-H. Dai and W. R. Dolbier, Jr.: J. Amer. Chem. Soc. *94*, 3946 (1972).
[47] A. Streitwieser, Jr., R.H. Jagow, R.C. Fahey and S. Suzuki: J. Amer. Chem. Soc. *80*, 2326 (1958).
[48] E.A. Halevi: "Secondary Isotope effects" in S.G. Cohen, A. Streitwieser, Jr. and R.W. Taft, Jr. *eds. Progress in Physical Organic Chemistry, vol. 1*. Wiley-Interscience, New York 1963.
[49] R.J. Crawford and D.M. Cameron: J. Amer. Chem. Soc. *88*, 2589 (1966).
[50] J.E. Baldwin and J.A. Kapecki: J. Amer. Chem. Soc. *91*, 3106 (1969); J. Amer. Chem. Soc. *92*, 4874 (1970).
[51] E. Koerner von Gustorf, D. V. White, J. Leitich and D. Henneberg: Tetrahedron Lett. *1969*, 3113.
[52] E. Koerner von Gustorf, D. V. White, B. Kim, D. Hess and J. Leitich: J. Org. Chem. *35*, 1155 (1970).
[53] E.I. Snyder: J. Org. Chem. *35*, 4287 (1970).
[54] N.S. Isaacs and B.G. Hatcher: J. Chem. Soc. Chem. Commun. *1974*, 593.
[55] R.W. Holder, N.A. Graf, E. Deusler and J.C. Moss: J. Amer. Chem. Soc. *105*, 2929, (1983).
[56] E.A. Halevi and D. Becker: *unpublished*. Having been denied publication in 1976 on grounds of being "speculative", the manuscript of this communication was distributed privately to several workers in the field. Its resubmission for publication was postponed *sine die* after the mechanism outlined in Fig. 6.7 was cited briefly, along with indirect experimental evidence in its support [57].
[57] D. Becker and N.C. Brodsky: J. Chem. Soc. Chem. Commun. *1978*, 237.
[58] R.G. Pearson: *Symmetry Rules for Chemical Reactions.* Wiley, New York 1976.
[59] R.W. Taft, Jr.: Chapter 13 in M.S. Newman (ed.) *Steric Effects in Organic Chemistry.* Wiley, New York, 1956.
[60] J. Shorter: *Correlation Analysis of Organic Reactivity.* Wiley, Chichester 1982.
[61] W.F. Bayne: Tetrahedron Lett. *1970*, 2263.
[62] T.J. Katz and R. Dessau: J. Amer. Chem. Soc. *85*, 2172 (1963)
[63] L.S. Bartell: J. Amer. Chem. Soc. *83*, 3567 (1961).

Chapter 7

Cycloadditions and Cycloreversions: II. Beyond [2+2]

There are two ways of going beyond [2+2]-cycloaddition: The straightforward way is to extend the length of one or both of the conjugated π systems of the reactants; the number of electrons involved in [n+m]-cycloaddition is then given by: $k = (n + m)$. It was shown in Chapter 5, however, in connection with the benzvalene-benzene and cubane-cyclooctatetraene interconversions, that just how many electrons are "involved" in a given reaction is a matter of interpretation. For present purposes, a reaction in which the presence of more than four electrons – bonding or non-bonding – cannot be safely ignored in the symmetry analysis will be considered under the heading "beyond [2+2]", even if it results in closure of a four membered ring.

7.1 $[_\pi 4 + {}_\pi 2]$-Cycloaddition; Anasymmetrization

Although the procedures illustrated so far are capable of dealing with all of the reactions to be discussed in this chapter, some economy – and perhaps a bit more insight – can be attained by the use of a formal symmetry-raising procedure, which might be called *anasymmetrization*[1], It has already been applied implicitly in the construction of Fig. 5.6 and will become increasingly useful in the analysis of sigmatropic rearrangements and certain photochemical reactions. It will be illustrated with the Diels-Alder Reaction.

7.1.1 The Diels-Alder Reaction

This prototypic $[_\pi 4 + {}_\pi 2]$-cycloaddition has been documented so thoroughly [1, 2][3, pp. 745ff.] that there is very little new that can be said about it. As is clear from the examples taken up in Chapter 1, there is universal agreement that the $[_\pi 4_s + {}_\pi 2_s]$ and $[_\pi 4_a + {}_\pi 2_a]$ pathways are both *allowed* on the ground-state surface. [4, pp. 70,78] This is true of OCAMS as well: The analysis of each pathway is carried out after mutually orienting the reactants as in (a) and (c) of Fig. 1.1 for the $[_\pi 4_s + {}_\pi 2_s]$ and $[_\pi 4_a + {}_\pi 2_a]$ pathways respectively. Two correspondence diagrams are then drawn, with *cis*-cyclohexene in $\mathbf{C}_s$ in the first case and with *trans*-cyclohexene in $\mathbf{C}_2$ in the second. [5]

[1] From *ana* = upward, in the sense of *building up*. (cf. anagram, anabolism, analepsis).

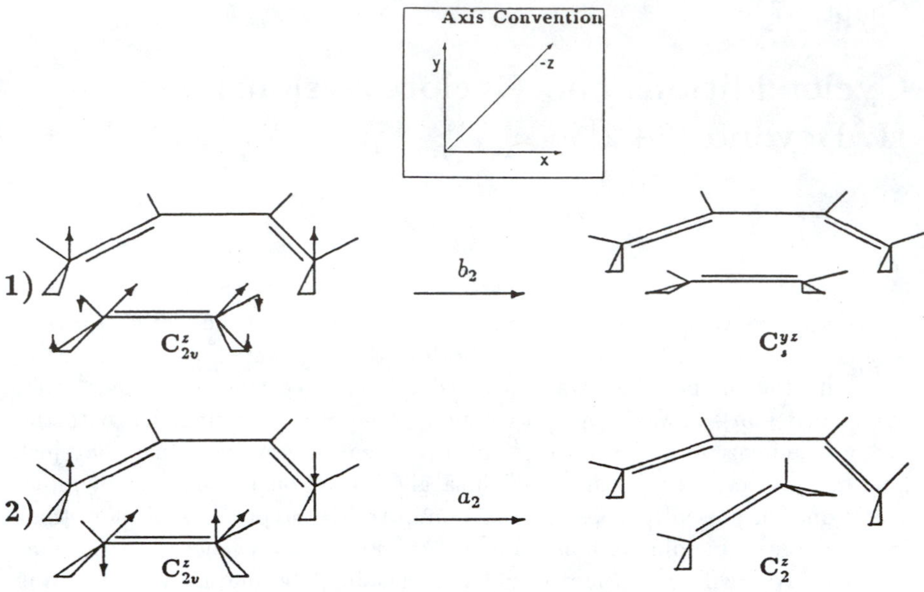

Figure 7.1. Desymmetrization of approach to Diels-Alder reaction: (a) $\mathbf{C}_{2v}^{z} \longrightarrow \mathbf{C}_{s}^{xz}$; (b) $\mathbf{C}_{2v}^{z} \longrightarrow \mathbf{C}_{2}^{z}$

The initial orientations for the two *WH-allowed* pathways for the reaction between butadiene and ethylene are shown at the right side of Fig. 7.1. In the chosen axis convention, the appropriate geometry for embarkation on the $[_\pi 4_s +_\pi 2_s]$ pathway is $\mathbf{C}_s^{zx}$ and that for $[_\pi 4_a +_\pi 2_a]$ is $\mathbf{C}_2^z$. These two groups are subgroups of $\mathbf{C}_{2v}^z$, which is therefore the *supergroup* that is chosen to represent them both. The coordinate that moves the ethylene molecule out of the yz plane has the irrep b_1 and desymmetrizes the reactant pair to $\mathbf{C}_s^{zx}$, whereas rotating it about the symmetry axis (a_2) reduces the symmetry to $\mathbf{C}_2^z$. If the correspondence diagram with the product cyclohexene – also constrained to $\mathbf{C}_{2v}^z$ – requires an a_2 displacement, the $[_\pi 4_a +_\pi 2_a]$ pathway is *allowed*; if a b_1 displacement is specified, the $[_\pi 4_a +_\pi 2_a]$ pathway is *allowed*. The correspondence diagram in $\mathbf{C}_{2v}^z$ is set up in Fig. 7.2.

7.1.1.1 Anasymmetrization

A glance at the illustration of "planar cyclohexene" in Fig. 7.2 triggers an immediate objection: The molecule is too strained to remain in plane, but would relax spontaneously from $\mathbf{C}_{2v}^z$ to either the *cis* ($\mathbf{C}_s^{zx}$) or the *trans* ($\mathbf{C}_2^z$) conformation. As in the case of Fig. 5.6, where neither the reactant nor the product is in its stable geometry, "planar cyclohexene" must be understood to be a formal construct representing both accessible conformations of the molecule. The procedure was formally justified in the paper outlining the theoretical basis and conceptual framework of *Orbital Correspondence Analysis in Maximum Symmetry.* [6]

OCAMS makes two tacit assumptions that are fundamental to all discussions of mechanism in terms of orbital symmetry:

a) The orbital approximation is adequate.[2]

b) The molecular orbitals are constructed as linear combinations of a well-defined atomic basis-set, generally a minimal basis-set consisting of the valence-shell *s* and *p* orbitals, augmented when necessary by *d* orbitals in the case of elements of the second-row and beyond.

Up to this point, whenever it was desired to analyze a reaction between a reactant R and product P that differ in their symmetry properties, the correspondence diagram was constructed in the point group of lower symmetry. If, for example, R belongs to **G** whereas the less symmetric P belongs to **H**, a subgroup of **G**, the symmetry R was reduced to to that of P by a slight displacement of the nuclei into **H**. *Anasymmetrization* is a procedure that formally raises the symmetry of P to that of R, and thus allows a correspondence diagram to be set up in **G**, the symmetry point group of the *more* symmetric of the two.

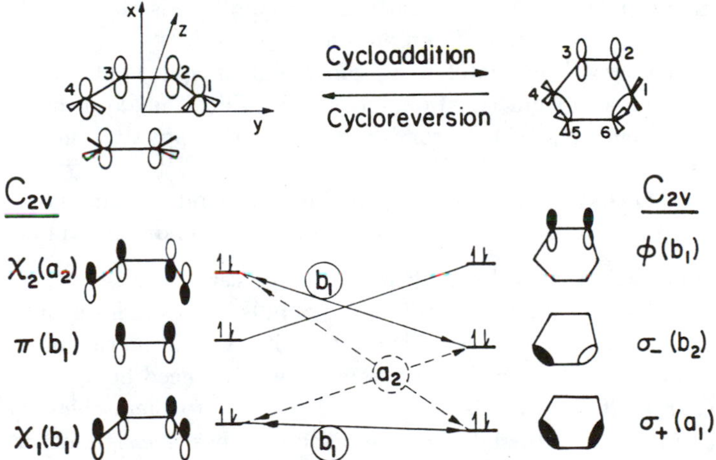

Figure 7.2. Correspondence diagram for [$_\pi 4 +_\pi 2$]-cycloaddition ($\mathbf{C}^z_{2v}$)

Assume that some sym-op S present in **G** is missing in **H**, and that **G** can be generated as the direct or semidirect product of **H** and the two-element group $\{E, S\}$. In the reaction under consideration, **G** is $\mathbf{C}^z_{2v}$, that can be generated as the direct product of either $\mathbf{C}^{zx}_s$ or $\mathbf{C}^z_2$ with $\mathbf{C}^{yz}_s$, which comprises the identity

[2] This assumption is valid for most molecules in their ground-states, though it breaks down whenever the description of the ground-state in terms of a single-electron configuration is particularly poor, as in the case of biradicals. [7, p. 63] Even then, as in the biradical mechanism for [2+2]-cycloaddition (Section 5.3.1) the MO approximation retains enough vestigial force to be qualitatively useful.

operation E and $\sigma(yz)$; the *anasymmetrizer*, S, is $\sigma(yz)$. It has been shown [6, p. 11] that each MO of P, ψ_i, can be relabeled in the higher group **G** by operating with S on $\psi_i(\mathbf{H})$ and taking the appropriate linear combination of $\psi_i(\mathbf{H})$ with $S\psi_i(\mathbf{H})$.

$$\psi_i(\mathbf{G}) = \psi_i(\mathbf{H}) \pm S\psi_i(\mathbf{H}) \tag{7.1}$$

in our example:

$$\psi_i(\mathbf{C}_{2v}^z) = \psi_i(\mathbf{C}_s^{zx}) \pm \sigma(yz)\psi_i(\mathbf{C}_s^{zx}) \tag{7.2}$$

or

$$\psi_i(\mathbf{C}_{2v}^z) = \psi_i(\mathbf{C}_2^z) \pm \sigma(yz)\psi_i(\mathbf{C}_2^z) \tag{7.3}$$

Since the number of MOs in **G** must be the same as that in **H**, only one of the two linear combinations can represent ψ_i in the higher group. The choice of the proper sign is made on the basis of a few simple rules,[3] that ensure continuity of the phases of the occupied orbitals along the reaction path, and guarantee that the result is independent of the particular sym-op used as anasymmetrizer:

1. If an atomic orbital that is included in any occupied MO is transformed in the higher group to itself, or to an AO that was symmetrically equivalent to it before anasymmetrization, it cannot be allowed to vanish in *all* of the anasymmetrized MOs. (A particularly useful corollary is that no MO can be allowed to disappear completely on anasymmetrization).
2. When two AOs centered on initially non-equivalent atoms become equivalent upon anasymmetrization, any MO in which they appear, $\psi_i(\mathbf{H})$, and its "mirror image", $S\psi_i(\mathbf{H})$ are combined once with a positive and once with a negative sign.
3. An AO that participates in a bonding-antibonding combination cannot be allowed to appear in one but disappear from the other on anasymmetrization.

The rules will have to be applied with some care in subsequent examples, but their application to Fig. 7.2, is quite simple: The diagram only includes bonding orbitals, so Rule 3 is irrelevant to it. As all six carbon atoms are reflected into themselves, Rule 1 is the only one that need be applied to them. Their $2s$, $2p_y$ and $2p_z$ orbitals are reflected by σ_{yz} into themselves, so the σ_{CC} MOs constructed from them have to be taken with positive sign; otherwise they would vanish. The mirror image of a p_x orbital in σ_{yz} is $-p_x$, so the π orbitals must be taken with negative sign in order for their component p_x orbitals not to vanish. Thus, for example: $\phi + \sigma_{yz}\phi = 0$ whereas $(\phi - \sigma_{yz}\phi)/2 = \phi$, which has the irrep b_1 in $\mathbf{C}_{2v}^z$ and is labeled accordingly in Fig. 7.2

The MOs involving the H atoms require the use of Rule 2. The two upper H atoms bonded to C_1 and C_4, for example, are equivalent to one another in $\mathbf{C}_s^{zx}$, the symmetry point group of *cis*-cyclohexene, and combine to a positive (a') and a negative (a'') combination. The same is true of the lower two H atoms,

[3] These rules establish the continuous one-to-one correspondence between the Born-Oppenheimer eigenstates and those of the anasymmetrized Hamiltonian which is required by theory [6, p. 12]. They are related to Goddard's [8] *phase continuity rule*, which, however, does not invoke symmetry explicitly.

that are interconvertible with one another but not with the upper ones. On anasymmetrization to C_{2v}^z, all four of the σ_{CH} bonds to these atoms become equivalent. Rule 2 requires that the two a' orbitals combine to a_1 and b_1, both of which are symmetric with respect to reflection in $\sigma(xz)$ – the mirror plane initially present, and that the two a'' orbitals form an a_2 and a b_2 combination.[4]

The eight σ_{CH} combinations are omitted from the correspondence diagram, in Fig. 7.2, not because they are unimportant but because their behavior is easily ascertained by analogy with Fig. 5.5. As in that diagram, the CH bonds on C_1 and C_4 of butadiene lie in the molecular plane and combine to two a_1 and two b_2 orbitals. In cyclohexene they are symmetrically equivalent, their combinations spanning the four irreps of C_{2v}^z. As a result, one a_1 and one b_2 σ_{CH}-bonding combination on the left correlates directly with two similar combinations on the right, whereas the remaining two have to be induced to correlate with one a_2 and one b_1 combination. As in Fig. 5.5, this can be accomplished by either a disrotation (b_1 in the axis convention of Fig 7.2) or a conrotation (a_2). The four CH bonds of ethylene, that combine similarly to two a_1 and two b_2 orbitals, become those bonded to C_5 and C_6 of "planar cyclohexene", which – like those to C_1 and C_4 – belong to the *regular representation* of C_{2v}^z that spans its four irreps. Therefore, either a disrotation or a conrotation will correlate these four σ_{CH}-bonding orbitals across the diagram as well.

It is clear from Fig. 7.2 that the CC-bonding orbitals can also be induced to correspond along either pathway, so *OCAMS* – like *WH-LHA* – finds no symmetry-imposed barrier to the reaction along either pathway. It is obvious from Fig. 7.1 why the disrotatory pathway is ordinarily taken in preferece to the conrotatory one. Starting from a coplanar geometry, both reaction coordinates incorporate a totally symmetric (a_1) coordinate, depicted in Fig. 7.1 by means of arrows indicating the least-motion approach of the ethylene molecule to its reaction parter. If, in addition, the latter moves out of plane along a relative translation of irrep b_1 that costs no energy, the reactants are ideally oriented for disrotatory cycloaddition. In order to attain the proper mutual orientation for conrotatory cycloaddition, two a_2 coordinates must be added to the totally symmetric component: the ethylene molecule has to rotate about the symmetry axis and the terminal methylene groups have to twist out of plane. The former motion costs no energy, but the latter disrupts the conjugated π system of butadiene before any advantage can be gained from formation of the new σ bonds. This factor, in addition to the more sterically strained transition state for conrotatory cycloaddition, selects the disrotatory mode as the preferred pathway.

---

[4] The reader who finds wading through the formal argument tedious may be content to recognize that the irreps of the anasymmetrized MOs are precisely those that would be obtained if the cyclohexene molecule were forced into plane. While this is often the case, it cannot be adopted as a general rule, since constraining a molecule to an unnatural conformation may change the energetic order of its MOs, perhaps interchanging occupied and unoccupied orbitals.

7.1.1.2 Changing the Initial Orientation

The initial orientation of the reactants in Fig. 7.2 was chosen because the *cis* and *trans* forms of the product belong to its non-trivial subgroups, $\mathbf{C}_s^{zx}$ and $\mathbf{C}_2^z$ respectively. However, the two orientations on the left and right sides of Fig. 7.3 are equally probable modes of approach and also have $\mathbf{C}_{2v}^z$ symmetry.

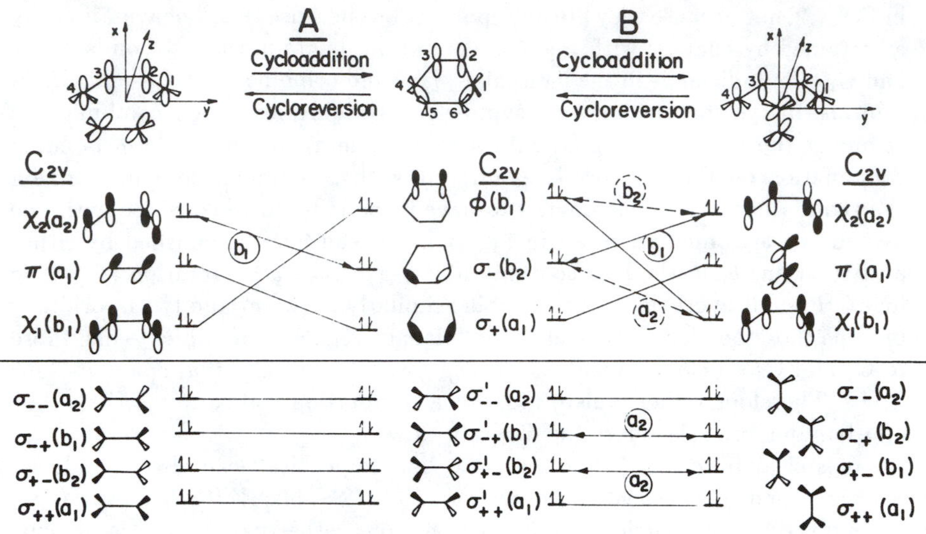

Figure 7.3. Alternative $\mathbf{C}_{2v}^z$ approaches for $[_\pi 4 +_\pi 2]$-cycloaddition

In both modes, the product hexadiene and the reactant butadiene are set up as in Fig. 7.2, and only the ethylene molecule is oriented differently. As before, CH-bonding MOs at C_1 and C_4 correlate under either a conrotation or a disrotation and can be left out of the diagram. In Approach (**A**) at the left of the figure, the π orbital of ethylene is in the yz plane and its four σ_{CH} orbitals correlate directly with those of the product. Raising it above the plane and rotating it about its own longitudinal axis are both b_1 displacements that cost no energy and bring it into the proper orientation for $[_\pi 4_s +_\pi 2_s]$-cyloaddition. Alternatively, rotation about the z axis (a_2) takes it into the appropriate geometry for $[_\pi 4_a +_\pi 2_s]$-cycloaddition. The b_1 correspondence between χ_2 of butadiene and σ_- of hexadiene affirms the "allowedness" of the former mode and asserts that the latter is *forbidden*.

At first sight, Approach (**B**) is only suitable for the $[_\pi 4_a +_\pi 2_s]$ mode; the required rotation about z (a_2), which allows the CH-bonding orbitals to retain their individual identities, costs no energy. Induction of the $\chi_2(a_2) \leftrightarrow \sigma_-(b_2)$ correspondence still calls for a b_1 displacement, so $[_\pi 4_a +_\pi 2_s]$-cycloaddition remains *forbidden*. Nevertheless, we can construct composite motions that regenerate the reaction coordinate for either the $[_\pi 4_s +_\pi 2_s]$ or $[_\pi 4_a +_\pi 2_a]$ modes as follows:

1. Rotation of ethylene by 90° along the a_2 coordinate brings it back to the initial orientation of Approach (**A**), from which an out-of-plane displacement (b_1) takes it onto the $[_\pi 4_s +_\pi 2_s]$ pathway.

2. We note that the a_2 perturbation induces the correspondence: $\chi_1(b_1) \leftrightarrow \sigma_-(b_2)$, leaving $\xi_2(a_2) \leftrightarrow \phi(b_1)$ to be induced by the imposition of a b_2 displacement. If a 90° rotation of the ethylene molecule about its longitudinal axis (here b_2) is superposed on its partial rotation about the symmetry axis (a_2), the reactants come into position for $[_\pi 4_a +_\pi 2_a]$-cycloaddition.

Having made an "unwise" choice of the initial orientation in Approach (**B**), we are obliged to pay for it with greater complexity; the mechanistic conclusions, however, are unaltered.

7.2 Reactions Related to $[_\pi 4 + {}_\pi 2]$-Cycloaddition

7.2.1 The Homo-Diels-Alder Reaction

This related reaction, [9] illustrated in Fig. 7.4, differs from $[_\pi 4 +_\pi 2]$-cycloaddition in that the π bonds of the diene are not conjugated, so it is more rigorously referred to as a $[_\pi 2 +_\pi 2 +_\pi 2]$-cycloaddition, that is also *WH-allowed* [4, p. 106].

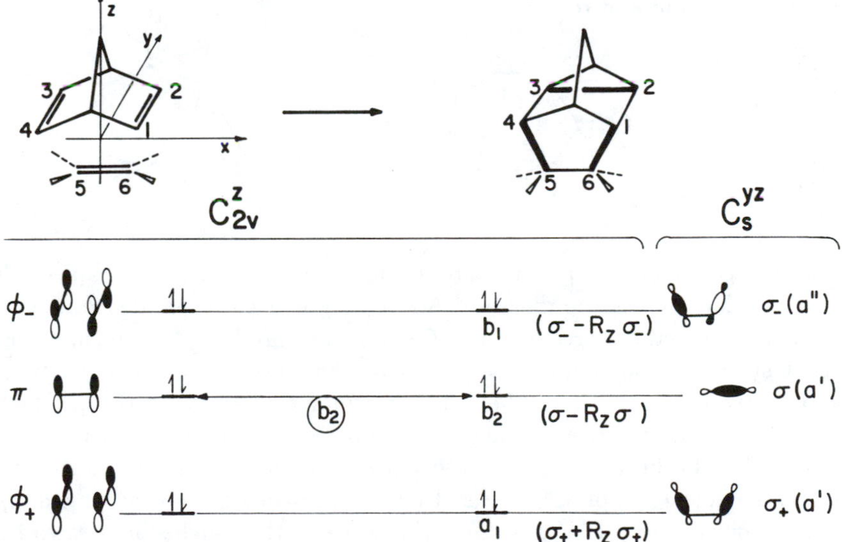

Figure 7.4. Correspondence diagram for the homo-Diels-Alder reaction

Only the three π bonds of the reactant and the three newly formed σ bonds of the product need be included in the diagram. Bicycloheptadiene and the dienophile, characteristically an electronegatively polysubstituted ethylene, are set up in $\mathbf{C}_{2v}^z$, the π orbital of the latter lying in the zx plane. This orbital is totally symmetric; so is ϕ_+, whereas ϕ_- has the irrep b_1. The symmetry point group of the product is $\mathbf{C}_s^{yz}$, which can be raised to $\mathbf{C}_{2v}^{z}$ using either $\sigma(zx)$ or $C_2(z)$ as the anasymmetrizer; for a change, let us use the latter. C_5 and C_6 are rotated into each other by $C_2(z)$, so $C_2(z)\sigma_+ = \sigma_+$ and $C_2(z)\sigma_- = -\sigma_-$. By the corollary to Rule 1, $\sigma_+ \rightarrow (\sigma_+ + C_2(z)\sigma_+)(a_1)$ and $\sigma_- \rightarrow (\sigma_- - C_2(z)\sigma_-)(b_1)$, because the alternative combination vanishes in both cases. Then, since the positive sign has necessarily been assigned to σ_+, Rule 2 insists that the second a' orbital, σ, be taken with the negative sign: $\sigma \rightarrow (\sigma - C_2(z)\sigma)(b_2)$.

The one mismatch in Fig. 7.4 is between $\pi(a_1)$ and $\sigma(b_2)$. Correspondence between them is induced by a b_2 displacement: motion of the dienophile parallel to the y axis so that it can bond concertedly either to C_1 and C_4 or to C_2 and C_3.

7.2.2 $n > 4$ and/or $m > 2$

The considerations developed so far are easily extended beyond $[_\pi 4 +_\pi 2]$-cycloaddition, and agree fully with the *WH-rules* when both reactants are hydrocarbons. As March points out, [3, p. 776] the conformational flexibility of the reactants makes formation of a ring with eight or more atoms difficult to achieve in a single step, but it is facilitated when one or both of the reactants are cyclic. Thus, the *WH-allowed* $[_\pi 6_s +_\pi 4_s]$ cycloaddition of cyclopentadiene to tropone is rapid and reversible [10]:

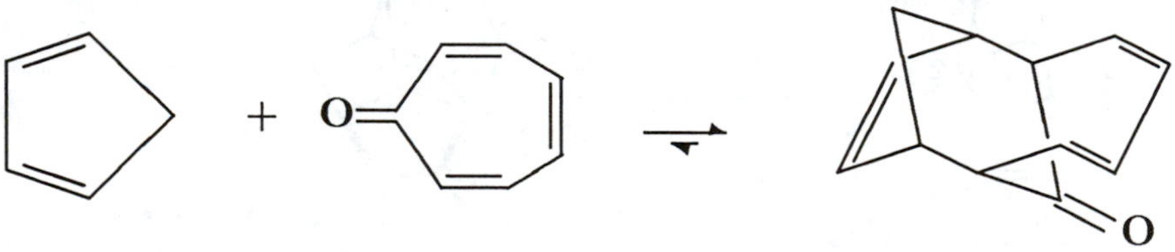

+ O=

O

If the reactants and product are set up in $\mathbf{C}_s$, all of the occupied MOs would correlate across the diagram. Alternatively, a correspondence diagram, in which the reactants are set up in $\mathbf{C}_{2v}$ and the product is anasymmetrized to that symmetry point group, would show that formal desymmetrization of the pathway to $\mathbf{C}_s$ – i.e. to the true molecular symmetry of the product – is called for. The methylene bridge of cyclopentadiene is innocuous; so, as it turns out, is the bridging carbonyl group in tropone. It will become evident from subsequent examples, however, that the presence of heteroatoms and/or multiple bonds can make a substantial difference to the conclusions drawn from an orbital symmetry analysis.

7.2.3 1,3-Dipolar Cycloaddition

This fully documented reaction [11] resembles $[_\pi 4 +_\pi 2]$-cycloaddition in that two σ bonds are formed between adjacent C atoms of an olefin and the termini of a chain of π-bonded atoms. It differs in that the chain is triatomic, most often includes one or more heteroatoms, and is characterized in the valence bond description by resonance between ionic or dipolar cannonical forms. These are of two main types, ordinarily distinguished by the nature of the contributing structures, [3, pp. 743–745] but it is more convenient for present purposes to adopt the MO formalism, and exemplify them by means of two symmetric parent reactants that differ in their geometry: azide ion and ozone. The former represents linear 1,3-dipoles of the *propargyl allenyl type* in Huisgen's nomenclature and the latter represents bent 1,3-dipoles of the *allyl type* [12].

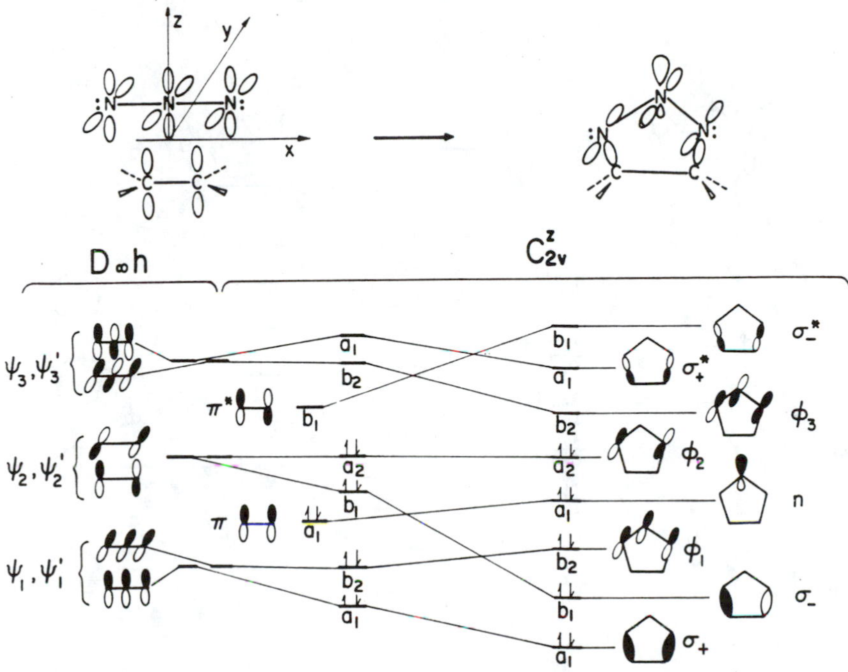

Figure 7.5. Correspondence diagram for the cycloaddition of N_3^- to ethylene

Fig. 7.5 shows the correspondence diagram for the cycloaddition of azide ion to ethylene. Each N atom has one $2s$ and three $2p$ orbitals; the 12 MOs constructed from them have to house the 16 valence electrons of N_3^-, but the two NN bonding orbitals and their antibonding counterparts, as well as the lone-pair orbitals[5] on the terminal atoms, retain their identity across the diagram and need not be considered explicitly. We are thus left with 6 MOs, three pairs of degenerate π orbitals, the lower four of which are occupied by eight electrons. When ethylene is brought up in the geometry adopted in the diagram, the cylindrical symmetry of the azide ion is reduced to C_{2v}^z and the degeneracy of

[5] One or both of the lone pairs may be replaced by σ bonds to substituent groups: e.g. azides or nitrile imines respectively.

its orbitals is split as shown. The left side of the diagram is completed by adding the π and π^* orbitals of ethylene and the two electrons that occupy the former.

The MOs of the cyclic product are set up in the conventional manner: bonding below non-bonding below antibonding, and σ below π, MOs of the same type being ordered according to the number of nodal planes. The five doubly-occupied MOs are seen to correlate smoothly across the diagram, so no reduction of symmetry below $\mathbf{C}_{2v}$ is called for.

The correspondence diagram for the cycloaddition of ozone and ethylene, which is the first step in the Criegee mechanism of the ozonolysis of olefins, [17], [3, pp. 1067–1070] is illustrated if Fig. 7.6.

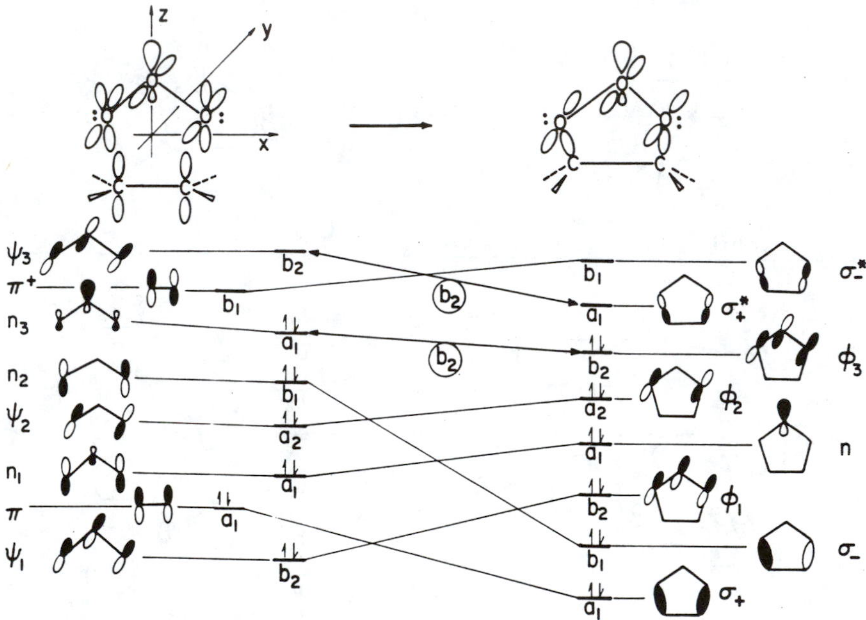

Figure 7.6. Correspondence diagram for cycloaddition of ozone to ethylene

The right side of Fig. 7.6 is identical with that of Fig. 7.5, except for the fact that $\phi_3(b_2)$ is doubly occupied. The MOs of ozone are easily related to those of N_3^-; the additional occupied MO, $n_3(a_1)$, is a non-bonding orbital largely localized on the central atom, that is derived from ψ_3', the in-plane member of N_3^-'s degenerate pair of π^* orbitals. The required $n_3(a_1) \leftrightarrow \phi_3(b_2)$ correspondence implies that the ethylene molecule has to move above or below the molecular plane of O_3 along a reaction path that can retain no more than $\mathbf{C}_s^{yz}$ symmetry.

Figs. 7.5 and 7.6 are consistent with the hypothesis, in support of which Huisgen [12, 13] has marshalled a great deal of evidence from solvent, substituent and isotope effects, that both types of 1,3-dipolar cycloaddition generally take place in a single step. Stepwise cycloaddition, advocated by Firestone [14, 15], has also been observed [16], but can compete successfully with the concerted process only when steric hindrance to bond formation at one end of the dipole is particularly severe.

The analogy with $[_\pi 4 +_\pi 2]$-cycloaddition is close, but not perfect: Fig. 7.5 shows the propargyl allenyl type of reaction to be *allowed* along the high symmetry, least-motion pathway; the alkene evidently reacts suprafacially,[6] but there is no way – experimentally or conceptually – of distinguishing between suprafacial and antarafacial bonding at its cylindrically symmetrical reaction partner. For 1,3-dipolar cycloaddition of the allyl type, Fig. 7.6 specifically selects desymmetrization to $\mathbf{C}_s^{yz}$, so both reactants can be said to be reacting suprafacially.

7.3 More Complex $[_\pi 2 +_\pi 2]$-Cycloadditions

Cycloaddition of species with triple bonds, which should logically be addressed at this point, will be postponed to later chapters. The reluctance of acetylene to dimerize to cyclobutadiene (CBD) on the ground-state surface follows directly from Fig. 6.2. It is sufficient to note that when two acetylene molecules approach one another in the plane-rectangular ($\mathbf{D}_{2h}$) orientation, the two additional π orbitals in acetylene are retained as such in CBD, so they cannot alleviate the "forbiddenness" of the $[_\pi 2_s + {}_\pi 2_s]$ pathway [5, Fig. 4]. Discussion of the reaction between dioxygen and acetylene to form 1,2-dioxetene and the cycloreversion of tetraalkyl-1,2-dioxetanes to two ketonic fragments has to be postponed until the relation between space and spin symmetry has been introduced in Chapter 9.

The rest of this chapter is devoted to an examination of how $[_\pi 2 + {}_\pi 2]$-cycloaddition and $[_\sigma 2 + {}_\sigma 2]$-cycloreversion are affected by the presence of additional multiple bonds that remain intact after the reaction has occurred, and of heteroatoms that – at first sight – play no apparent rôle in the reaction.

7.3.1 Dimerization of Cyclobutadiene

After CBD has been generated photochemically in an argon matrix at low temperature and the matrix is thawed at 35° K, it dimerizes to *syn*-tricyclo-[4.2.0.0^{2,5}]octa-3,7-diene (TCOD), [18] which calculations indicate to be less stable than its *trans* isomer [19, Table 1]:

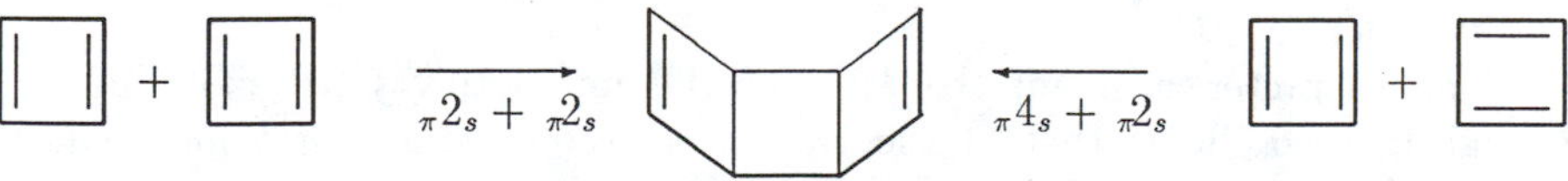

[6] If the diagram is set up – unreasonably – with the ethylene molecule rotated by 90° about its long axis (x), the irrep of π becomes b_2 and correspondence with the product has to be induced by incorporating a b_2 dispacement in the reaction coordinate, i.e. rotating it back into its orientation in Fig. 7.5.

Woodward and Hoffmann [4, p. 147] characterize the reaction as an *llowed* $[_\pi 4_s + {}_\pi 2_s]$-cycloaddition, in which *syn*-TCOD is formed rather than its *anti* isomer as a result of secondary orbital interactions. The reaction was analysed by OCAMS [5, p. 598] for a nearly-coplanar axial approach of the two CBD molecules, leading to an impossibly strained dimer with axial symmetry and a puckered central ring. The "axial dimer", however, is not intended to depict a real molecule, but is a purely formal model produced by anasymmetrization to $\mathbf{D}_2$ of either the *syn* ($\mathbf{C}_s$) or the *anti* ($\mathbf{C}_2$) isomer, and can thus represent them both. The correspondence diagram [5, Fig. 7] then shows that relaxation to the *syn* isomer involves a lower investment in distortional energy, in agreement with experiment.

In order to ascertain once more that the result of a symmetry analysis does not depend on how the reactant and product were set up in the diagram, let us begin by asking why cyclobutadiene does not dimerize to cubane when the CBD molecules are brought up face to face. The answer that "$[_\pi 4_s + {}_\pi 4_s]$-cycloaddition is *forbidden*" is hardly satisfying, in view of the conclusion reached in Chapter 5, that there is no symmetry-imposed barrier to the interconversion of COT and cubane *via* two simultaneous *forbidden* $[_\pi 2_s + {}_\pi 2_s]$-cycloadditions. The attempt to answer this question will also serve as an example of how a correspondence diagram in one symmetry point group suggests an alternative pathway that has to be analyzed in a second. The second analysis, in turn, may point to a third pathway that had not been considered at first, and so on. If the method is reliable, the sequence of analyses will lead inexorably to a single mechanistic conclusion, which – it is to be hoped – will agree with experiment.

7.3.1.1 The (Non)-Dimerization of CBD to Cubane

In the correspondence diagram drawn in (**A**) of Fig. 7.7, the reactant molecules are arranged in $\mathbf{D}_{2h}$, with the π orbitals directed towards one another. The symmetry of cubane can be reduced from $\mathbf{O}_h$ to $\mathbf{D}_{4h}$ by a slight elongation of the σ bonds that are formed in the reaction, and then desymmetrized further to $\mathbf{D}_{2h}$ by a minimal shortening of the bonds between atoms that were doubly-bonded in CBD. More simply, it is sufficient to recognize that when the molecules are fixed in the orientation adopted in the diagram, only the eight p_z AOs change their bonding functions during the reaction. Therefore, only four bonding combinations need be constructed on either side of the diagram and their irreps specified in $\mathbf{D}_{2h}$, the highest symmetry point group common to the reactant and product.

The characterization of the MOs of the interacting CBD molecules is straightforward. So is that of the two *linear combinations of bond orbitals (LCBOs)* of cubane, labeled σ_{++} and σ_{--}; those labeled σ_{-+} and σ_{+-} are interconvertible by a 90° rotation about z, and are therefore degenerate in $\mathbf{T}_d$ and in any of its subgroups that include the sym-op $C_4(z)$. $\mathbf{D}_{2h}$ does not, so they split formally to b_{2u} and b_{3u}. Two pairs of orbitals are in direct correspondence, and the non-correlating pairs of orbitals can be induced to correspond in either of two ways:

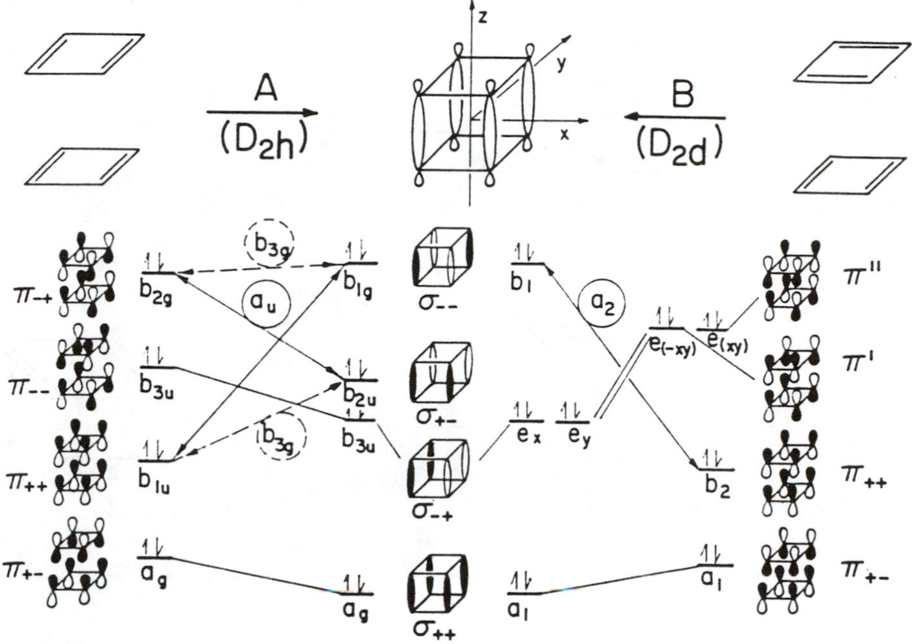

Figure 7.7. Correspondence diagram for the (non-)dimerization of CBD to Cubane: (**A**) $\mathbf{D}_{2h}$ orientation; (**B**) $\mathbf{D}_{2d}$ orientation

$$\pi_{++}(b_{1u}) \leftrightarrow \sigma_{--}(b_{1g}) \quad and \quad \pi_{-+}(b_{2g}) \leftrightarrow \sigma_{+-}(b_{2u}),$$

or

$$\pi_{++}(b_{1u}) \leftrightarrow \sigma_{+-}(b_{2u}) \quad and \quad \pi_{-+}(b_{2g}) \leftrightarrow \sigma_{--}(b_{1g}).$$

The a_u twist that induces the first pair of correspondences and the b_{3g} glide that induces the second pair are illustrated in (**A**) of Fig. 7.8. Motion along either coordinate takes the system away from the geometry of cubane; this is *prima facie* evidence that the dimerization is *forbidden*.

On second thought, we recognize that if the twist is continued beyond $\mathbf{D}_2$, the kernel of a_u, the system is taken into $\mathbf{D}_{2d}$, in which the π bonds of the two are at 90° to each other. Evidently, dimerization to cubane cannot be firmly disallowed before the reaction has also been analyzed in $\mathbf{D}_{2d}$; this is done in (**B**) of Fig. 7.7. The two lower MOs of the reactant pair retain their identity and their irreps are determined easily: π_{+-} is totally symmetric, and π_{++}, being antisymmetric to S_4 and to rotation about the two C_2 axes perpendicular to z,[7] is assigned to b_2. The two upper π orbitals are degenerate; although $\mathbf{D}_{2d}$ does not include C_4, it does include S_4, that also interconverts x and y. As to

[7] In the convention adopted in Fig. 7.7, these axes lie along the diagonals rather than along x and y. As a result, the mirror planes ordinarily denoted by σ_d (diagonal) lie in the yz and zx planes. Conveniently, the labeling of MOs by the irreps of $\mathbf{D}_{2d}$ is independent of the axis convention.

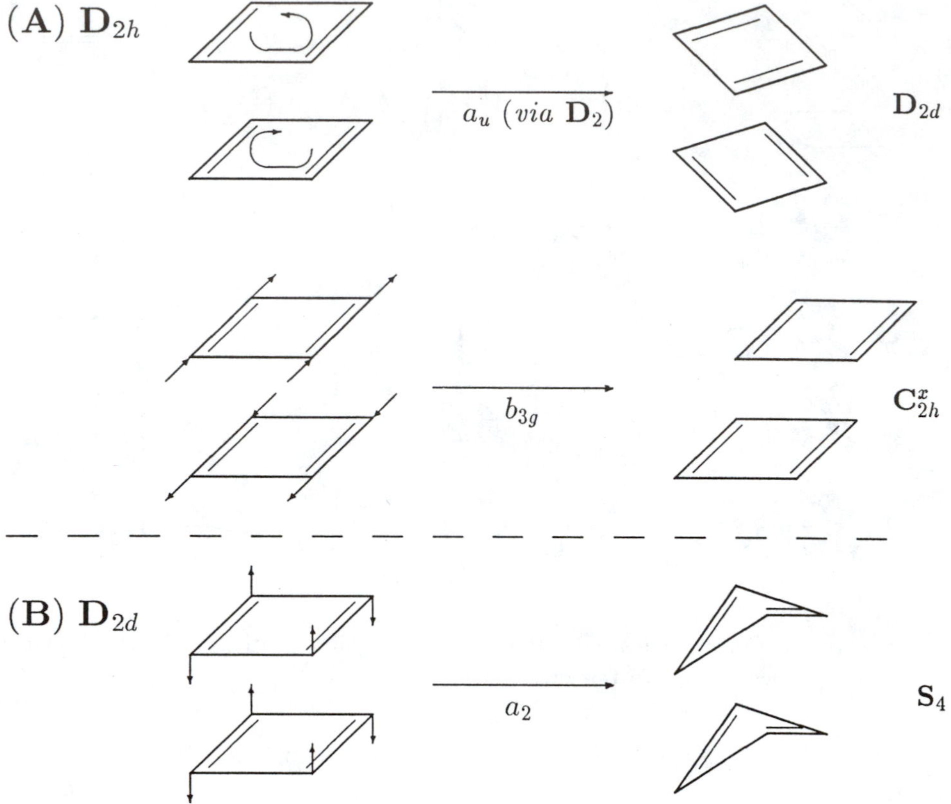

Figure 7.8. Prescribed symmetry coordinates for dimerization of Cyclobutadiene to Cubane. (**a**) $\mathbf{D}_{2h}$ orientation; (**b**) $\mathbf{D}_{2d}$ orientation.

the product: σ_{++} is totally symmetric and σ_{--}, being symmetric with respect to rotation about the C_2 axes, is b_1, whereas the two interconvertible MOs, σ_{-+} and σ_{+-}, are formally degenerate, as they are in the full $\mathbf{T}_d$ symmetry of cubane. There is one non-correlating pair of MOs, $\sigma_{--}(b_1)$ and $\pi_{++}(b_2)$; they can be induced to correspond if an a_2 symmetry coordinate is incorporated in the reaction coordinate. The only skeletal symmetry coordinate of that irrep is shown in (**b**) of Fig. 7.8; it describes a vibrational mode in which both CBD rings are puckered simultaneously and is therefore opposed by a substantial restoring force.

Dimerization of CBD to cubane thus appears to be genuinely *forbidden*, but the b_{3g} displacement prescribed in (**a**) of Fig. 7.8 leads to a mutual orientation of the CBD molecules that looks suitable for dimerization to *anti*-tricyclooctadiene. The disposition of the π orbitals is, moreover, different from that in the original analysis [5, Fig. 7], which found formation of *syn*-TCOD to be energetically less costly. Evidently, this mode of dimerization has to be thought through once more.

7.3.1.2 Dimerization of Cylobutadiene to Tricyclooctatriene

In Fig. 7.9 the analysis is carried out twice in $\mathbf{D}_{2h}$. The CBD molecules are brought up in the xz plane: in (**A**) the π orbitals are placed in the intuitively more reasonable mutual orientation for $[_\pi 2 +_\pi 2]$-cycloaddition; in (**B**) they are put into the arrangement suggested by Fig. 7.8, that would appear to favor an electrocyclic process in which two of the four π bonds shift while the other two are converted to the σ bonds that complete the central ring.

The dimer can be anasymmetrized to $\mathbf{D}_{2h}$ from $\mathbf{C}_{2v}^{y}$, the subgroup of *syn*-TCOD or from $\mathbf{C}_{2h}^{x}$, that of its *anti* isomer. It is easily confirmed that the irreps of the four occupied MOs are just those that would be assigned to the molecule if either isomer were forced into planarity. The two lower orbitals of the reactant pair, which are the same on both sides of Fig. 7.9, correlate directly with the two upper orbitals of the product, and two non-correlating pairs of orbitals remain in both cases. Despite the apparently different nature of the bonding processes

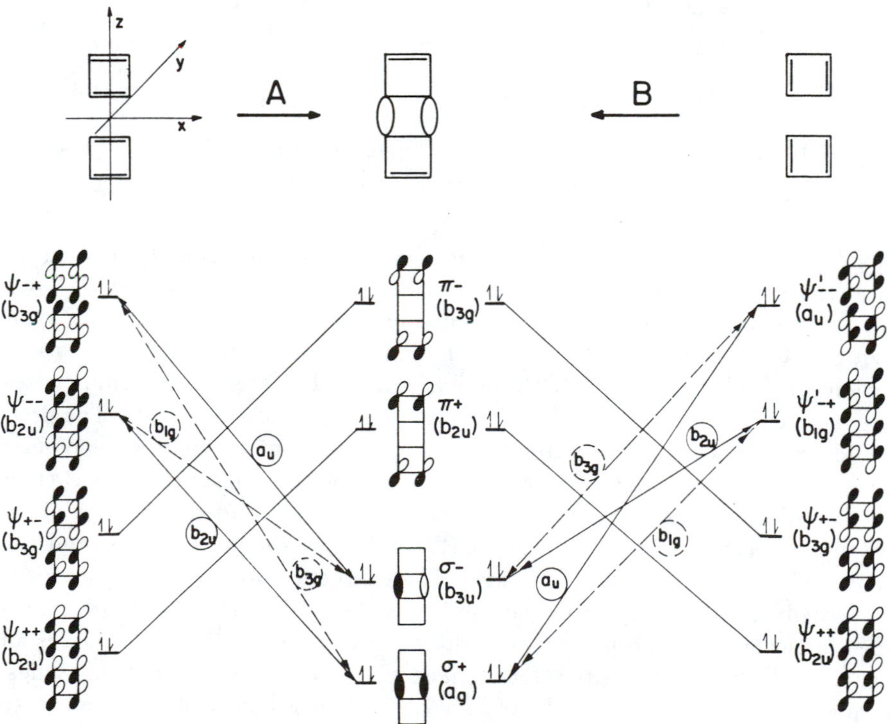

Figure 7.9. Correspondence diagram for dimerization of CBD to Tricyclo-$[4.2.0.0^{2,5}]$octadiene-3,7 (TCOD): Coplanar approach

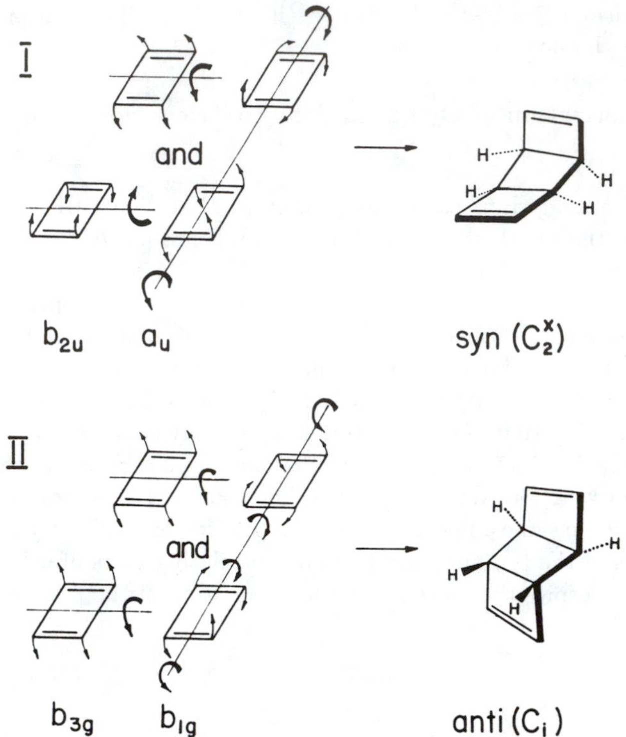

Figure 7.10. Symmetry coordinates for the dimerization of CBD to Tricyclooctadiene. (See (**A**) in Fig. 7.9.) (**I**) $b_{2u} \oplus a_u \longleftrightarrow \mathbf{C}_{2v}^y$; (**II**) $b_{3g} \oplus b_{1g} \longleftrightarrow \mathbf{C}_{2h}^x$

involved in pathways (**A**) and (**B**), the two necessary correspondences can be induced in both cases by either of two composite motions: $b_{3g} \oplus b_{1g}$ or $b_{2u} \oplus a_u$.

It is clear from Fig. 7.10 that motion along a b_{3g} coordinate lowers the symmetry of the system to $\mathbf{C}_{2h}^x$ and takes the reactants into the appropriate orientation for producing *anti*-TCOD. Similarly, a b_{2u} displacement desymmetrizes the reaction path to $\mathbf{C}_{2v}^y$, which is suited to generation of *syn*-TCOD. The preference for formation of the latter necessarily arises from the different energetic consequences of the second displacement that must be incorporated into each of the two reaction coordinates: The a_u twist that has to be added to the *syn* pathway merely rotates the approaching CBD molecules about the z axis, and the central ring is puckered as it is formed. The energetic cost of this displacement is negligible; after all, cyclobutane itself is puckered. [20] In contrast, displacement along the b_{1g} coordinate that facilitates production of the *anti* dimer is an energetically costly concerted out-of-plane distortion of both CBD molecules, similar to that illustrated in (**b**) of Fig. 7.8, that is geometrically irrelevant to the dimerization.

A final objection might be raised: Interaction between the approaching CBD molecules is weaker in the coplanar approach illustrated in Fig. 7.9 than when the reactant molecules approach each other face-to-face; why then should the dimerization to tricyclooctadiene not be analyzed for the approach illustrated in Fig. 7.7, which was found not to lead to the formation of cubane? The two correspondence diagrams in Fig. 7.11 reach precisely the same conclusion as those in Fig. 7.9: The $\mathbf{C}_2$ pathway leading to *syn*-TCOD is entered by a "soft" puckering displacement, so it is taken in preference to a $\mathbf{C}_i$ pathway to the *anti* dimer, for which orbital symmetry conservation prescribes the energetically more costly distortion of the cyclobutene rings.

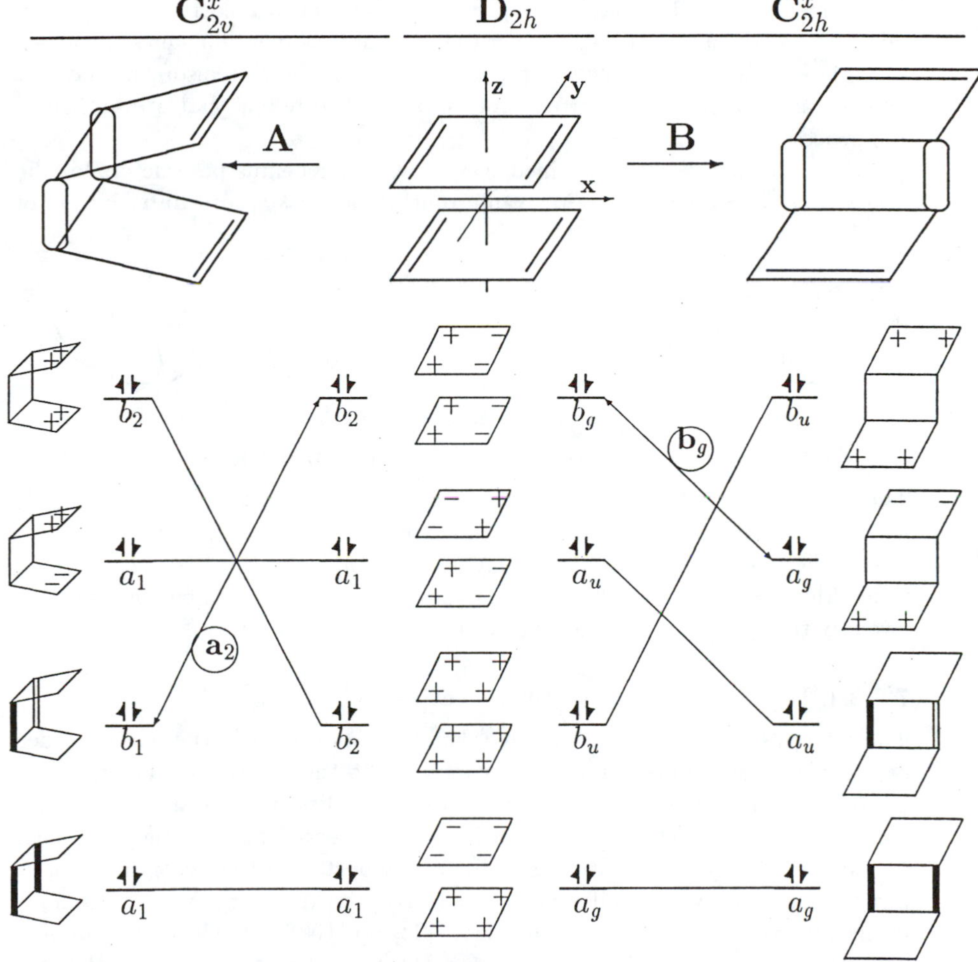

Figure 7.11 . Correspondence diagram for the dimerization of CBD to TCOD, Face-to-face approach: (**A**) $\mathbf{D}_{2h} \rightarrow \mathbf{C}^x_{2v}$; (**B**) $\mathbf{D}_{2h} \rightarrow \mathbf{C}^x_{2h}$. p_z is represented by + and $-p_z$ by −. (For the irreps of the MOs in $\mathbf{D}_{2h}$ see Fig. 7.7)

7.3.2 [2 + 2]-Cycloreversion of o,o′-Benzene-dimer

In a sense, the thermal decomposition of benzene-dimer can be regarded as the converse of the dimerization of CBD, in which two molecules of an antiaromatic cyclic polyene combine rapidly to the thermodynamically more stable dimer. The thermolysis of o,o'-benzene-dimer exemplifies the relatively facile fragmentation of a strained polycyclic dimer to two molecules of its thermodynamically much more stable aromatic monomer. In the former case, the stereoselectivity of the dimerization was shown to be consistent with the greater ease of the two diplacements from $\mathbf{D}_{2h}$ that are required to induce orbital correspondence between the reactant and the *syn* rather than the *anti* isomer. The question arises whether here too the symmetry properties of one of the stereoisomers may not be such as to facilitate concerted rupture of the two bonds that are broken in the process.

Yang [21] has recently obtained the following Arrhenius parameters for the thermal decomposition in cyclohexane solution of the *syn* and *anti* isomers of o,o'-benzene-dimer:

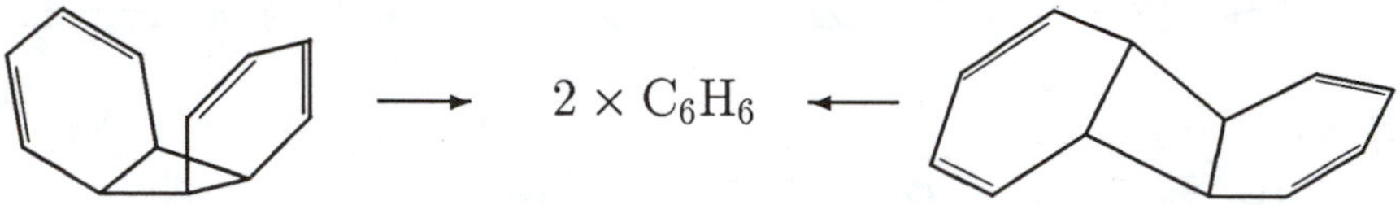

$$\longrightarrow \quad 2 \times C_6H_6 \quad \longleftarrow$$

$\Delta H^{\ddagger}$:	22.5 kcal/mol	24.9 kcal/mol
$\Delta S^{\ddagger}$:	−14.0 cal/mol°K	0. cal/mol°K

The two stereoisomers decompose thermally with nearly the same activation enthalpy, which – although the reaction is nominally a $[_{\sigma}2_s +_{\sigma}2_s]$ cycloreversion – is in the range conventionally assigned to *symmetry-allowed, concerted* reactions. Moreover, the markedly different values of the activation entropy suggest strongly that the fragmentation proceeds by different pathways.

7.3.2.1 Digression on Entropy of Activation

The entropy of activation is generally taken to be a measure of the relative order or disorder of the transition state as compared to the reactants. In the case of unimolecular reactions, a positive value of $\Delta S^{\ddagger}$ implies that the transition state is looser than the reactant and a negative value suggests that it is tighter. Thus, a positive entropy of activation is ordinarily expected in the decomposition of a ring compound, whereas it should be negative in the decomposition of a noncyclic reactant *via* a cyclic intermediate. [22, pp. 109ff.] Arene–arene adducts are polycyclic, so a substantial positive entropy of activation for their thermal decomposition implies that at least one of the two bonds between the moieties is fully ruptured at the transition state, or nearly so. In contrast, a negative $\Delta S^{\ddagger}$, particularly when accompanied by a relatively low $\Delta H^{\ddagger}$, indicates an otherwise energetically favorable pathway *via* a sterically congested transition state for concerted rupture of both bonds. The point can be illustrated by a comparison of the activation parameters reported by Yang et al. for the thermolysis of the *ortho* ([4+2]) and *para*([4+4]) adducts of benzene to anthracene [24]:

[4 + 2]: $\Delta H^\ddagger = 24.3$ kcal/mol; $\Delta S^\ddagger = -3.$ cal/mol °K.
[4 + 4]: $\Delta H^\ddagger = 33.$ kcal/mol; $\Delta S^\ddagger = +16.4$ cal/mol °K.

In both adducts, anthracene is bonded across its central ring, so it does not gain much resonance energy when the bonds are broken. The cycloreversion of the [4+2] adduct, in which the benzene is bonded at its *ortho* positions, is a *retro*-Diels-Alder reaction, to which there are no symmetry restrictions. Since the reactant is much higher in energy than the product, the reaction can be presumed to have an *early, reactant-like* transition state [23], that is perhaps somewhat more highly congested than the reactant by virtue of the tendency of the moieties to flatten as a result of their incipient aromaticity. Accordingly, it has a rather low enthalpy of activation and the small negative entropy of activation characteristic of the retro-Diels-Alder reactions of arene–arene adducts [26].

We need not take on faith the *Rule* that [4+4]-cycloreversion is *forbidden* in order to be able to conclude that concerted rupture of the bonds to the *para* positions of benzene in the [4+4] adduct is opposed by a symmetry-imposed barrier: The dimer has $\mathbf{C}_{2v}$ symmetry; choosing z as the C_2 axis, which passes through the center of both moieties, and x parallel to the long axis of anthracene, the electron configurations of reactant and products are easily worked out. The three doubly-occupied π orbitals of benzene have the irreps a_1, b_1 and b_2; the seven occupied π-orbitals of anthracene [25, p. 125] have the $\mathbf{C}_{2v}$ configuration $[2\times a_1^2,\ a_2^2,\ 2\times b_1^2,\ 2\times b_2^2]$. The MOs of the reactant are labeled as follows: The two σ-orbitals that hold the dimer together are a_1 and b_2; the two π-bonding orbitals in the benzene moiety are a_1 and b_1; the six combinations of the occupied "benzene MOs" in the external rings include two each of irreps a_1 and b_1 and one each of a_2 and b_2.[8] The resulting non-correlation is:

$$[4\times a_1^2,\ a_2^2,\ 3\times b_1^2,\ 2\times b_2^2] \ \not\Leftrightarrow\ [3\times a_1^2,\ a_2^2,\ 3\times b_1^2,\ 3\times b_2^2]$$

Correspondence between the mismatched a_1 and b_2 orbitals is induced by incorporating in the reaction coordinate a b_2 displacement that lengthens one of the σ bonds more than the other. As a result, the reaction takes a stepwise course, in which the bonds are ruptured one by one and the increased enthalpy of activation is compensated by a large positive activation entropy.

Returning to the thermolysis of *o,o′*-benzene dimer, we could set up two correlation diagrams, in $\mathbf{C}_{2v}$ for the *syn* dimer and in $\mathbf{C}_{2h}$ for the *anti*. Alternatively, a single correspondence diagram can be constructed in $\mathbf{D}_{2h}$ between two coplanar benzene molecules and the anasymmetrized dimer, as was done in Figure 7.9 for the dimerization of CBD. The same mechanistic conclusions can be drawn in yet a third way, by comparing the electron configurations of the isomeric reactants and the product; this is done in Figure 7.12.

[8] Our axis convention differs from that of Heilbronner and Bock [25] by an interchange of x and y.

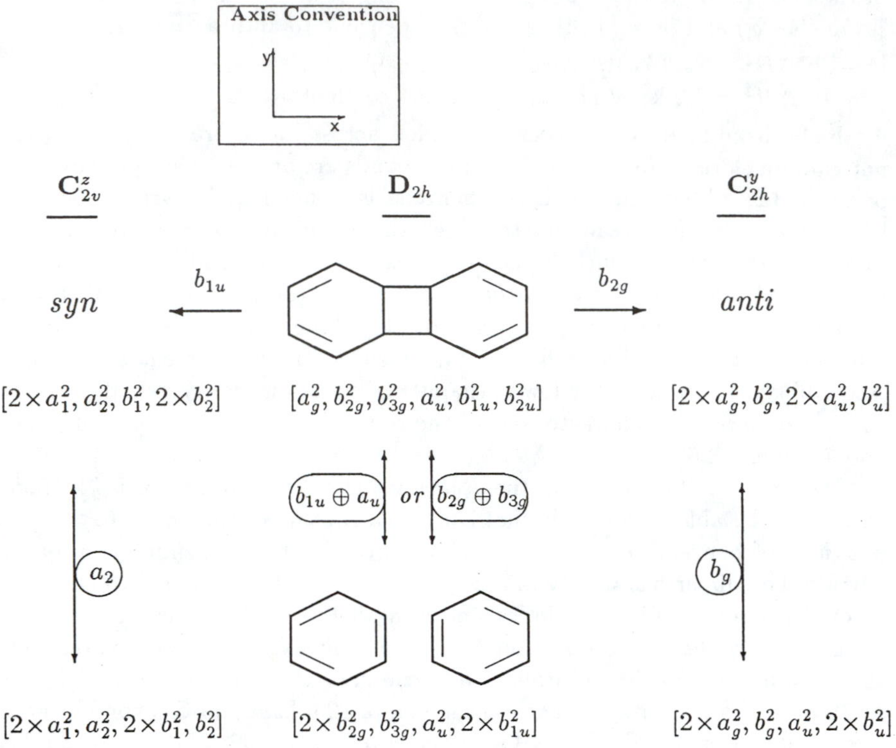

Figure 7.12 . Relationships between electron configurations relevant to the fragmentation of o, o'-Benzene-dimer. (Note the change of axis convention from that of Figs. 7.9–7.11)

The anasymmetrized "planar" dimer can be induced to correspond in $\mathbf{D}_{2h}$ with two coplanar benzene molecules by either of two composite motions: $b_{1u} \oplus a_u$ or $b_{2g} \oplus b_{3g}$. The first component of each pair takes the dimer into one of its isomers: the former to the *syn*-, the latter to the *anti* dimer. The *syn* dimer can undergo fragmentation under the influence of a puckering motion of the central ring $a_2(\mathbf{C}_{2v}^{z}) \rightarrow \mathbf{C}_2^{z}$. Thermolysis of the *anti* dimer requires incorporation of a motion that retains inversion as the only remaining sym-op: $b_g(\mathbf{C}_{2h}^{y}) \rightarrow \mathbf{C}_i$. Formally, this can be be a lateral displacement of the benzene rings, distorting the central ring to a parallelogram, or a simultaneous out-of-plane distortion of both benzene rings,[9] both of which are opposed by a much larger restoring force than the puckering a_2 displacement.

It follows from the analysis that the *syn* dimer is the isomer that is more likely to break both bonds of the four-membered ring at once, for precisely the

[9] The latter appears to be the more appropriate, since it is derived from b_{3g} rather than b_{1g} of $\mathbf{D}_{2h}$ in the axis convention of Fig. 12.

same reason that CBD prefers to dimerize to its *syn* dimer: the relative ease of puckering the central cyclobutane ring. The transition state for such a reaction is expected to be tight and the entropy of activation to take on a large negative value. A comparison with the thermolysis of o,p′-benzene-dimer by a classsical retro-Diels-Alder reaction [26] is instructive:

$\longrightarrow$ $2 \times C_6H_6$

$$\Delta H^{\ddagger} = 14.3 \text{ kcal/mol}; \quad \Delta S^{\ddagger} = -6. \text{ cal/mol}\,^\circ\text{K}$$

No reduction of symmetry below $\mathbf{C}_s$ is formally called for, so the transition state is less sterically strained than that of the *syn-o,o′*-dimer. Accordingly the activation entalpy, $\Delta H^{\ddagger} = 14.3$ kcal/mol, is appreciably lower and the entropy of activation, $\Delta S^{\ddagger} = -6.$ cal/mol °K, is less negative, but both sets of activation parameters are consistent with simultaneous rupture of both bonds.

The *anti* isomer cannot conveniently break both bonds as once, but it has an alternative stepwise pathway at its disposal:

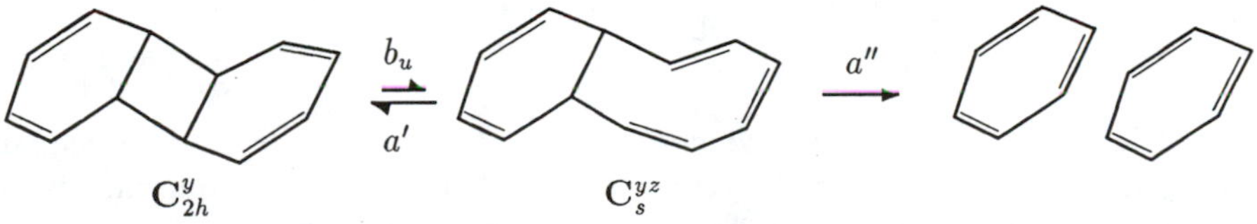

b_u / a' a''

$\mathbf{C}_{2h}^{y}$ $\mathbf{C}_{s}^{yz}$

The first step is a disrotatory "cyclohexadiene-hexatriene" isomerization. Its product, *cis*-dihydrobenzocyclooctatetraene, is less stable than the *trans* dimer and is known to isomerize to it, [27] the isomerization presumably taking place *via* an a'' displacement that reduces symmetry to $\mathbf{C}_1$, in which no reaction is *forbidden*. At the higher temperatures at which fragmentation occurs, the first product should be in equilibrium with the reactant, and its eight-membered ring is sufficiently flexible that a similar desymmetrization would allow it to serve as an unstable intermediate. The activation parameters cited above, which – for the postulated mechanism – measure the enthalpy and entropy differences between the reactant and the transition state of the second step, are not inconsistent with concerted electrocyclic rupture of both bonds *via* a relatively unconstrained transition state.

7.3.3 Dimerization of Cyclomonoalkenes

Let us return for a moment to (**A**) of Fig. 7.9 and note that the reaction appears to be a simple [$_\pi 2 +_\pi 2$]-cycloaddition of the proximal π bonds. Similarly, Fig. 7.12 suggests that the butadiene units in the two benzene moieties remain intact, except for growing resonance stabilization, as fragmentation proceeds.

If this can be interpreted as implying that it is only the interconversion of two π bonds and two σ bonds that determines the symmetry requirements of the reaction, it would appear that a composite "bending *cum* puckering" motion like that shown in (**I**) of Fig. 7.10 should suffice to "allow" concerted $[_\pi 2 + {_\pi}2]$-cycloaddition, provided that the reacting olefins are suitably substituted for producing a *cis*-joined four-membered ring.

7.3.3.1 Cyclopropene

An obvious test reaction is the dimerization of cyclopropene to tricyclo[3.1.0.0^{2,4}]-hexane (TCH) and its reversal, the fragmentation of TCH to cyclopropene. Formally, the analysis is easily seen to hold here as well, but it does not follow that the energetic advantages of this perturbation are sufficient for concerted cycloaddition to *syn*-TCH to overcome the normal preference for formation of a *trans*-biradicaloid species, illustrated profusely in the preceding chapter.

The experimental evidence is suggestive, but inconclusive. Although cyclopropene is a highly strained molecule, with an estimated strain energy of 26 kcal/mol,[28] it does not dimerize spontaneously. The *anti*-isomer of TCH, the *bis-gem*-dimethyl derivative of which has been formed by catalytic dimerization of *gem*-dimethyl cyclopropene,[29] does not revert thermally to two monomer molecules. Instead it isomerizes to vibrationally excited cyclohexadiene [30]:

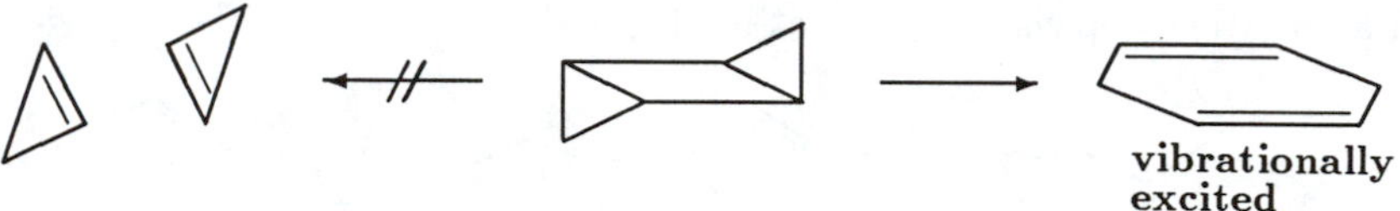

The four-membered ring evidently chooses to break in the alternative manner, producing a pair of endocyclic π bonds. The symmetry requirements are the same as those of the fragmentation. The fact that the product molecule is vibrationally excited confirms the conclusion reached above, that concerted bond rupture of the central ring of the *anti* isomer of a tricyclic molecule requires the gratuitous excitation of vibrational motion. We might expect the analogous isomerization of *syn*-TCH to proceed more readily to an unexcited product molecule, but the prediction cannot be checked experimentally: *syn*-TCH has yet to be prepared.

The course of the reverse reaction, thermal dimerization of cyclopropene, was followed computationally [31]. The molecules were mutually oriented in $\mathbf{C}_{2v}$; as they were brought together the symmetry was relaxed to $\mathbf{C}_s$, so as to bias the reaction path in favor of concerted closure of the four-membered ring to *syn*-TCH, in analogy with the behaviour of CBD. As illustrated in Fig. 7.13, the dimerization follows the conventional stepwise path instead, bonding across a diagonal to form a transoid biradical that is more stable thermodynamically than the reactants and can close – if at all – only to the *anti* isomer.

In a very recent experimental study [32], dimerization of cyclopropene was found to yield a dienic product with two three-membered rings, that could be

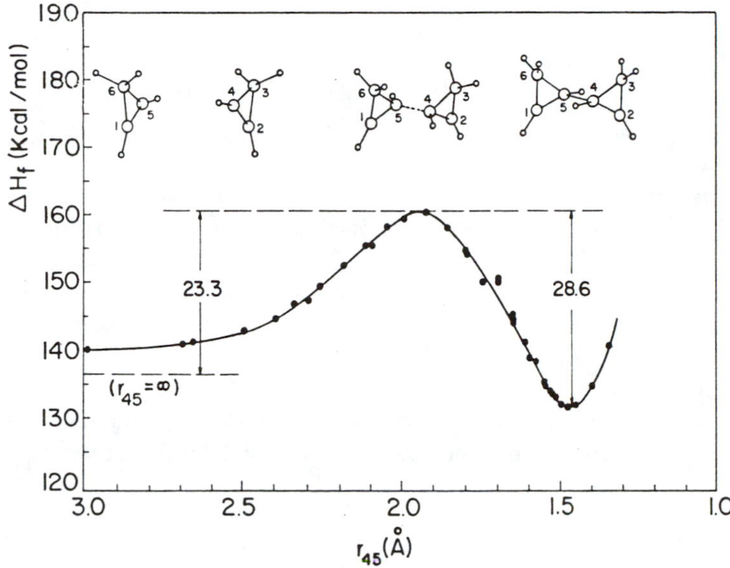

Figure 7.13. Computed $\mathbf{C_2}$ reaction path for dimerization of cyclopropene.The numbering of the atoms follows that of Chemical Abstracts for tricyclo[3.1.0.0^{2,4}]hexane. The slight computational instability near r_{45} = 1.7Å is due to a discontinuity in the limited CI procedure used.

trapped by a dienophile. Whether the biradical drawn in Fig. 7.13 might be a precursor of the observed product has not yet been determined at this writing.

7.3.3.2 Dimerization of Silacyclopropenes

This closely analogous reaction takes a surprisingly different course. [33] At low temperatures, silylenes react with acetylenes to form silacyclopropenes, but at higher temperatures, the usual products are 1,4-disilacyclohexa-2,5-dienes, which are presumably formed by dimerization of the low-temperature products.

It had been suggested [35] that the dimerization proceeds stepwise, *via* a silatricyclohexane intermediate. The formation of the *syn* isomer of silatricyclohexane should be considerably more facile than that of the *anti* isomer,[10] but the substitution pattern of the dimerization product speaks against the incursion of either *syn*- or *anti*-TCH as an intermediate. [36] When dimethylsilylene is generated in a mixture of 2-butyne and diphenylacetylene, the first step produces two symmetrically substituted molecules: $R_2SiC_2A_2$ and $R_2SiC_2B_2$ (A = CH_3; B = C_6H_5). Dimerization of either across the double bond would produce a mixed dimer of disila-TCH – and eventually of disilacyclohexadiene – in which one Si atom is flanked by two A substituents and the other by two of B. This isomer is not observed; in the mixed isomer that is observed, each Si atom is flanked by one A and one B.

[10] A correspondence diagram for the stepwise sequence is shown in Reference [34, Fig. 6].

The simplest mechanism consistent with these findings involves direct formation of the disilahexadiene by a *W.-H. forbidden* $[_\sigma 2_s + {}_\sigma 2_s]$-cycloaddition across the CSi single bonds. As is clear from the correspondence diagram for this reaction, displayed in Fig. 7.14, the single orbital mismatch is removed by a b_{1g} displacement that takes the reactants into $\mathbf{C}^z_{2h}$, thus bypassing the barrier imposed by orbital symmetry in $\mathbf{D}_{2h}$.

Construction of an orbital correlation digram in $\mathbf{C}^z_{2h}$ will illustrate once more the distinction between a correlation diagram and a correspondence diagram: The detailed pairwise connections beween pairs of orbitals across the diagram

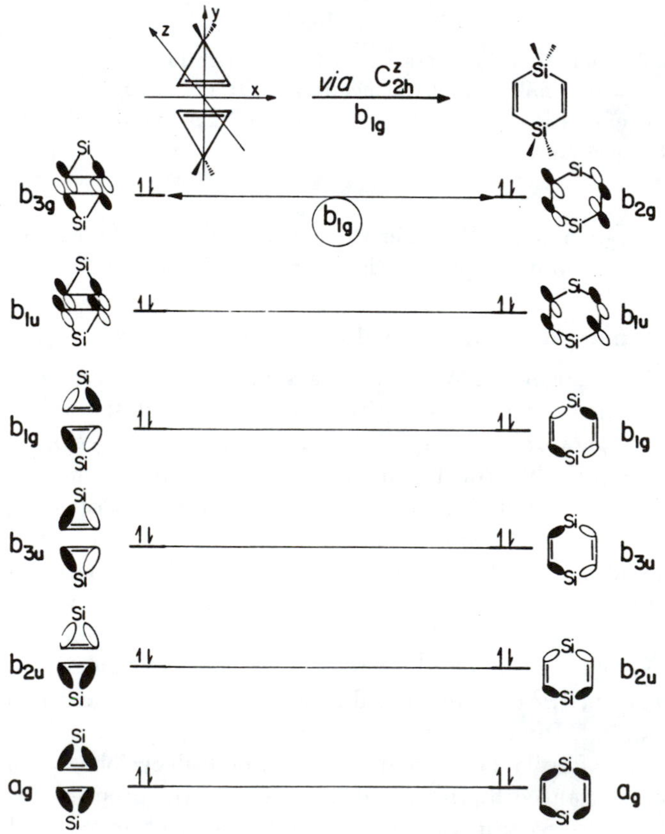

Figure 7.14. Correspondence diagram for dimerization of Silacyclopropene ($\mathbf{D}_{2h}$). (The asymmetry introduced by the substitutents is ignored)

will differ from those in Fig. 14, but the electronic cofigurations of the reactant pair and the product will correlate in the subgroup as predicted. The analogy with the dimerization of silaethylene discussed in Section 6.2.1.1 is striking. Again the dispacement is an in-plane reorientation of the reactants, but here it is not a simple glide but rather a relative rotation of the reactant molecules that brings them into position for head-to-tail cycloaddition.

$\mathbf{D}_{2h}$ $\xrightarrow{b_{1g}}$ $\mathbf{C}_{2h}^{z}$

A comparison of the three dimerizations discussed in the latter part of this chapter illustrates nicely the interplay between symmetry and energy: The presence of an additional π bond in cyclobutadiene offers enough energetic advantage to concerted closure of the four-membered ring to *syn*-TCOD for it to take precedence over the more general – and no less *allowed* – stepwise pathway *via* a *transoid* biradical. Cyclopropene, with just the one π-bond, behaves like a normal alkene and finds the latter pathway more convenient. Silacyclopropene starts off along a similar pathway, but makes use of the relative weakness of the CSi bonds to react in an entirely different manner, but one that is still consistent with the requirements of orbital symmetry conservation.

7.4 References

[1] R. Huisgen, R. Grashey and J. Sauer: In S. Patai (ed.) *The Chemistry of Alkenes, vol. 1.* Interscience, New York 1964, pp. 469ff.

[2] A. Wasserman: *Diels-Alder Reactions.* Elsevier, Amsterdam 1965.

[3] J. March: *Advanced Organic Chemistry, Third Edition.* Wiley, New York 1985.

[4] R.B. Woodward and R. Hoffmann: *The Conservation of Orbital Symmetry.* Verlag Chemie, Weinheim and Academic Press, New York 1970.

[5] E.A. Halevi: Angew. Chem. *88*, 664 (1976); Angew. Chem. Internat Ed. (English) *15*, 593 (1976).

[6] J. Katriel and E.A. Halevi: Theoret. Chim. Acta *40*, 1 (1975).

[7] L. Salem: *Electrons in Chemical Reactions.* Wiley-Interscience, New York 1982.

[8] W.H. Goddard III: J. Amer. Chem. Soc. *97*, 793 (1975).

[9] G.N. Fickes and T.E. Metz: J. Org. Chem. *43*, 4057 (1978) and references cited therein.

[10] R.C. Cookson, B.V. Drake: J. Hudec and A. Morrison: Chem. Comm. *1966*, 15.

[11] A. Padwa: *1,3-Dipolar Cycloaddition Chemistry.* Wiley, New York 1984.

[12] R. Huisgen: J. Org. Chem. *41*, 403 (1976).

[13] R. Huisgen: Chapter 1 in reference [11].

[14] R.A. Firestone: J. Org. Chem. *37*, 2181 (1972).

[15] R.A. Firestone: Tetrahedron *33*, 3009 (1977), and refernces cited therein.
[16] R. Huisgen, G. Mloston and E. Langhals: J. Amer. Chem. Soc., *108*, 640 (1986).
[17] R. Criegee: Angew. Chem. *87*, 765 (1975); Angew. Chem. Internat Ed. (English) *14*, 745 (1975).
[18] O.L. Chapman, C.L. McIntosh and J. Pacansky: J. Amer. Chem. Soc. *95*, 793 (1973).
[19] K. Hassenrück, H.-D. Martin and R. Walsh: Chem. Revs. *89*,1125 (1989).
[20] S. Meiboom and S. Snyder: J. Chem. Phys. *52*, 3857 (1970).
[21] N.C. Yang, T. Noh, H. Gan, S. Halfon and B.J. Hrnjez: J. Amer. Chem. Soc. *110*, 5919 (1988).
[22] A.A. Frost and R.G. Pearson: *Kinetics and Mechanism, Second edition.* Wiley, New York 1961.
[23] G.S. Hammond: J. Amer. Chem. Soc. *77*, 334 (1955).
[24] N.C. Yang, M.J. Chen and P. Chen: J. Amer. Chem. Soc. *106*, 7310 (1984).
[25] E. Heilbronner and H. Bock: *The HMO-Model and its Applications, vol. 3.* Verlag Chemie, Weinheim 1970.
[26] R. Braun, M. Kummer, H.D. Martin and M.B. Rubin: Angew. Chem. *97*, 1054 (1985); Angew. Chem. Internat. Ed. (English) *14*, 1059 (1985).
[27] H. Röttele, W. Martin, J.F.M. Oth and G. Schröder: Chem. Ber. *102*, 3985 (1969).
[28] P. von R, Schleyer, J.E. Williams and K.R. Blanchard: J. Amer. Chem. Soc. *92*, 2377 (1970).
[29] P. Binger, J. McMeeking and U. Schuchardt: Chem. Ber. *113*, 2372 (1980).
[30] J.A. Baldwin and J. Ollerenshaw: Tetrahedron Lett. 3757 (1972).
[31] E.A. Halevi: *Unpublished computations.*
[32] Y. Apeloig and E. Matzner: Results presented at the 10th International Conference on Physical Organic Chemistry, Haifa, 1990.
[33] For background literature, see ref. [34].
[34] E.A. Halevi and R. West: J. Organometall. Chem. *240*, 129 (1982).
[35] H. Gilman, S.G. Cottis and W.H. Atwell: J. Am. Chem. Soc. *85*, 1596 (1964).
[36] W.H. Atwell and D.R. Weyenberg: J. Am. Chem. Soc. *81*, 485 (1969); Angew. Chem. *97*, 1054 (1985); Angew. Chem. Internat. Ed. (English) *8*, 469 (1969).

Chapter 8

Degenerate Rearrangements

In their discussion of sigmatropic rearrangements, Woodward and Hoffmann state: "For the analysis of these reactions correlation diagrams are not relevant since it is only the transition state and not the reactants or products which may possess molecular symmetry elements."[1, p. 114] This is something of an overstatement. For example, they could hardly have meant it to apply to 1,5-hexadiene, which is no less symmetrical than any transition state that can be assumed for its Cope rearrangement.

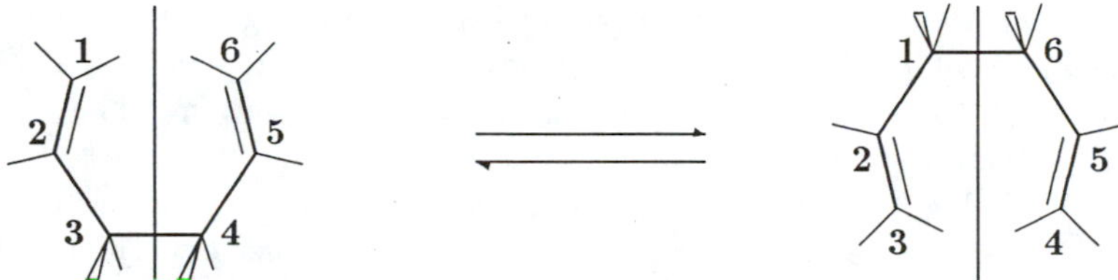

The *syn* and *anti* isomers of hexadiene have a mirror plane and a two-fold rotational axis respectively, each of which bisects the σ bonds broken and formed in the rearrangement. Woodward and Hoffmann themselves set up a correlation diagram in C_{2v} between hexadiene – treated as if it were planar – and an infinitely separated pair of allyl radicals [1, p. 149], in an attempt to rationalize the stereochemical course of the reaction. Dewar [2] rejects their attempt as "forced", adding: "It is clear that no simple interpretation is possible in terms of orbital correlations."

The Cope rearrangement will be dealt with in detail later in this chapter. At this point, it will be used to illustrate a general feature of *degenerate rearrangements*, in which the molecules of reactant and product are mirror images of one another. In the present example, the inapplicability of an orbital correlation diagram cannot be ascribed to the absence of symmetry elements that bisect bonds broken and formed in the reaction. The three "relevant" orbitals, one σ_{CC} orbital and two π combinations, have the same symmetry labels on both sides of the diagram and correlate formally in C_{2v}, but there can be no pathway along which C_{2v} symmetry is retained! A diagram limited to these three MOs necessarily ignores the fact that the in-plane methylene groups have to rotate out-of-plane and *vice versa*. When the methylene CH-bonding orbitals are taken into account, it becomes obvious that the highest symmetry that can be retained by the reaction coordinate is either C_2 along a *conrotatory* pathway or C_s along a *disrotatory* one. The three CC-bonding orbitals correlate in C_{2v} and in all of its subgroups, whereas the CH-bonding orbitals can be induced to correspond under either a conrotation or a disrotation. Therefore, neither a pair of correlation diagrams between the reactant and product in C_2 and C_s nor a correspondence diagram between them in C_{2v} can choose between the two pathways.

8.1 Correspondence Between Reactant and/or Product and Transition Structure

The difficulty can be circumvented as follows: Since the reactant and product of a degenerate rearrangement are identical or at least enantiomeric, an orbital correlation or correspondence diagram is set up between the MOs of either – or, when necessary, a superposition of both – and those of a postulated *transition structure (TS)*, the most symmetrical structure along the reaction coordinate.[1]

Before returning to the Cope rearrangement and going on to sigmatropic rearrangements of less symmetrical molecules, the procedure will be introduced by applying it to two particularly simple degenerate reactions. The first, in which the reactant and one of two postulated transition structures have the same relatively high symmetry, is a newly discovered reaction that – at the time of writing – has not yet been thoroughly investigated either experimentally or computationally. The second example, in which the TS is more symmetrical than the reactant, is not a rearrangement, but a venerable degenerate bimolecular reaction that has long served as a cornerstone of Physical Organic Chemistry.

8.1.1 1,2-Rearrangement of Tetraaryldisilenes

West and his coworkers [3, 4] have followed the isomerization of the *gem* tetrasubstituted disilene, $Ar_2Si{=}SiAr'_2$, in which Ar and Ar′ are distinguishable aryl groups with similar steric and electronic properties, e.g. Ar = mesityl and Ar'_2 = 2,6-xylyl. The two 1,2-isosubstituted isomers Z and E are formed at different rates. Having provisionally identified the first product as the Z isomer, the authors propose that the 1,1-isosubstituted molecule equilibriates with its 1,2-isomers by a stepwise mechanism: An initial *dyotropic shift*[2] *via* a bicyclobutane-like transition state to the Z isomer is followed by slower conversion to the E isomer, presumably by rotation about the SiSi double bond.

---

[1] For the purposes of the symmetry analysis, it is immaterial whether the TS is a genuine transition state or an unstable intermediate between two less symmetrical transition states that are mirror images of one another.

[2] A pericyclic valence isomerization in which two σ bonds migrate intramolecularly [5].

$$\mathrm{Ar_2Si{=}SiAr'_2}\ (gem) \xrightarrow{\ 1\ } \mathrm{ArAr'Si{=}SiAr'Ar}\ (Z) \xrightarrow{\ 2\ } \mathrm{ArAr'Si{=}SiAr'Ar}\ (E)$$

Tetraaryldisilenes are essentially planar molecules [6, 7] that distort out-of-plane with great ease to a *trans* conformation [8]. Ignoring the asymmetry imposed by the substituents, the molecule can be regarded as having $\mathbf{D}_{2h}$ symmetry. When it is placed in the xy plane with the SiSi bond along x, the "soft" out-of-plane displacement is assigned to b_{2g} and desymmetrizes the molecule to $\mathbf{C}_{2h}^{y}$, a motion that can play no essential rôle in the isomerization.[3]

The suggested reaction sequence is not the only way in which rearrangement can occur without instantaneous scrambling of the substituents. It is one of two alternative mechanisms, each proceeding by a dyotropic shift *via* a transition structure with a bicyclobutane-like ring. The TS preferred by the authors, leading to the Z isomer, has $\mathbf{D}_{2h}$ symmetry like the reactant. All four substituents move cyclically to adjacent positions; the highest symmetry that can be retained along the pathway is $\mathbf{C}_{2h}^{z}$, the kernel of b_{1g}. The alternative mechanism begins with an interchange of two *cis*-situated substituents between the silicon atoms, the other two remaining in place, and leads to the E isomer as the first product. If the in-plane C_2 axis that bisects the SiSi bond is labeled y, this TS has $\mathbf{C}_{2v}^{y}$ symmetry and the reaction coordinate (a_2) can retain $\mathbf{C}_{2}^{y}$ at most as the interchanging aryl groups "rotate" about the symmetry axis.

The diagrams in Fig. 8.1 are purely schematic, no attempt being made to reproduce the separation between the various MOs or even their order. Each substituent is represented by its σ bond to silicon, on the assumption that the aryl groups remain intact during the isomerization; the MOs of the disilene thus have the same irreps as ethylene. The only new element in Fig. 8.1 is the presence in each of the transition structures of two linear combinations of three-center bonds. Both of these are necessarily symmetric with respect to $\sigma(yz)$: in (**I**) they are labeled a_g and b_{2u} and in (**II**) – a_1 and b_1. Because the diagram is so schematic, the direct correspondences were drawn without strict adherence to the non-crossing rule, in order to emphasize the geometric relation between corresponding orbitals; the result of the analysis is unaffected: The induced correspondence in each case is the geometrically appropriate one, respectively $b_{1g} \to \mathbf{C}_{2h}^{z}$ and $a_2 \to \mathbf{C}_{2}^{y}$.

The mechanistic conclusion that can be drawn from Fig. 8.1 is that the pathways for rearrangement of the *gem* molecule to its E and Z isomers are equally consistent with the conservation of orbital symmetry. Since the authors

[3] Carried to an extreme, it would exchange both Ar subsituents on one silicon atom with the Ar′ groups bonded to the other, and merely regenerate the *gem* isomer.

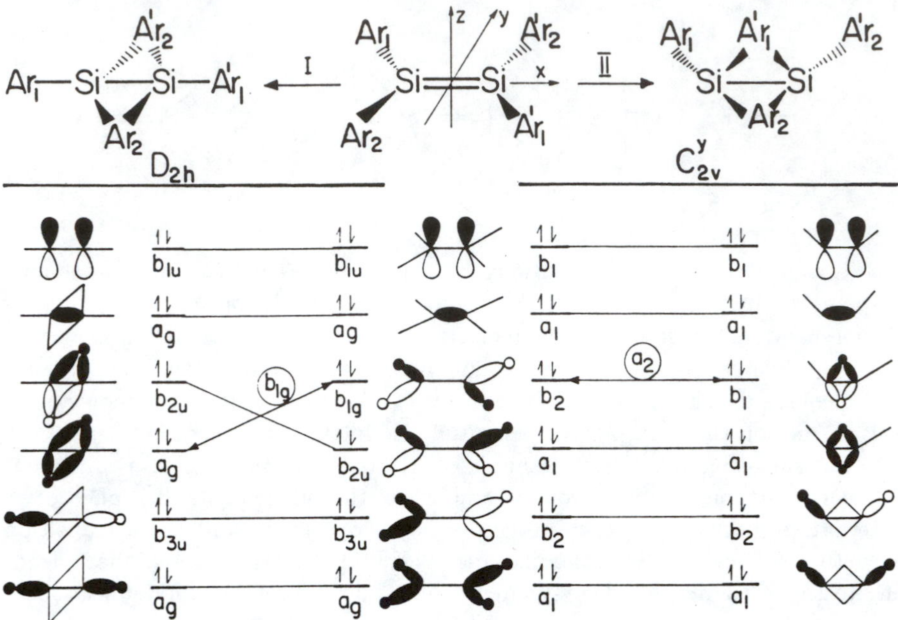

Figure 8.1. Schematic correspondence diagrams for rearrangement of tetrasubstituted disilenes. (**I**): Towards Z ($\mathbf{D}_{2h}$) ; (**II**): Towards E ($\mathbf{C}_{2v}^{y}$). (The asymmetry introduced by the substituents is ignored)

do not claim to have identified the first-formed isomer conclusively as Z, the sequence: *gem* $\rightarrow E \rightarrow Z$ can be regarded as being a no less feasible isomerization pathway than: *gem* $\rightarrow Z \rightarrow E$. It might also be worth investigating kinetically and/or computationally whether the two 1,2-isosubstituted disilenes may not be generated in parallel rather than in series.

8.1.2 Digression: Degenerate X^--Ion Substitution in CH_3X

Consider the bimolecular nucleophilic substitution, S_N2 in the familiar Ingold notation and $A_N D_N$ in Guthrie's [9] more recent scheme, that has been officially adopted by IUPAC [10]. A CH_3X molecule belongs to $\mathbf{C}_{3v}$; if the attacking halide ion approaches and the departing one recedes along the symmetry axis, the reaction coordinate is totally symmetric along the entire pathway and $\mathbf{C}_{3v}$ symmetry is retained throughout. In the special case where the entering and leaving group are identical except for an isotopic label, as in the classic radiochemical studies of Hughes and his coworkers [11], the reaction coordinate passes through a transition structure – here a genuine transition state – with $\mathbf{D}_{3h}$ symmetry.

$$X^{\ominus} \quad \mathrm{H_3C{-}X} \rightleftharpoons \left[\mathrm{X{-}CH_3{-}X} \right]^{\ominus} \rightleftharpoons \mathrm{X{-}CH_3} \quad X^{\ominus}$$

Halide ions are closed shell species with the same symmetry properties as H^-. X_A and X_B are therefore depicted in Fig. 8.2 as bearing just the two valence electrons that bond to carbon to form the new covalent bond, or are detached with the leaving group as the bond is broken.[4]

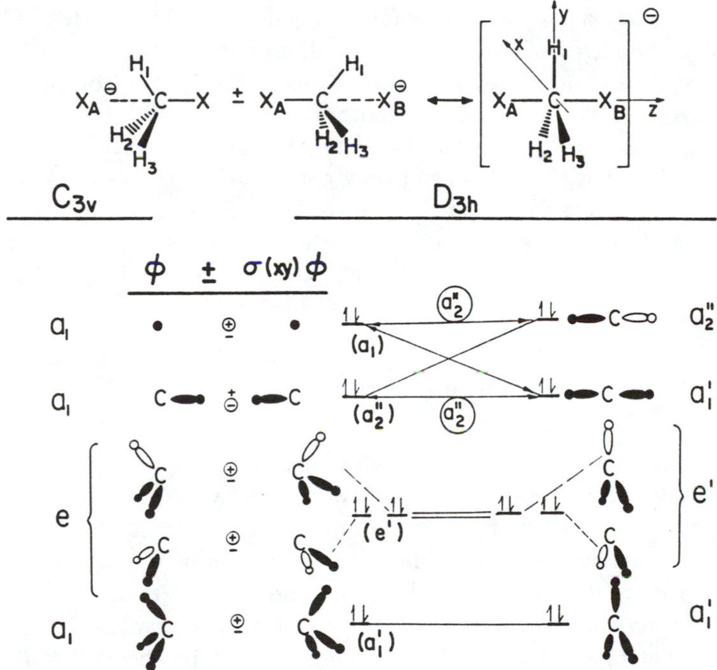

Figure 8.2. Degenerate bimolecular nucleophilic substitution: $\mathbf{C}^z_{3v} \leftrightarrow \mathbf{D}_{3h} \leftrightarrow \mathbf{C}^z_{3v}$

The three CH-bonding orbitals combine in the $\mathbf{D}_{3h}$ symmetry of the TS to one totally symmetric MO and a degenerate pair with the irrep e'; one of the two CX-bonding orbitals is also a_1' and the other is a_2''. Following the rules set out in Section 7.1.1.1, the MOs of the reactant are anasymmetrized to $\mathbf{D}_{3h}$ as follows by reflection in the xy plane:

[4] It makes no difference to the analysis whether the conventional picture of nucleophilic displacement as a two-electron process is retained or whether it is reinterpreted as a synchronous or asynchronous pair of one-electron transfers [12].

Since all three H atoms are transformed into themselves by $\sigma(xy)$, the first of their three linear combinations,[5] $(1s_1 + 1s_2 + 1s_3)$ has the irrep a_1', and the other two, $(2\times 1s_1 - 1s_2 - 1s_3)$ and $(1s_2 - 1s_3)$, comprise an e' pair. The same is true of the s, p_x, and p_y AOs of the carbon atom, so the three σ_{CH} orbitals – and the σ^*_{CH} orbitals associated with them – must be assigned to a_1' and e'. Both X atoms have a_1 symmetry in $\mathbf{C}_{3v}$, and are transformed by $\sigma(xy)$ into one another, so they are combined once with a positive and once with a negative sign to a_1' and a_2'' respectively. Only the second can combine in bonding and antibonding phase with the remaining AO of carbon, p_z, which similarly changes sign on reflection in $\sigma(xy)$; the a_1' combination remains to represent the closed shell of the entering and leaving group, X^-.

The five orbitals of the reactant correlate with those of the transition state in $\mathbf{D}_{3h}$, so they certainly correlate in its subgroup $\mathbf{C}_{3v}$, the symmetry point group retained along the highest-symmetry pathway. Note too that $\mathbf{C}_{3v}$ is the kernel of a_2'', the irrep of the reaction coordinate at the TS.[6] On either side of the symmetrical structure both a_1' and a_2'' map onto a_1, the totally symmetric irrep of $\mathbf{C}_{2v}$, and can mix with one another.

While it was hardly necessary to show that the conventional S_N2 mechanism is consistent with orbital symmetry conservation, the demonstration that anasymmetrization can be applied successfully to this simplest of degenerate reactions should lend credence to its use in the analysis of the more complicated sigmatropic rearrangements that follow.

8.2 The Cope Rearrangement

Interest in the mechanism and stereochemistry of the Cope rearrangement, which attracted considerable attention twenty-odd years ago [13], has recently been rekindled after several years of comparative neglect, and is presently the subject of a considerable amount of critical discussion, with particular emphasis on the *synchronicity*[7] of the bond-breaking and bond-forming processes [17].

The prototypical degenerate rearrangement of 1,5-hexadiene (HD) is a classic [3,3]-sigmatropic rearrangement, which is *allowed* by the *WH-Rules* to proceed on the ground-state potential energy surface along either a $[3s + 3s]$ or a $[3a + 3a]$ pathway [1, p. 125ff.]. Neither specification suffices to define the stereochemistry of the reaction: The sterically more convenient $[3s + 3s]$ mode, for example, can pass through either a *boat*-like ($\mathbf{C}_{2v}$) or a *chair*-like ($\mathbf{C}_{2h}$) transition state.

---

[5] Normalization of the linear combinations is implied.

[6] If the TS is a genuine transition state rather than an unstable intermediate, the reaction coordinate at the top of the barrier becomes the asymmetric CX-stretching coordinate with a negative force constant.

[7] There is some inconsistency in the way the terms *synchronous* and *concerted* are commonly used; see reference [16] for the recommended usage. As noted in footnote 1, synchronicity is irrelevant in the present context.

The experimental evidence indicates that in both the Cope rearrangement [18] and the isoelectronic Claisen rearrangement [19], in which one methylene group is replaced by an oxygen atom, reaction *via* the *chair* transition state is favored over that *via* the *boat*, unless the former is prohibited by steric constraints [13]. Although the unconstrained hexadiene molecule can take up either a *syn* ($\mathbf{C}_{2v}$) or an *anti* ($\mathbf{C}_{2h}$) conformation, its preference for reaction *via* the *chair* TS has been repeatedly confirmed by a variety of computational methods [2, 20, 21, 22, 23, 24, 25].

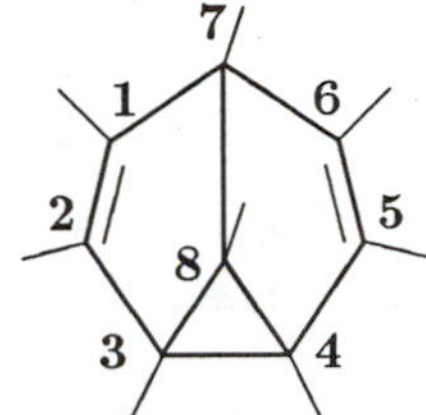

Semibullvalene

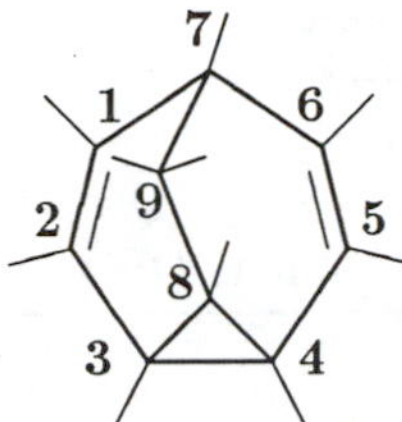

Barbaralane

In contrast, σ-bridged hexadienes like semibullvalene and barbaralane rearrange with great ease, despite the fact that the pathway is constrained to pass through a *boat*-like transition state [13]; the former hydrocarbon is cited to have "the lowest barrier of any known compound undergoing the Cope rearrangement" [26]. Computational studies using different semi-empirical methods [27, 28] confirm the remarkably low barrier to isomerization of semibulvalene, and ascribe it to the fact that the transition state resembles a pair of interacting allyl radicals. However, a recent exhaustive *ab initio* investigation [24] has shown that both the *chair TS* and the *boat TS* for rearrangement of the parent hexadiene are closed shell species with very little biradicaloid character. A detailed set of semiempirical computations [25] has confirmed this finding not only for hexadiene itself but – a fortiori – for semibullvalene and barbaralane as well. This being so, the Cope rearrangement *via* either pathway should be amenable to an orbital symmetry analysis.

8.2.1 Symmetry Analysis of the Cope Rearrangement

The following analysis is abstracted from a recent publication [25], in which its assumptions and conclusions were then checked computationally.[8] As noted above, the highest symmetry that can be retained along the pathway via the *chair* TS is $\mathbf{C}_2$ and that via the *boat* is $\mathbf{C}_s$. The former is conrotatory and the latter disrotatory but there are two variants of each, depending on the relative sense of rotation of the four methylene groups. The four reaction paths are il-

[8] The reader is assured that the qualitative symmetry analysis preceded its computational confirmation.

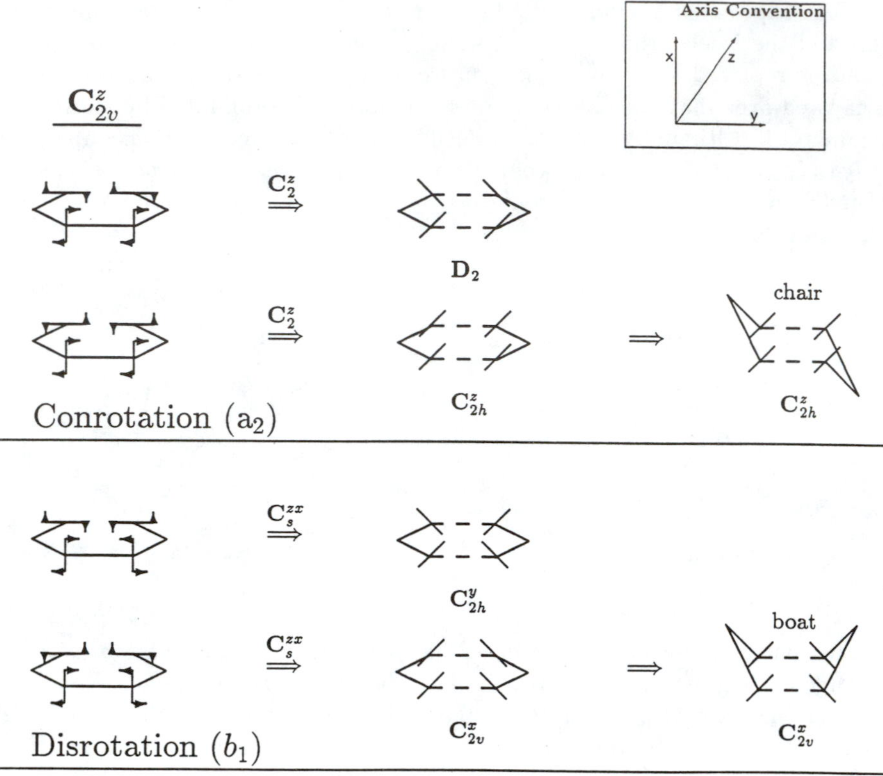

Figure 8.3. Reaction pathways for the Cope rearrangement of hexa-1,5-diene

lustrated in Fig. 8.3. In both conrotatory pathways, the two front methylene groups rotate in the same sense, as do the two rear methylenes; the only symop retained is C_2^z. In one, all four rotate in the same sense; in the other, the two in front rotate in an opposite sense from those at the rear. At the transition structure, half way between the reactant and product, the symmetry has momentarily become higher. The first goes to $\mathbf{D}_2$, acquiring two additional rotational axes; the second picks up a mirror plane and a center of inversion, and can relax in the ensuing $\mathbf{C}_{2h}^{z}$ symmetry to the *chair* conformation. Both disrotatory pathways retain $\sigma(zx)$, but only the second – in which the symmetry of the TS is momentarily raised to $\mathbf{C}_{2v}^{x}$ – allows the carbon skeleton to relax out-of-plane, this time to a *boat* conformation.

Both of the pathways that allow out-of-plane relaxation of the σ frame are $[3s + 3s]$, and are therefore *allowed* by the *Rules*. So are the other two, which are $[3a + 3a]$, but these can be rejected as sterically unfavorable. Instead of setting up two separate orbital correlation diagrams, between the reactant and the *chair* and *boat* transition structures respectively, we make do with a single correspondence diagram in $\mathbf{D}_{2h}$. For this purpose, the *syn* and *anti* conformers of the reactant are anasymmetrized to $\mathbf{D}_{2h}$, in analogy to the superposition of the reactant and product in nucleophilic substitution, that was discussed in the preceding section. Here, however, we also construct a formal *superposition of transition structures (STS)* in the same symmetry point group. The choice between the two transition structures is left to the correspondence diagram in Fig. 8.4, just as a planar model was adopted in Fig. 7.9 to represent both the *syn* and *anti* isomers of tricyclooctadiene, and the correspondence diagram was allowed to choose the isomer most likely to be formed by dimerization of cyclobutadiene.

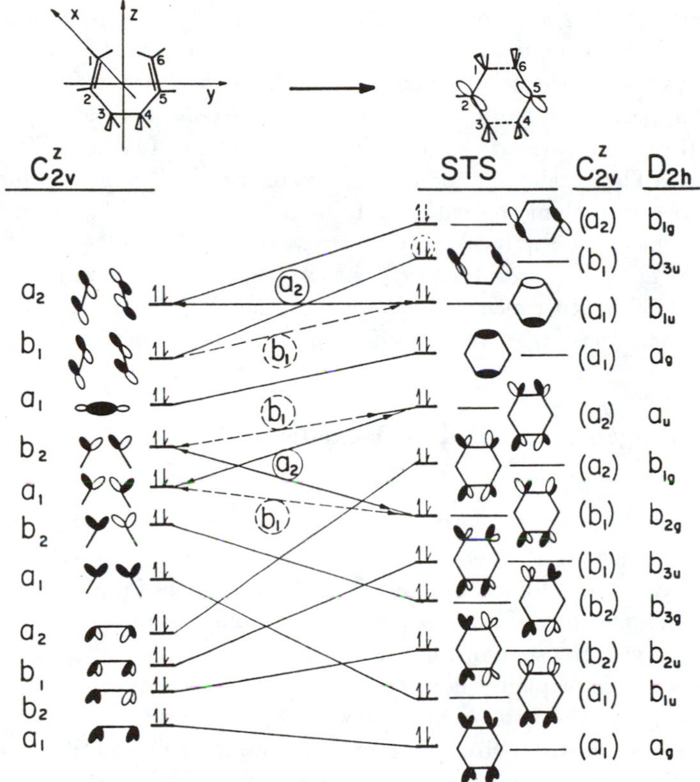

Figure 8.4. Correspondence diagram for the Cope rearrangement of hexadiene

The two stable conformations of hexadiene, *syn* and *anti*, are nearly isoenergetic and are separated by a sufficiently low barrier ($\approx$ 20 kcal/mol) that they can be assumed to be in rapid equilibrium at the temperatures required for the rearrangement. It was confirmed computationally that their electron configurations correlate smoothly with that of the planar molecule, so both conformers are represented on the left side of Fig. 8.4 by their superposition in C_{2v}^z. The four out-of-plane CH orbitals span all four *irreps* of C_{2v}^z, whereas two of the four in-plane combinations are a_1 and the other two are b_2.

On the right we set up a formal *superposition* in $\mathbf{D}_{2h}$ of the two *transition structures* (*STS*). The only moot point with regard to its electron configuration is the energetic order of the symmetric and antisymmetric p_x combinations. There can be no π-σ interaction in the planar conformation , so it was assumed that through-space interaction would stabilize the former (b_{3u}) and destabilize the latter (b_{1g}). These two MOs would then become the respective HOMO and LUMO of the STS, though the possibility of an inversion of their order in either the *chair* or *boat* TS has to be kept in mind. The eight CH bonds are interconvertible and their linear combinations span the eight irreps of $\mathbf{D}_{2h}$. Since the reactant has only been anasymmetrized to $\mathbf{C}_{2v}^{z}$, the orbitals of the STS are *desymmetrized* to this subgroup by lengthening slightly the bond between C_1 and C_6, and thus moving towards the reactant from whichever TS will be shown by the analysis to be the correct one. The labels of the MOs in $\mathbf{C}_{2v}^{z}$ are easily confirmed; in particular, two CH-bonding orbitals go into each of its irreps.

The CH orbitals correlate along either a conrotatory (a_2) or a disrotatory (b_1) pathway. The totally symmetric σ_{CC} correlates with one a_1 orbital and the symmetric π combination with the HOMO (b_1), but the antisymmetric a_2 combination has to be induced to correlate with the second $a_1\,\sigma$ orbital by displacement along a conrotatory (a_2) pathway *via* the *chair* transition state, in agreement with experiment. It follows that the HOMO and LUMO in the *chair* TS remain those that were specified in Fig. 8.4 for the STS .

8.2.2 Rearrangement of Bridged Hexadienes

In semibullvalene, C_1 and C_6 of hexadiene are bonded to one carbon atom, C_3 and C_4 are bonded to another, and the two additional C-atoms (C_7 and C_8 respectively) are linked to one another. Barbaralane differs from it only in that C_7 and C_8 are linked by a third aliphatic carbon atom, C_9. Both molecules have $\mathbf{C}_s$ symmetry, with the sole mirror plane passing through C_7 and C_8, whereas a perpendicular mirror plane and a two-fold rotational axis appear at the TS, that necessarily has the *boat* ($\mathbf{C}_{2v}^{x}$) geometry.

The σ frame and the CH-bonding orbitals retain their identity across the diagram in Fig. 8.4, so only three occupied orbitals on each side of the diagram need be considered explicitly. On the left they are the three uppermost MOs: the σ_{CC} orbital of the bond broken (a' in $\mathbf{C}_s^{zx}$), and the symmetric (a') and antisymmetric (a'') combinations of the π orbitals. As was confirmed computationally, the orbitals corresponding to the two σ_{CC} combinations at the right of Fig. 8.4 are occupied; both are symmetric with respect to the mirror plane retained along the *boat* pathway (a'). Therefore, if the great facility with which the rearrangement takes place implies that it does not have to overcome a symmetry-imposed barrier, the third occupied orbital cannot be the symmetric combination of p_x orbitals on C_2 and C_5, which also has a' symmetry in $\mathbf{C}_s^{zx}$, but has to be the antisymmetric one (a''). This must also be the case in the rearrangement of the parent hydrocarbon via the *boat TS* which, though less favored by some 11. kcal/mol than the *chair* pathway [29], is thermally accessible.

At first sight it may appear that since the symmetric combination is the more stable of the two in $\mathbf{D}_{2h}$, bending it into $\mathbf{C}_{2v}^{x}$ should favor it still more as the co-facial lobes of the p_x orbitals are brought into proximity. It was confirmed computationally [25], however, that π-σ interaction predominates to invert the HOMO-LUMO order in the *boat* transition structures for the rearrangement of the bridged hexadienes, and that its electron configuration indeed correlates with that of the reactant in both cases. The ease with which the polycyclic dienes rearrange is also quite easy to understand: In contrast to both the *chair* and *boat* pathways for the isomerization of hexa-1,5-diene and its unbridged analogs, the geometric reorganization that occurs on going from semibullvalene or barbaralane to the corresponding *boat TS* is minimal. Specifically, there is little or no increase in steric compression of the σ frame on going from the reactant to the transition state, with the consequence that $\Delta H^{\ddagger}$ decreases and $\Delta S^{\ddagger}$ is less negative.

8.3 [1, j]-Sigmatropic Rearrangements

The primary exposition of the *Woodward-Hoffmann Rules* for sigmatropic rearrangements [1, p. 114] was based on this homologous series of degenerate rearrangements, in which the shift of a hydrogen atom from atom j to atom 1 in a molecule with $(j-1)/2$ intervening conjugated π bonds is accompanied by migration of the π bonds in the reverse direction. When $j = 3, 7, \ldots (4n-1)$, the thermal rearrangement is *allowed* along an antarafacial and *forbidden* along a suprafacial pathway; when $j = 5, 9, \ldots (4n+1)$, the *Rules* are reversed: suprafacially *allowed* and antarafacially *forbidden*. Their analysis was based on hydrogen shifts in linear mono- and polyalkenes and its conclusions were then extended to shifts of hydrogen and carbon atoms in cyclic systems. We will find it convenient to deal with each type of rearrangement separately.

8.3.1 [1, j]-Hydrogen Shifts in Non-Cyclic Molecules

No sym-op at all is retained along the reaction path of these rearrangements, some degree of symmetry being momentarily attained only at its midpoint: $\mathbf{C}_2$ at the transition structure for the antarafacial pathway or $\mathbf{C}_s$ at the suprafacial TS. The first two members of the series, the [1,3]-sigmatropic rearrangement of propylene ($j = 3$) and the [1,5]-sigmatropic rearrangement of s-*cis*-pentadiene ($j = 5$), are analysed below in $\mathbf{C}_{2v}$ the supergroup of $\mathbf{C}_2$ and $\mathbf{C}_s$. The correspondence diagram is set up between a superposition of the reactant and product in $\mathbf{C}_{2v}$ and a superposition of the transition structures in the same symmetry point group. The preferred TS is selected by the subgroup of $\mathbf{C}_{2v}$ into which the STS relaxes: the one on the antarafacial pathway if it is desymmetrized to $\mathbf{C}_2$ or that on the suprafacial pathway if it chooses to go to $\mathbf{C}_s$.

8.3.1.1 [1,3]-Sigmatropic Rearrangement of Propylene

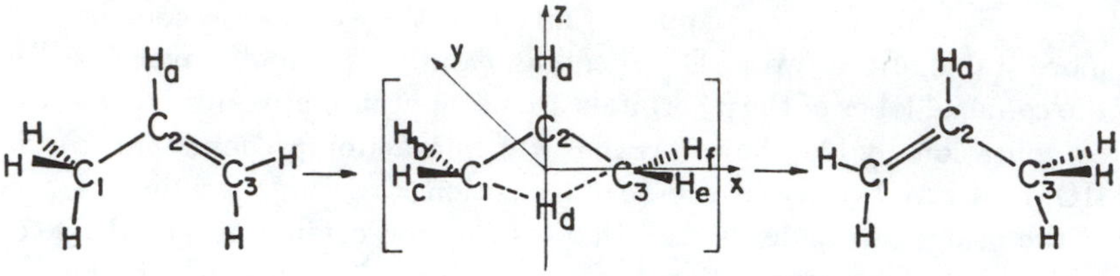

Figure 8.5. Degenerate [1,3]-sigmatropic rearrangement of propylene

This reaction will be analysed in somewhat greater detail than has been done up to now. As noted in Section 7.1.1.1, a common assumption made whenever reaction path computations are carried out is that the same atomic orbital basis set is adequate for following the reaction path. If the reactive system is made up entirely of carbon and hydrogen atoms, the electron and orbital count is particularly simple: Each C atom brings four electrons and each H atom provides one; the minimal basis set comprises a $1s$ AO for each H atom and a set of four – one $2s$ and three $2p$ AOs – for each C atom. The number of independent linear combinations that can be constructed from a system with N_{C} carbon and N_{H} hydrogen atoms is therefore $(4N_{\mathrm{C}} + N_{\mathrm{H}})$, half of which will be occupied and half unoccupied in the closed shell ground-state of all the molecular species concerned.

Table 8.1. [1,3]-Sigmatropic rearrangement of propylene: AO combinations in the STS ($\mathbf{C}_{2v}$). The axis covention and numbering of the atoms is shown in Fig. 8.5

Orbitals in minimal basis set: 18 ; Valence electrons: 18

Symmetry species of the atomic and group orbitals in $\mathbf{C}_{2v}$:

$$C_2 \quad [2s, 2p_z(a_1)\ ;\ 2p_y(b_2)\ ;\ 2p_x(b_1)]$$

$$C_{1,3} \quad [2s^+, 2p_z^+, 2p_x^-(a_1);\ 2s^-, 2p_z^-, 2p_x^+(b_1);\ 2p_y^+(b_2)\ ;\ 2p_y^-(a_2)]$$

$$H \quad [1s_a,\ 1s_d,\ (1s_b + 1s_c + 1s_e + 1s_f)(a_1)\ (1s_b + 1s_c - 1s_e - 1s_f)(b_1);\ (1s_b - 1s_c + 1s_e - 1s_f)(a_2);\ (1s_b - 1s_c - 1s_e + 1s_f)(b_2)];$$

Therefore, for the STS:

$$\Gamma_{\text{orbitals}} = [8\times a_1 + 2\times a_2 + 5\times b_1 + 3\times b_2]$$

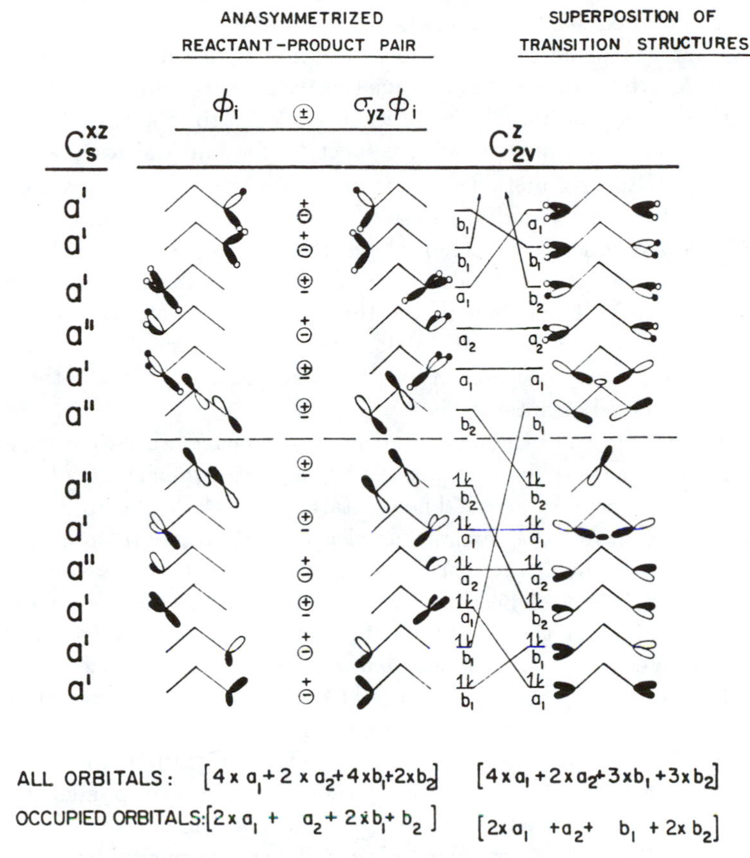

Figure 8.6. Correspondence diagram for [1,3]-sigmatropic rearrangement of propylene

In propylene and in the STS for its rearrangement $N_{\mathrm{C}} = 3$ and $N_{\mathrm{H}} = 6$, so the 18 AOs are combined to 18 MOs. We begin with the STS, assigning each of its MOs to the appropriate irrep of $\mathbf{C}_{2v}$. This is done in two stages: Linear combinations of the carbon and hydrogen AOs are labeled by symmetry; they are listed in Table 8.1. Combinations with the same symmetry label are then allowed to interact with one another in bonding and antibonding phase, and the resulting MOs are stacked by energy on the right side of Figure 8.6. Only twelve of the MOs are included in the diagram, because the three σ bonds to the central carbon atom stay in place during the rearrangement and need not be considered explicitly. It is easy to see that the six electrons involved in them occupy two a_1 and one b_1 orbitals. Omitting these and their antibonding partners, we end up with 12 electrons in the six lowest of twelve orbitals, distributed: $[4 \times a_1 + 2 \times a_2 + 3 \times b_1 + 3 \times b_2]$. The characterization of the orbitals and their stacking by energy in Fig. 8.6 is straightforward.

If the sigmatropic rearrangement is to take place without encountering a symmetry-imposed barrier along its pathway, the electronic configuration of the reactant has to be brought into correlation with that of the STS in $\mathbf{C}_{2v}^{z}$ or one of its subgroups, preferably not the trivial subgroup $\mathbf{C}_1$. The reactant, which has $\mathbf{C}_s^{zx}$ symmetry; all four of its methylene orbitals, two of them σ and two σ^*, are a', i.e. symmetric with respect to the only mirror plane present, $\sigma(zx)$. Four of the six methyl orbitals are also a' and two are antisymmetric to $\sigma(zx)$ (a''), as are π and π^*. In all: $[8\times a' + 4\times a'']$. Recalling that a_1 and b_1 of $\mathbf{C}_{2v}^{z}$ are symmetric to $\sigma(zx)$ whereas a_2 and b_2 are antisymmetric with repect to it, we see that the electron configuration of the STS maps onto $\mathbf{C}_s^{zx}$ as $[7\times a' + 5\times a'']$. It does not match that of the reactant, so symmetry with respect to $\sigma(zx)$ cannot be retained along the pathway and the TS cannot have the full $\mathbf{C}_{2v}$ symmetry of the STS but, at most, *either* $\mathbf{C}_s^{yz}$ or $\mathbf{C}_2^{z}$: the pathway is either suprafacial or antarafacial.

To find out which of these two pathways is preferred, if either, we anasymmetrize, in each case taking the positive or negative combination of the orbital with its mirror image in $\sigma(yz)$. The orbitals that include the $1s$ orbital of the mobile H atom have to be symmetric with respect to $\sigma(yz)$, so do the π and $\pi*$ orbitals, which involve p_y of the central carbon atom; otherwise these AOs would vanish (*Rule 1* of Section 7.1.1.1). The CH-bonding and antibonding orbitals, which include the p_y orbitals of the end atoms as well, have to be combined with a negative sign (*Rules 2 and 3*). Focusing on the occupied orbitals in Fig. 8.6, we see that there is a mismatch between the number of b_1 and b_2 orbitals, which can be overcome by imposing an a_2 displacement. The "true" TS is thus obtained by desymmetrizing the STS to $\mathbf{C}_2^{z}$, the kernel of a_2.

There are no surprises: The pathway chosen by the correspondence diagram is indeed the one that goes through the antarafacial transition state, as predicted by the *WH-Rules*. To be sure, there is very little experimental evidence for the occurrence of thermal [1,3]-sigmatropic rearrangements, but their rarity has been reasonably ascribed to the high steric strain [14, p. 1016] and poor orbital overlap [30, p. 99] in the *allowed* antarafacial transition state.

8.3.1.2 [1,5]-Hydrogen Shift in s-*cis*-Pentadiene

This reaction can be analysed by means of a simple extension of the analysis of the [1,3]-shift just discussed. Pentadiene has two carbon and two hydrogen atoms more than propylene; the total number of orbitals is thus raised by ten to twenty-eight, as is the number of valence electrons. However, after the four σ_{CC} and three "fixed" σ_{CH} bonds are combined to form seven localized molecular orbitals and their antibonding counterparts to seven more, we are left with 14 mobile electrons in 7 doubly occupied MOs. There is no need to draw a new correspondence diagram for the reaction; to a slight amplification of Fig. 8.6 is all that is required.

The only change that has to be made in the hypothetical STS at the extreme right of Fig. 8.6 is the replacement of the non-bonding $p_y(b_2)$ orbital by three allyl π orbitals: a bonding b_2, a non-bonding a_2 and an antibonding b_2 orbital;

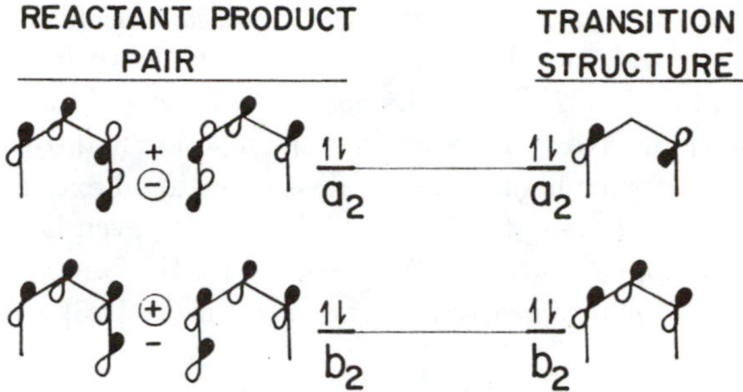

Figure 8.7. Upper occupied MOs for [1,5]-sigmatropic rearrangement of s-*cis*-pentadiene

the first two, which are doubly-occupied, are shown in Fig. 8.7. The six occupied orbitals of the STS in Fig. 8.6 have thus merely been augmented by a seventh, of a_2 symmetry.

Now to the reactant: The occupied π orbital on the left side of Fig. 8.6 has been replaced by the two depicted in Fig. 8.7: one in which all of the four p_y orbitals are in phase and the other with a nodal surface between adjacent bonding pairs. We take the positive combination of the former and its mirror image in order to ensure that the p_y orbital of the central carbon atom does not vanish. Now, however, since the p_y orbital of the leftmost carbon atom is involved in CH bonding in the reactant and in π_{CC} bonding in the product, *Rule 2* requires that the negative combination of the second bonding π-MO and its mirror image be taken. As a result, the uppermost occupied b_2 orbital on the left of Figure 8.6 has been replaced by one of the same symmetry species and another of a_2 symmetry. Therefore, one doubly-occupied MO of a_2 symmetry has been added on the left of the diagram, just as on the right. These two orbitals can correlate along either pathway, so the orbital correspondence that has to be induced ($b_1 \leftrightarrow b_2$) is the same as in the [1,3]-rearrangement, and calls for a transition state that retains the C_2 axis. Our analysis thus favors the *antarafacial* pathway for the [1, 5]-sigmatropic rearrangement of s-*cis*-pentadiene!

The conclusion reached above is not only in flagrant violation of the *WH-Rules*, but also appears to contravene the general view that [1,5]-sigmatropic rearrangements are common, and are "well known" to proceed concertedly along a suprafacial pathway. For example, Gajewsky [31, p. 106] points out that the energy of activation for rearrangement of the parent pentadiene, 36–38 kcal/mol, is "very low" as compared to the dissociation energy of a CH bond. It is nevertheless considerably higher than for suprafacial [1,5]-shifts of hydrogen, as well as carbon, in cyclic dienes, which will be taken up very shortly.[9] The one investigation routinely cited [14, p. 1016], [31, p. 106] as providing compelling

[9] The greater flexibility of the linear dienes should – and does – show up in a negative entropy of activation, but should not raise the energy of activation unless the transition state is quite strained; that for a suprafacial hydrogen shift in pentadiene is not.

evidence in support for rearrangement *via* an *allowed* suprafacial pathway is the classical stereochemical and kinetic investigation of substituted pentadienes by Roth, Koenig and Stein [32], in which that pathway was shown to be preferred by at least 8 kcal/mole. Note, however, that the reaction had to be carried out at 250°C and above. At such high temperatures, vibrational excitation can permit the reaction to proceed along the suprafacial pathway, even if it is *forbidden* at the orbital level of approximation, particularly since the competing antarafacial transition state is probably nearly as strained as in the [1,3]-rearrangement.

8.3.1.3 A Word About [1,7]-Hydrogen Shifts

OCAMS agrees with the *WH-Rules* with regard to [1,3]-hydrogen shifts, disagrees with them on [1,5]-shifts, and the two methods converge once more in predicting the antarafacial pathway to be favored for [1,7]-hydrogen shifts. The relative ease of the isomerization of substituted heptatrienes [14, p. 1017], [31, p. 223], that are flexible enough to react antarafacially, is in agreement with the prediction. In Havinga's much cited investigation [33], a linear cycloheptatriene, sterically constrained to transfer the mobile hydrogen atom antarafacially by incorporation of the two terminal double bonds in cyclohexene rings, rearranges with the activation parameters: $\Delta H^{\ddagger} = 21.$ kcal/mol; $\Delta S^{\ddagger} = -17.$ kcal/mol °K. The fact that $\Delta H^{\ddagger}$ is so much lower than that for the corresponding rearrangement of pentadiene, for which the suprafacial pathway is geometrically convenient, supports our conclusion that the suprafacial rearrangement of the latter, though it is undoubtedly the preferred pathway at elevated temperatures – and is probably concerted though perhaps not synchronous – is inconsistent with orbital symmetry conservation.

8.3.2 Circumambulatory Rearrangements

When atoms 1 and j are linked to one another in a j-membered ring, the transfer of the substituent from atom j to atom 1 with a concomitant shift of the conjugated π-system, regenerates one of j product molecules that are identical with the reactant. The substituent can simply *walk* around the ring; hence the name *circumambulatory-* or *walk rearrangement*. This fascinating family of reactions has been thoroughly investigated and extensively reviewed [34, 35], with particular concern for the applicability to it of orbital symmetry considerations. We will here limit the discussion to two representative reactions, both in electrically neutral carbocyclic molecules. In the first, the shift involves a single migratory atom, which can be chosen for simplicity to be hydrogen. In the second, the migrating C atom is part of an [n.1.0] bicyclic system.

Note first that the shift of a substituent from atom j to atom 1 – or equivalently to atom $(j-1)$, can be described as either a $[1,2]$- or a $[1,j]$-shift; it is universally regarded to be the latter. Moreover, unless the ring is extremely large, only the suprafacial pathway need be considered. The same *Rules* are conventionally applied to it as to the linear $[1,j]$-rearrangement, so the thermal rearrangement of a single migratory atom is *forbidden* when $j = 3, 7, \dots (4n-1)$, and *allowed* when $j = 5, 9, \ldots (4n+1)$. The *Rules* for bicyclic molecules depend on the stereochemistry at the migrating carbon atom, as will be discussed below.

8.3.2.1 [1,5]-Hydrogen Shift in Cyclopentadiene

In contrast to the [1,5]-sigmatropic rearrangement of pentadienes, that of cyclopentadienes proceeds at an appreciable rate at room temperature along a suprafacial pathway [36, 37]. For example, the enthalpy of activation for the hydrogen shift in 5-methylcyclopentadiene, is 20. kcal/mol [38] and that in cyclopentadiene itself, which has been proven isotopically to be intramolecular, is 24. kcal/mol [39]. The latter reaction is illustrated in Fig. 8.8, in which the circled hydrogen atom is depicted as migrating from C_1 to C_2.

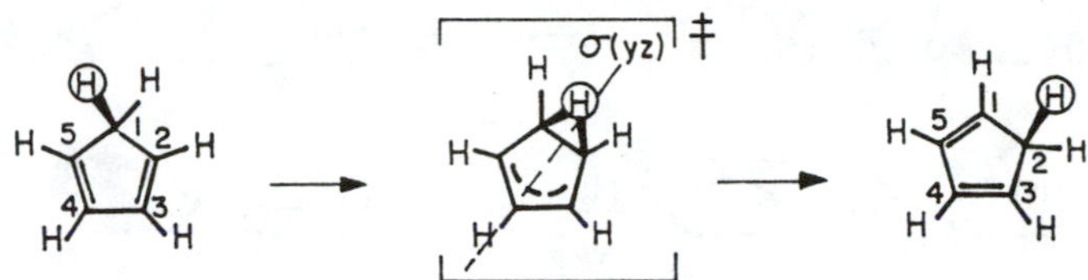

Figure 8.8. Suprafacial [1,5]-rearrangement of cyclopentadiene ($\mathbf{C}_s^{yz}$)

The σ frame is retained throughout and, for qualitative purposes, we can consider C_1 to be using an in-plane sp^2 hybrid to bond the non-migrating H atom and its $2p_y$ orbital to bond the mobile one. We need therefore only consider three doubly-occupied orbitals in the reactant: the σ-bonding orbital to the migrating H atom and the two π orbitals of the butadiene moiety. The suprafacial transition structure has a mirror plane, $\sigma(yz)$, passing through this atom and C_4. As can be seen in Fig. 8.9, the orbital that bonds the shifting atom equivalently to C_1 and C_2 is symmetric with respect to $\sigma(yz)$ (a'), whereas the two pairs of π-electrons occupy the bonding a' and non-bonding a'' allyl orbitals.

Anasymmetrization of the occupied orbitals of the reactant-product pair is trivially simple: In order for the $1s$ AO of the migrating H atom not to vanish, the σ(CH) orbital and its mirror image must be combined in-phase. Having thus taken the positive combination of the p_y orbitals of C_1 and C_2 in the lowest MO, we are obliged by *Rule 2* to take the negative combination in at least one of the two π orbitals, say the upper one, so that these two AOs appear in it with opposite sign. Now, however, the lower π orbital has to be taken with positive sign in order ensure the survival of the p_y AO of C_4 in at least one of of the bonding MOs.[10]

The orbitals on both sides of the correspondence diagram are $[2 \times a' + a'']$ and the rearrangement *via* the suprafacial transition state in clearly *allowed*, in agreement with both the *Woodward-Hoffmann Rules* and experiment. It is

[10] The alternative choice would exchange the symmetry labels of the two anasymmetrized π orbitals, but the electron configuration would remain the same.

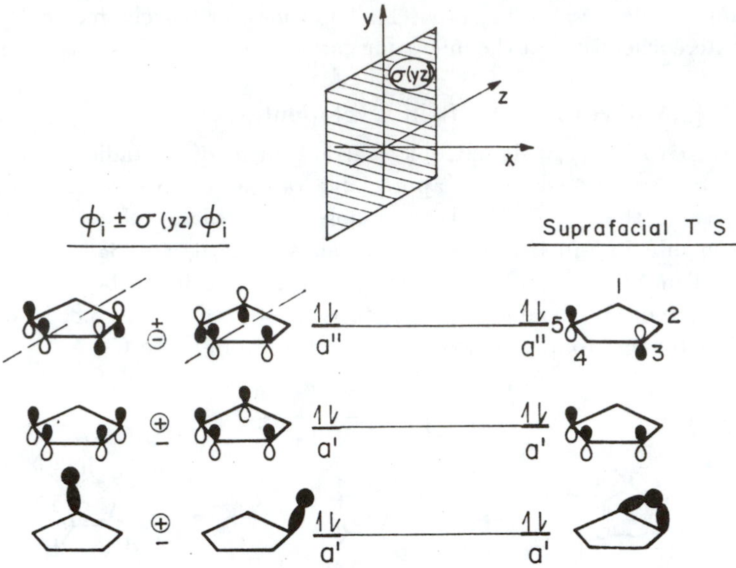

Figure 8.9 . Correspondence diagram for [1,5]-rearrangement of cyclopentadiene ($\mathbf{C}_s^{yz}$)

noteworthy that the present analysis, which takes all of the occupied orbitals into account, finds the rapid suprafacial [1,5]-sigmatropic rearrangement of cyclopentadiene to be *allowed* at the orbital level of approximation but the much less facile corresponding proceess in the open chain pentadiene to be *forbidden*, whereas the *Rules*, which are based on the properties of the frontier orbitals alone, characterize both of them as *allowed.*

8.3.2.2 The "Norcaradiene Walk" Rearrangement

The strongest presumptive evidence for facile sigmatropic shifts that obey the *Woodward–Hoffmann Rules* is cited for molecules with a cyclic σ frame in which the migrating bond is to carbon rather than to hydrogen [30, pp. 98–103]. A particularly intriguing example, discovered by Berson and Wilcott some 25 years ago [40], is illustrated in Fig. 8.10.

It is one of a large series of similar rearrangements that has been reviewed more recently by Klärner [35]. In them, a methylene group, exocyclically bonded to two adjacent atoms of a $(j+1)$-membered ring containing $(j-1)/2$ conjugated π bonds, "walks" around the ring. Here too we are dealing with a suprafacial $[1,j]$-sigmatropic shift, but it differs from the H-atom shifts in that the migrating carbon atom detaches a $2p$, rather than a $1s$, orbital from one of the ring atoms and can therefore reattach itself to the ring atom that flanks it on the other side with either the same or the opposite lobe. In the former case, reaction occurs with retention of stereochemistry at the migrating C atom, and the *WH-Rules* are the same as for the suprafacial H-atom shift: *forbidden* for $j = 3, 7, \ldots (4n-1)$

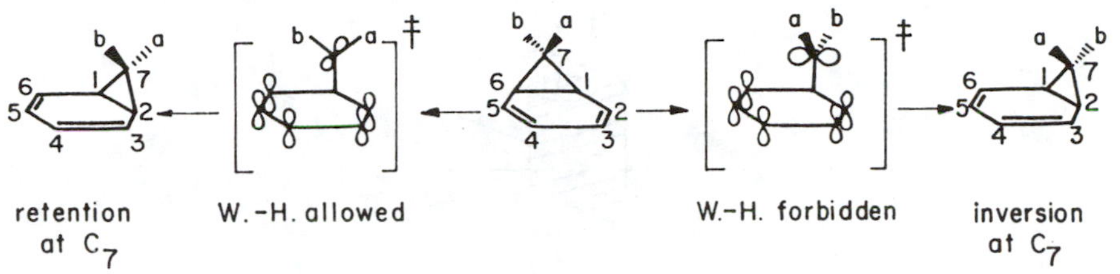

Figure 8.10. "Norcaradiene Walk" rearrangement

and *allowed* for $j = 5, 9, 13, \dots (4n + 1)$. In the latter case, the reaction occurs with inversion and the rules are reversed [1, p. 114].

Accordingly, the "norcaradiene walk" rearrangement ($j = 5$) was assumed to proceed with retention at C_7, along the *WH-allowed* pathway [41] until 1974, when Klärner and his coworkers showed that it occurs with inversion [42]. Since this pathway had been characterized as *forbidden* on the ground-state surface Klärner [43] initially discussed it in terms of biradical transition states, but the one-step nature of the reaction became evident when it was found that that there is no one-center epimerization at C_7 [44]. Rearrangement with inversion is indeed generally observed. For example, it occurs in bicyclopentene ($j = 3$) [35, p. 9] and *cis*-bicyclo[6.1.0]nona-2,4,6-triene ($j = 7$) [35, p. 22] as well. In these cases it was simply characterized as *allowed*, whereas the "forbiddenness" of the inversion pathway for the "norcaradiene walk" was assumed [35, p. 15] to be mitigated by *subjacent orbital interaction* [45, 46].

Schoeller [47] calculated both pathways semi-empirically and found the inversion pathway to be favored over that with retention by some 1.4 kcal/mole. At the temperature at which the experiments were carried out, this energetic difference is large enough to account for 95% stereoselectivity, but it can hardly serve as a basis for a qualitative distinction between an *allowed* and a *forbidden* reaction. It is difficult to contest Childs' conclusion [34, p. 597]: "[Since in] cyclopropyl circumambulations about 4, 5, 6, 7 and 8-membered rings ... the preferred migration pathway is that involving inversion at the migrating carbon, ... orbital symmetry is clearly not the all-determining factor." What remains to be seen, however, is whether orbital symmetry makes any stereochemical prediction at all when the full complement of occupied orbitals are taken into acount, rather than just the HOMO and LUMO.

Both reaction paths are analysed in Fig. 8.11: The reaction involves 12 mobile electrons; assuming that neither TS is a biradical, all we need consider are six doubly occupied orbitals: four constructed from the σ bonds to the migrating C atom and two π orbitals. Both transition structures have a mirror plane passing through C_7, C_1 and C_4, and the symmetry of their occupied MOs is characterized accordingly in the diagram on either side of the anasymmetrized reactant-product pair.

The *WH-allowed* and *forbidden* transition states differ in only one feature. In the former, the antisymmetric CH-bonding orbital is a'' and the p orbital

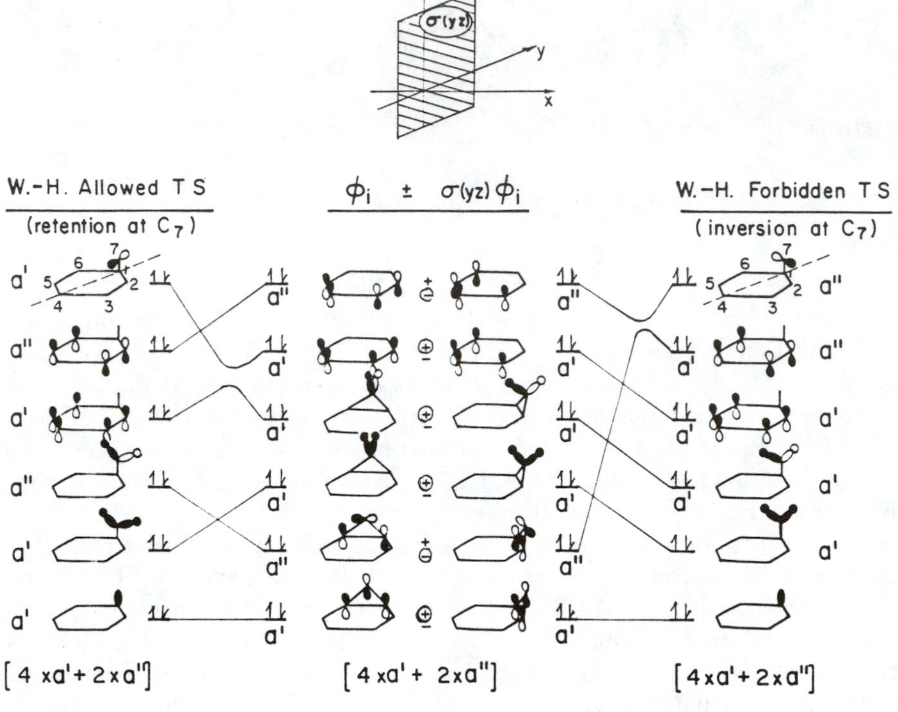

Figure 8.11. Correspondence diagram for the "Norcaradiene Walk" (C_s^{yz})

of C_7, which bonds it simultaneously to C_2 and C_6, is a'; in the latter the two are reversed. Both TS configurations are identically $[4\times a' + 2\times a'']$, so if one of them correlates with the reactant-product pair so will the other. It follows that when all of the MOs are taken into account – not just the HOMO and LUMO – either both pathways are *allowed* or both are *forbidden*.

Anasymmetrization of the occupied orbitals of norcaradiene is straightforward. The two bonds forming the cyclopropane ring involve the p_z and p_x orbitals of C_7 respectively; in order for them not to vanish, the first MO must be taken as a' and the second as a''. The bonds to the substituent H atoms (a and b) both lie in the yz plane, so their positive and negative combinations both have to be a'. C_3 and C_5 are reflected into each other; if their p_z AOs are in-phase in the lower π orbital they have to be taken out-of-phase in the upper. The occupied orbitals are thus anasymmetrized to $[4\times a' + 2\times a'']$. As can be seen, both transition states correlate equally well with the anasymmetrized reactant-product pair, so both have to be characterized as *allowed.* The choice between them is evidently made by energetic factors that are not related to orbital symmetry, perhaps the better overlap of C_7's p_x than its p_z with the p_z AOs of C_2 and C_6.

8.4 Fluxional Isomerization of Cyclobutadiene

Because of its presumed antiaromaticity, cyclobutadiene and its derivatives have generated an enormous amount of interest [48], which gained new impetus after the parent CBD was isolated [49, 50] and shown experimentally to be a rectangular molecule with a singlet ground-state [51, 52]. Understandably, considerable attention is paid in recent comprehensive reviews [53, 54] to the nature of the transition state for interconversion of the two rectangular structures . It is sufficient to recognize for our present purpose that, although it is an electrocyclic shift of two π bonds to adjacent positions, it is no less clearly a degenerate rearrangement. As in the case of the Cope rearrangement, the reactant and product have the same symmetry, here $\mathbf{D}_{2h}$, which contains sym-ops that bisect the bonds made and broken in the reaction. They are mirror images of one another – but less obviously so, because the mirror plane that interconverts them is diagonal to the existing planes of symmetry rather than perpendicular to them. This reaction was one of the earliest analysed by OCAMS [55], before it was realized that a correspondence diagram between the reactant and product of a degenerate rearrangement can be misleading. We begin by describing the early analysis[11] before going on to show how its erroneous conclusions can be avoided by adopting the procedure described earlier in this chapter.

8.4.1 Correspondence Between Reactant and Product

The correspondence diagram between the two rectangular forms of CBD appears in Fig. 8.12. The four π orbitals, labeled by their irreps in $\mathbf{D}_{2h}$, are identical on both sides of the diagram; the only difference is in their order, the HOMO and LUMO on the left being interchanged on the right. Correspondence between the two HOMOs, b_{3g} and b_{2g} is induced by incorporating a b_{1g} displacement into the reaction coordinate for the isomerization. The obvious choice is a distortion of the "least motion" square-planar transition state as shown at the bottom of Fig. 8.12. The energetic advantage to be gained from such a displacement was confirmed [55] by a Hückel-Hubbard calculation [56], which makes approximate allowance for electron interaction .

The implication was that, if the transition state is provisionally assumed to have $\mathbf{D}_{4h}$ symmetry, it should be stabilized by relaxation along a b_{2g} coordinate (in the axis convention of Fig. 8.12) towards rhomboidal ($\mathbf{D}_{2h}^{xy}$) geometry.[12] Three multi-configurational computations appeared soon thereafter [57, 58, 59], all indicating that the transition state is indeed square-planar, though the restoring force to the rhombic displacement was extremely small; [59] in one

[11] "Then spake the chief butler unto Pharaoh, saying: I do remember my faults this day." – Genesis, ch.41-v.9.

[12] The implication is reasonable, but it does not follow inexorably from the correspondence diagram. In principle, the designated b_{1g} displacement can be incorporated sufficiently early in the reaction path for a lower energy rhomboidal transition state to exist that is separated by a potential energy barrier from the square planar structure.

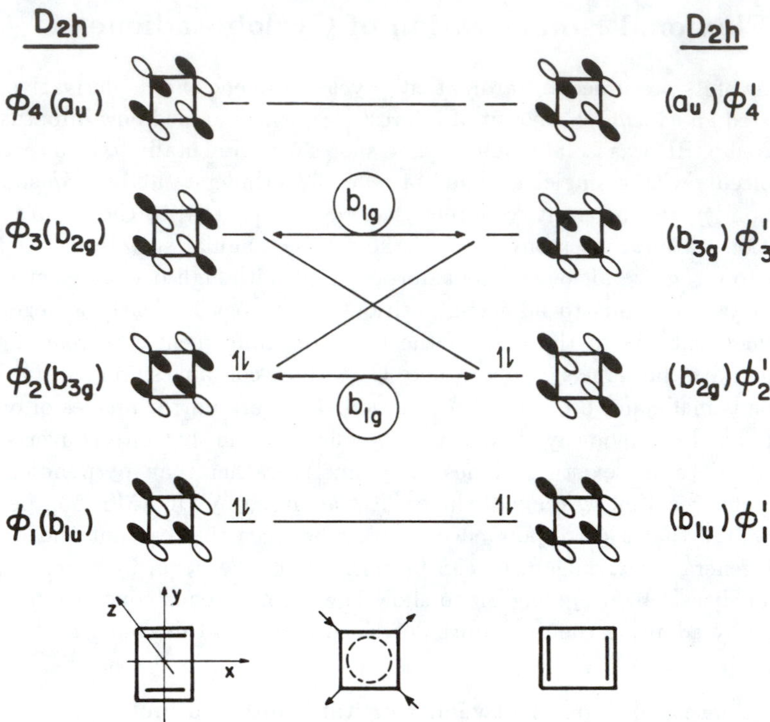

Figure 8.12. Correspondence diagram between the rectangular isomers of CBD ($\mathbf{D}_{2h}$). (Adapted from ref. [55])

study [58] its quadratic force constant was actually found to vanish. Very recently, a full vibrational analysis of the square-planar structure was carried out by Janoschek and Kalcher [60], proving incontrovertibly that it is a true transition state. It has only one imaginary frequency (b_{1g} in $\mathbf{D}_{4h}$, a_g in $\mathbf{D}_{2h}$), whereas the frequency of the normal mode for displacement to the rhombus (b_{2g} in $\mathbf{D}_{4h}$, b_{1g} in $\mathbf{D}_{2h}$) is real and appreciable.

The failure of Fig. 8.12 to predict the correct transition state symmetry arises from a breakdown of the molecular orbital approximation at the geometry of the transition state, where the HOMO and LUMO cross. In $\mathbf{D}_{4h}$ the two orbitals depicted at the left of Fig. 8.13, become degenerate components of the irrep E_g, so any orthogonal combination of ϕ_2 and ϕ_3 is as good an *a priori* choice as the original pair. As a basis for an approximate computation, the positive and negative combinations at the right of the figure are actually better: A pair of electrons in a strictly closed shell configuration can keep farther apart and reduce electron repulsion, and would presumably be stabilized further by deformation to the rhombus. However, as a result of the orbital degeneracy, the open shell, biradical configuration – in which each electron is localized on a diagonally disposed pair of atoms – is more stable still and the transition state remains square-planar.

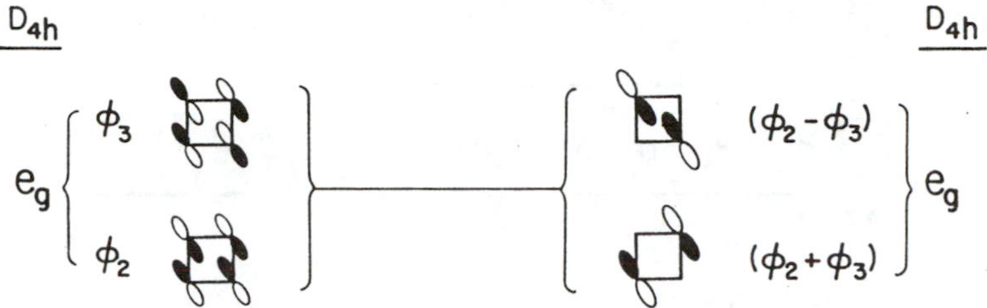

Figure 8.13. Alternative choices of the degenerate π orbitals in square-planar ($\mathbf{D}_{4h}$) CBD

8.4.2 Correspondence Between Anasymmetrized Reactant and Transition Structure

The breakdown of the MO approximation, and – as a consequence – of the analysis in Fig. 8.12, becomes immediately obvious when the reaction is treated like the degenerate rearrangement that it is. The four orbitals of the reactant at the extreme left of Fig. 8.14 are anasymmetrized to $\mathbf{D}_{4h}$ with one of the two diagonal mirror planes, σ_d: The lowest occupied and highest unoccupied MOs evidently have to be taken in their positive combinations or they would vanish. If the positive sign is taken for the combination of the occupied b_{3g} orbital with its mirror image, the negative sign has to be used in that of the unoccupied b_{2g} orbital, as the two positive – or negative – combinations would be identical. An important, if rather subtle, point must be kept in mind: Though these two combinations appear to be a degenerate e_g-pair, the "degeneracy" is purely formal; the true symmetry of the reactant molecule is still $\mathbf{D}_{2h}$. The mirror image of the reactant's HOMO is the HOMO of the product, so the energy of their superposition is simply that of the HOMO; similarly the energy of the LUMO combination remains the same as in the rectangular molecule, and is substantialy higher than that of the HOMO, despite their formal degeneracy.

At the TS, here presumably a genuine transition state, the two e_g orbitals are identical in form with the two combinations on the left and are truly degenerate. Therefore, as the square-planar geometry is approached along the reaction coordinate and the open shell biradical configuration becomes the most stable one, a slight perturbation – such as is always present in real systems – suffices to produce the singlet biradical. Beyond the TS, as rectangular geometry is restored, CBD returns to its stable closed shell configuration.

A final point might be raised. The rectangular structures isomerize much more rapidly than warranted by the computed potential barrier ($\approx$ 10 kcal/mol) when the classical Arrhenius equation is used. This fact and an excessively large negative entropy of activation led Carpenter [61] to propose that the reaction proceeds largely by means of *heavy atom tunneling*. After some initial skepticism, his interpretation has been widely accepted [54], but its general acceptance may be premature. If the rather soft b_{2g} vibration has substantial anharmonic cross-terms with the b_{1g} reaction coordinate, tunneling rates computed with

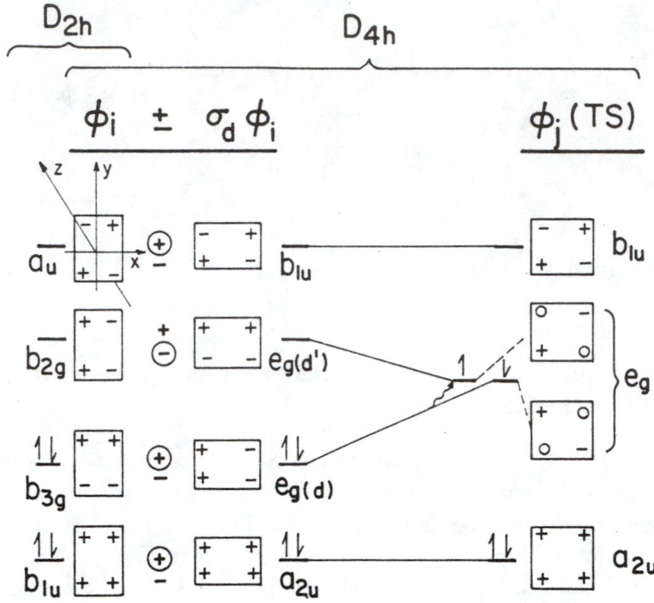

Figure 8.14. Correspondence diagram between the anasymmetrized isomers of CBD and the square-planar TS($\mathbf{D}_{2h}$)

a one-dimensional model [13] become unreliable [62, pp. 136–138, 166–167]. In that case, the deviation of the rate of CBD isomerization from that predicted by the classical Arrhenius equation may not necessarily be evidence of heavy atom tunneling but may be due instead to anharmonic coupling of the reaction coordinate with the rather low-frequency "rhombic" b_{2g} vibration. If this eventually turns out to be so, Fig. 8.12 will have served a useful purpose, however erroneous its original interpretation [55] may have been.

8.5 References

[1] R.B. Woodward and R. Hoffmann: *The Conservation of Orbital Symmetry.* Verlag Chemie, Weinheim and Academic Press, New York 1970.

[2] A. Brown, M.J.S. Dewar and W.W. Schoeller: *J. Amer. Chem. Soc.* **92**, 5376 (1970).

[3] H.A. Yokelson, J. Maxka, D.A. Siegel and R. West: *J. Amer. Chem. Soc.* **108**, 4239 (1986).

[4] H.A. Yokelson, D.A. Siegel, A.J. Millevolte, J. Maxka and R. West: Organometallics *9*, (1990) (in press)

[13] At this writing, even the proper model to use within the one-dimensional framework is in dispute; as a result,the tunneling rates calculated by different authors vary widely [61, 63, 64, 65].

[5] M.T Reetz: Adv. Organomet. Chem. *16*, 33 (1977).
[6] B.D. Sheperd, D.R. Powell and R. West: Organometallics *8*, 2664 (1989).
[7] B.D. Sheperd, D.R. Powell and R. West: J. Hetroatom. Chem., *1*, 1 (1990).
[8] D.E. Goldberg, P.B. Hitchcock, M.T. Lappert, K.M. Thomas, A.J. Thorne, T. Fjeldberg, A. Haaland and B.E.R. Schilling: J. Chem. Soc. Dalton Trans. *1986*, 2387.
[9] R.D. Guthrie: J. Org. Chem. *40*, 402 (1975); R.D. Guthrie and W.P. Jencks: Accts. Chem. Resarch *22*, 343 (1989).
[10] IUPAC Commission on Physical Organic Chemistry: Pure Appl. Chem. *61*, 23 (1989).
[11] E.D. Hughes, F. Juliusberger, S. Masterman, B. Topley and J. Weiss: J. Chem. Soc. *1935*, 1525; E.D. Hughes, F. Juliusberger, A.D. Scott, B. Topley and J. Weiss: J. Chem. Soc. *1936*, 1173; W.A. Cowdrey, E.D. Hughes, F. Juliusberger, T.P. Nevell and C.L. Wilson: J. Chem. Soc. *1938*, 209.
[12] A. Pross and S.S. Shaik: Accts. Chem. Research *16*, 363 (1983).
[13] For citations of the early work, see, for example, References [14, pp. 1021–1032] and [15, pp. 220–230].
[14] J. March: *Advanced Organic Chemistry, Third Edition*. Wiley, New York 1985.
[15] T.L. Gilchrist and R.C. Storr: *Organic Reactions and Orbital Symmetry*. Cambridge University Press, Cambridge 1972.
[16] V. Gold (Compiler and Editor): *Glossary of Terms Used in Physical Organic Chemistry*, Pure and Appl. Chem. *55*, 1281ff. (1983).
[17] W.T. Borden, R.J. Loncharich and K.N. Houk: Ann. Revs. Phys. Chem. *39*, 215–24 (1988).
[18] W. v.E. Doering and W.R. Roth: Tetrahedron *18*, 67 (1962).
[19] G. Frolin, A. Habich, H.-J. Hansen and H. Schmid: Helvet. Chim. Acta *52*, 335 (1969).
[20] M.J.S. Dewar, G.P. Ford, M.L. McKee, H.R. Rzepa and L.E. Wade, Jr.: J. Amer. Chem. Soc. *99*, 5069 (1977).
[21] M.J.S. Dewar: J. Amer. Chem. Soc. *106*, 7127 (1984).
[22] Y. Osamura, S. Kato, K. Morokuma, D. Feller, E.R. Davidson, and W.T. Borden: J. Amer. Chem. Soc. *106*, 3363 (1984).
[23] M.J.S. Dewar and Caoxian Jie: J. Amer. Chem. Soc. *109*, 5893 (1987).
[24] K. Morokuma, W.T. Borden and D.A. Hrovat: J. Amer. Chem. Soc. *110*, 4474 (1988).
[25] E.A. Halevi and R. Rom: Israel J. Chem. *29*, 311 (1989).
[26] A.K. Cheng, F.A.L. Anet: J. Mindusky and J. Meinwald, J. Amer. Chem. Soc. *96*, 2887 (1974).
[27] R. Hoffmann and W.-D.Stohrer: J. Amer. Chem. Soc. *93*, 6941 (1971).
[28] M.J.S. Dewar and D.H. Lo: J. Amer. Chem. Soc. *93*, 7201 (1971).
[29] W.v.E. Doering, V.G. Toscano, G.H. Beasley: Tetrahedron *27*, 299 (1971).
[30] I. Fleming: *Frontier Orbitals and Organic Chemical Reactions*. Wiley-Interscience, Chichester 1976.
[31] J.J. Gajewsky: *Hydrocarbon Thermal Isomerizations*. Academic Press, New York 1981.
[32] W.R. Roth, J. König and K. Stein: Chem. Ber. *103*, 426 (1970).
[33] J.L.M.A. Schlatmann: J. Pot and E. Havinga, Rec. Trav. Chim. Pays Bas *83*, 1173 (1964).
[34] R.F. Childs: Tetrahedron *38*, 567 (1982).

[35] F.-G. Klärner: Chapter 1 in E.J. Eliel, S.H. Wilen and N.L. Allinger (eds.) *Topics in Stereochemistry, vol. 15*. Wiley, New York 1984.
[36] K. Alder and H.-J.Ache: Chem. Ber. *95*, 503, 511 (1962)
[37] V.A.M. Mironov, E.V. Sobolev, A.N. Elizarovna: Tetrahedron *19*, 1939 (1963).
[38] S. McLean and P. Haynes: Tetrahedron *21*, 2329 (1965).
[39] W.R. Roth: Tetrahedron Lett., 1009 (1964).
[40] J.A. Berson and M.R. Wilcott III: J. Amer. Chem. Soc. *87*, 2751, 2752 (1965).
[41] J.A. Berson: Accts. Chem. Res. 1, 152 (1968).
[42] F.-G. Klaerner: Angew. Chem. *86*, 270 (1974); Angew. Chem. Internat. Ed. (English) *13*, 268 (1974).
[43] F.-G. Klaerner, S. Yaslak and M. Wette: Chem. Ber. *112*, 1168 (1979).
[44] F.-G. Klaerner and B. Brassel: J. Amer. Chem. Soc. *102*, 2469 (1980).
[45] J.A. Berson and L. Salem: J. Amer. Chem. Soc. *94*, 8917 (1972).
[46] J.A. Berson: Accts. Chem. Res. *5*, 406 (1972).
[47] W.W. Schoeller: J. Amer. Chem. Soc. *97*, 1978 (1975).
[48] M.P. Cava and M.J. Mitchell: *Cyclobutadiene and Related Compounds*. Academic Press, New York 1967.
[49] O.L. Chapman, P. De la Cruz, R. Roth and J. Pacansky: J. Amer. Chem. Soc. *95*, 1337 (1973).
[50] A. Krantz, C.Y. Lin and M. D. Newton, J. Amer. Chem. Soc. *97*, 2744 (1975).
[51] S. Masamune, F.A. Soto-Bachiller, T. Machiguchi and J.E. Bertie: J. Amer. Chem. Soc. *100*, 4889 (1978).
[52] B.A. Hess, Jr., P. Carsky and L.J. Schaad: J. Amer. Chem. Soc. *105*, 695 (1983).
[53] T. Bally and S. Masamune: Tetrahedron *36*, 343 (1980).
[54] G. Maier: Angew. Chem. *100*, 317 (1988); Angew. Chem. Internat. Ed. (English) *27*, 309 (1988).
[55] E.A. Halevi, F.A. Matsen and T.L. Welsher: J. Amer. Chem. Soc. *98*, 7088 (1976).
[56] F.A. Matsen: Internat. J. Quantum Chem. *10*, 511 (1976).
[57] H. Kollmar and V. Staemmler: J. Amer. Chem. Soc. *99*, 3583 (1977).
[58] W.T. Borden, E.R. Davidson and P. Hart: J. Amer. Chem. Soc. *100*, 388 (1978).
[59] J.H Jafri and M.D. Newton: J. Amer. Chem. Soc. *100*, 7088 (1978).
[60] R. Janoschek and J. Kalcher: Internat. J. Quantum Chem. *37* (1990) (in press)
[61] B.K. Carpenter: J. Amer. Chem. Soc. *105*, 1700 (1983).
[62] H.R. Johnston: Large Tunneling Corrections in Chemical Reaction Rates. In: Adv. Chem. Phys. *3*, 131 (1961).
[63] M.J. Huang and M. Wolfsberg: J. Amer. Chem. Soc. *106*, 4039 (1984).
[64] M.J.S. Dewar, K.M. Merz and J.J.R. Stewart: J. Amer. Chem. Soc. *106*, 4040 (1984).
[65] R. Lefebvre and N. Moiseyev: J. Amer. Chem. Soc. *112*, 5052 (1990).

Part IV

Spin and Photochemistry

Chapter 9

Electron Spin

Up to this point we have managed to sidestep the issue of spin [1, 2] entirely, because the reactions treated so far have all involved closed shell singlet reactants and products. Even in those few cases where the transition state is essentially open shell, as in the fluxional isomerization of cyclobutadiene (Section 8.4), its singlet state lies sufficiently far below the corresponding triplet [3, p. 363] that electron spin can be ignored. This is no longer the case in photochemical reactions, several of which will be dealt with in the following chapter, or in the less common – but by no means rare – thermal reactions in which the spin state of the product differs from that of the reactant.

The classical model of the spinning electron is reminiscent of that of the electron in a p orbital, referred to briefly in Chapter 2. In analogy to the decription of a p_{+1} electron as rotating in one sense about an axis (z) aligned along an external magnetic field and one in p_{-1} as rotating about it in the other, an electron in an α spin state is said to be spinning about its axis in one sense and a β electron in the other, so that one is stabilized and the other is destabilized when the axis of spin lines up as best it can with that of an external magnetic field. The analogy breaks down in that there is no counterpart to the p_0 state, which has no net angular momentum about z and consequently does not interact with the field. For molecules composed of light atoms, spin- and orbital-angular momentum are very nearly independent of one another and can be treated separately. [2, p. 10] Moreover, the energy of magnetic interactions is negligible in the absence of very strong magnetic fields, so the principal effect of spin on chemical reactivity manifests itself through symmetry-based selection rules.

9.1 Spin and Symmetry

The symmetry of molecular systems with non-zero spin will be taken up in three stages: We begin with an elementary discussion of the symmetry properties of spinning electrons and follow it with a few words about how the net spin of the electrons in a molecule affects its state symmetry. Then, in preparation for the analysis of thermal reactions in which electron spin is not conserved, we will consider how the *overall symmetry* of a reacting system can be retained by compensatory changes in spin- and orbital-angular momentum. [2, Chaps. 3–5]

9.1.1 The Symmetry of Spinning Electrons

Let us return to the analogy between spin- and orbital-angular momentum. It is basic common knowledge that the unit of angular momentum is $\frac{h}{2\pi}$ and that an s state is characterized by the quantum number $l = 0$, a p state by $l = 1$, d by $l = 2$, f by $l = 3$, and so on. Each of these has $(2l+1)$ substates with the same total angular momentum $\sqrt{l(l+1)}\frac{h}{2\pi}$ variously aligned with respect to an external magnetic field, no more than $\pm l\frac{h}{2\pi}$ of which can be oriented parallel or antiparallel to z. Of the $(2l+1)$ substates, l are stabilized by the field, l are destabilized by it, and the remaining one is unaffected. The spinning electron has quantum number $s = \frac{1}{2}$, so it has two substates with angular momentum $\frac{\sqrt{3}}{2}\frac{h}{2\pi}$, of which $\frac{1}{2}\frac{h}{2\pi}$ is aligned parallel and $-\frac{1}{2}\frac{h}{2\pi}$ antiparallel to z. A spin-flip from the first orientation (α) to the second (β) or vice versa thus implies a change by one unit ($\frac{h}{2\pi}$) in the z component of the spin-angular momentum.

9.1.1.1 A Single Spinning Electron

The symmetry properties of the spin wave functions, α and β, are not covered by the Character Tables in Appendix A, but require the expansion of the usual symmetry point groups to *double groups* [4, p. 308ff.]. The imperfect analogy beween orbital- and spin-angular momentum can be drawn once more in order to provide an intuitive rationale for double groups, and to explain at the same time why we can do without them in the context of this book.

Equation 2.1 reminds us that the p_x and p_y orbitals are two orthogonal combinations of p_{+1} and p_{-1}, representing rotation of the electron about the z axis with one unit of angular momentum. Similarly, d_{xy} and $d_{x^2-y^2}$ (Fig. 2.10) are combinations of d orbitals with two units of orbital-angular momentum aligned along z. Rotation of p_x by 90°, i.e. $\frac{1}{4}(2\pi)$, converts it to p_y, rotation by $\frac{1}{2}(2\pi)$ takes it to $-p_x$, rotation by $\frac{3}{4}(2\pi)$ transforms it to $-p_y$, and a full 360° rotation regenerates p_x. In contrast, $d_{x^2-y^2}$ is transformed to d_{xy} under a rotation of 45°, a rotation of 90° changes its sign, and one of $\frac{1}{2}(2\pi)$ (180°) suffices to regenerate it. In general, the minimal rotation about z that can take an orbital with angular quantum number l into itself is $\frac{1}{l}(2\pi)$. It follows that in order to transform an α electron ($s = +\frac{1}{2}$) to itself it has to be rotated by 4π – two full 360° rotations – about z. The *identity operator* for a spinning electron is thus not C_1, as in all of the symmetry point groups, but rather $C_{\frac{1}{2}}$, or rotation by 4π (720°); hence the need for a special kind of symmetry group to deal with it.

9.1.1.2 Two Spinning Electrons

Fortunately, all of the reactions discussed in this book involve singlets, with no electron spin ($S = 0$), and triplets, in which two electrons with parallel spin combine to one full unit of spin-angular momentum ($S = 1$) and the net spin wave function has three components, each with the symmetry properties of a p orbital. The simultaneous spin-flip of more than one electron in a molecule is a rare event, so the spin change between reactant and product involves at most a single unit of angular momentum.[1] These spin-non-conservative reactions can therefore be discussed without going beyond the regular symmetry point groups. [2, p. 201ff.]

There are four ways in which the angular momentum of two independently spinning electrons can be combined:

$$\Lambda_{+1} : \alpha(1)\alpha(2) \tag{9.1}$$
$$\Lambda_{-1} : \beta(1)\beta(2) \tag{9.2}$$
$$\Lambda_0^{\alpha\beta} : \alpha(1)\beta(2) \tag{9.3}$$
$$\Lambda_0^{\beta\alpha} : \beta(1)\alpha(2) \tag{9.4}$$

In analogy to the orbital motion of an electron in p_{+1} or p_{-1}, a pair of electrons in Λ_{+1} or Λ_{-1} can be loosely regarded as spinning around axes parallel to z, with one unit of angular momentum aligned to positive and negative z respectively. In $\Lambda_0^{\alpha\beta}$ and $\Lambda_0^{\beta\alpha}$ the z-components of the spin-angular momenta of electrons 1 and 2 cancel, so they can be expected to behave in a magnetic field like an electron in a state with magnetic quantum number $m = 0$, such as $2s$ or $2p_0$. The two Λ_0 spin-functions defined in Equations 9.3 and 9.4 are unacceptable, because they specify which electron is α and which is β, ignoring the fact that electrons are indistinguishable. They therefore have to be replaced by their negative and positive combinations; the first changes sign when electrons 1 and 2 are interchanged and the second does not.

$$\Lambda_0^- : (\alpha(1)\beta(2) - \beta(1)\alpha(2))/\sqrt{2} \tag{9.5}$$
$$\Lambda_0^+ : (\alpha(1)\beta(2) + \beta(1)\alpha(2))/\sqrt{2} \tag{9.6}$$

Λ_0^- represents complete cancellation of spin, analogous to to an atomic S state, which has zero orbital angular-momentum. In Λ_0^+ the electron spins add to one another, but – since their z components cancel – the spin state corresponds to p_0, with one unit of angular momentum about some undetermined axis in the xy plane.

In symmetry point groups in which x, y and z belong to separate non-degenerate representations, a naive extension of the analogy with orbital motion might suggest that the three spin functions associated with the triplet (Λ_{+1}, Λ_{-1}, Λ_0^+) should combine similarly to functions that transform like x, y and z, but this is not the case. Due to the unique nature of spin, the three spin functions Λ_x, Λ_y and Λ_z have to be assigned to the irreps of R_x, R_y and R_z respectively [2, p. 206].

A simplistic rationalization might run as follows: The symmetry point group is determined by the electrostatic field set up by the nuclei; therefore the electron's charge density and – consequently – its potential energy vary in differently

---

[1] This is no less true of a single spin-flip in molecules with other multiplicities, e.g. a doublet to a quartet.

oriented p orbitals (Section 2.2). In contrast, the spin of an electron cannot interact directly with the electrostatic field, but is affected by it indirectly *via* magnetic coupling of the spin-angular momentum with that of the electron's orbital motion. In the absence of an external magnetic field, each electron can be thought of as spinning about an axis parallel to that of the orbital magnetic moment [1, p. 17]. In a symmetry point group that distinguishes x, y and z , the orbital-angular momentum of a P state is oriented along one of the Cartesian axes, so the three spin components behave like rotations about them.

9.1.2 Space-, Spin- and Overall Symmetry

Let us return to the isomerization of CBD in Section 8.4, and compare the symmetry properties of the orbital products that can be constructed from ϕ_2 and ϕ_3 in $\mathbf{D}_{2h}$ and in $\mathbf{D}_{4h}$. We recall from Section 3.2.2 that the irrep of a product of orbitals is the product of the irreps of the occupied orbitals, each taken once for every occupying electron.

9.1.2.1 Example 1: Rectangular Cyclobutadiene ($\mathbf{D}_{2h}$)

In $\mathbf{D}_{2h}$ ϕ_2 is more stable than ϕ_3, so the orbital products are – in order of increasing energy:

$$\Psi_1(A_g) : \phi_2(b_{3g})^2 \tag{9.7}$$

$$\Psi_2(B_{1g}) : \phi_2(b_{3g})^1\, \phi_3(b_{2g})^1 \tag{9.8}$$

$$\Psi_3(A_g) : \phi_3(b_{2g})^2 \tag{9.9}$$

The two totally symmetric orbital products can only be occupied by a pair of electrons with opposite spin, so – since all of the other electrons in the molecule are also paired – they represent closed shell singlet states.[2] The B_{1g} configuration gives rise to two open shell states, ${}^1B_{1g}$ and ${}^3B_{1g}$, that have the same energy at the orbital level of approximation but split when electron interaction is taken into account.

The Pauli exclusion principle, which we have not yet had occasion to apply explicitly,[3] must now be taken into account. It requires the overall wave function, expressed as the *space wave function*, Ψ_i – represented here by its dominant orbital product – multiplied by the spin function Λ_i, to be antisymmetric. Since the singlet and triplet spin functions are respectively antisymmetric and symmetric, the space functions associated with them have to be of opposite parity. The two totally symmetric closed shell space functions are obviously symmetric to interchange of the two electrons in a single orbital. Ψ_2 in Equation 9.8 stands for two equivalent open shell alternatives with the orbital product $\phi_2\,\phi_3$: $\phi_2(b_{3g})(1)\ \phi_3(b_{2g})(2)$, in which the first electron is in ϕ_2 and the second in ϕ_3,

---

[2] Configurations with the same symmetry label interact with one another as described for singlet dioxygen (Equation 3.11). It is therefore implicitly understood that Ψ_1 contains a stabilizing contribution from Ψ_3 whereas Ψ_3 includes a destablizing admixture of Ψ_1.

[3] See footnote 8 in Chapter 3.

and $\phi_2(b_{3g})(2)\ \phi_3(b_{2g})(1)$, in which they have been interchanged. They are neither symmetric nor antisymmetric to electron interchange, so their normalized symmetric and antisymmetric combinations have to be taken:

$$\Psi_2^+ : (\phi_2(1)\phi_3(2) + \phi_3(1)\phi_2(2))/\sqrt{2} \tag{9.10}$$

$$\Psi_2^- : (\phi_2(1)\phi_3(2) - \phi_3(1)\phi_2(2))/\sqrt{2} \tag{9.11}$$

Each of the three symmetric combinations, $\Psi_1(A_g)$, $\Psi_2^+(B_{1g})$ and $\Psi_3(A_g)$ is therefore associated with the antisymmetric spin function Λ_0^- whereas $\Psi_2^-(B_{1g})$ can be combined with each of the symmetric spin functions: Λ_x, Λ_y and Λ_z.

We take note of the fact that the two closed shell singlets differ in energy and are not interconvertible; both are totally symmetric, so the lower is labeled 1^1A_g and the upper 2^1A_g. Since ϕ_2 transforms like yz and ϕ_3 like xz, their product, like xyz^2 (or xy) belongs to the irrep B_{1g}; so do the sum and difference, which transform like $y(1)x(2) \pm y(2)x(1)$. The open shell singlet and triplet thus have the same *space symmetry*, and are labeled $^1B_{1g}$ and $^3B_{1g}$ respectively.

9.1.2.2 Example 2: Square-Planar Cyclobutadiene ($\mathbf{D}_{4h}$)

In $\mathbf{D}_{4h}$, ϕ_3 and ϕ_2 are degenerate; like x and y they belong to the doubly degenerate irrep E_g. The four space functions that can be formed from them have therefore to be assigned to irreps included in the direct product $E_g \otimes E_g$, which is evaluated in Table 9.1.

Table 9.1. The direct product of E_g with itself ($\mathbf{D}_{4h}$)

$\mathbf{D}_{4h}$	E	$2C_4$	C_2	$2C_2'$	$2C_2''$	i	$2S_4$	σ_h	$2\sigma_v$	$2\sigma_d$	
E_g	2	0	-2	0	0	2	0	-2	0	0	$(\phi_3, \phi_2), (d_{zx}, d_{yz})$
$E_g \otimes E_g$	4	0	4	0	0	4	0	4	0	0	
A_{1g}	1	1	1	1	1	1	1	1	1	1	z^2, (x^2+y^2)
A_{2g}	1	1	1	-1	-1	1	1	1	-1	-1	R_z
B_{1g}	1	-1	1	1	-1	1	-1	1	1	-1	(x^2-y^2)
B_{2g}	1	-1	1	-1	1	1	-1	1	-1	1	xy

$$E_g \otimes E_g = A_{1g} \oplus A_{2g} \oplus B_{1g} \oplus B_{2g}$$

These are evidently the four one-dimensional irreps that are symmetric with respect to E, C_2, i and σ_h; one of the four space functions is assigned to each of them.

Taking the e_g orbitals at the right of Fig. 8.13 as our basis and renaming ϕ_+ and ϕ_-, the four space functions corresponding to Equations 9.7, 9.9, 9.10 and 9.11 become:

$$\Psi_1(A_{1g}) \ : \ (\phi_+(1)\phi_+(2) + \phi_-(1)\phi_-(2))/\sqrt{2} \tag{9.12}$$

$$\Psi_2^+(B_{1g}) \ : \ (\phi_+(1)\phi_-(2) + \phi_-(1)\phi_+(2))/\sqrt{2} \tag{9.13}$$

$$\Psi_2^-(A_{2g}) \ : \ (\phi_+(1)\phi_-(2) - \phi_-(1)\phi_+(2))/\sqrt{2} \tag{9.14}$$

$$\Psi_3(B_{2g}) \ : \ (\phi_+(1)\phi_+(2) - \phi_-(1)\phi_-(2))/\sqrt{2} \tag{9.15}$$

The assignments can be confirmed as follows: Since $\phi_2 \sim yz$ and $\phi_3 \sim xz$,[4] $\phi_+ \sim (y+x)z$ and $\phi_- \sim (y-x)z$. It follows that the four space functions transform:

$$\Psi_1, \Psi_3 \sim z_1(y_1+x_1)\cdot z_2(y_2+x_2) \pm z_1(y_1-x_1)\cdot z_2(y_2-x_2) \quad (9.16)$$
$$\Psi_2^+, \Psi_2^- \sim z_1(y_1+x_1)\cdot z_2(y_2-x_2) \pm z_1(y_1-x_1)\cdot z_2(y_2+x_2) \quad (9.17)$$

Expansion and simplification leads to the assignments[5]:

$$\Psi_1(A_{1g}) \sim z_1z_2(y_1y_2+x_1x_2) \sim z^2(y^2+x^2) \quad (9.18)$$
$$\Psi_2^+(B_{1g}) \sim z_1z_2(y_1y_2-x_1x_2) \sim z^2(y^2-x^2) \quad (9.19)$$
$$\Psi_2^-(A_{2g}) \sim z_1z_2(-y_1x_2+x_1y_2) \sim R_z \quad (9.20)$$
$$\Psi_3(B_{2g}) \sim z_1z_2(y_1x_2+x_1y_2) \sim z^2xy \quad (9.21)$$

As in rectangular CBD, the three space functions that are symmetric to electron interchange are assigned to the singlet and $\Psi_2^-(A_{2g})$, that changes its sign when the indices 1 and 2 are permuted, is associated with the triplet. When interaction with the other electrons in the molecule is ignored, and square-planar CBD is characterized solely in terms of two electrons restricted to the four space functions in Table 9.1, it can be thought of as a *perfect biradical* [8], the lowest state of which is invariably the triplet. It turns out, however, that $^1B_{1g}$ lies some 10 kcal/mol below $^3A_{2g}$ [3, p. 363]. Needless to say, the relative thermodynamic stability of these two open shell states could only be computed by methods that not only go beyond the simple MO approximation but include interaction with many more space functions than the four that can be formed from ϕ_2 and ϕ_3 [5, 6, 7].

9.1.2.3 Overall Symmetry

The *overall symmetry* of a state is the direct product of the irreps of its space and spin wave functions [2, p. 206]:

$$\Gamma_{\text{overall}} = \Gamma_\Psi \otimes \Gamma_\Lambda \quad (9.22)$$

The singlet spin function is always totally symmetric, so we are justified in identifying the overall symmetry of a singlet state with that of its space function:

$$\Gamma_{\text{overall}}(\mathbf{S}) = \Gamma_\Psi(\mathbf{S}) \quad (9.23)$$

The overall symmetry of the three components of a triplet, provided that R_x, R_y and R_z belong to non-degenerate irreps,[6] are:

---

[4] The similarity sign($\sim$) is being used for conciseness to mean "behaves similarly to" or "has the same symmetry properties as".

[5] The assignment of the three symmetric space functions is straightforward. The antisymmetric one can be assigned by default to the remaining irrep, A_{2g}, or it might be recognized that $(y_1x_2 - x_1y_2)$ is similar in form to the explicit expression for rotation about z: $y\frac{\partial}{\partial x} - x\frac{\partial}{\partial y}$.

[6] The same considerations can be extended with care to cases in which two of them belong to degenerate representations.

$$\Gamma_{\text{overall}}(\mathbf{T}) = \Gamma(R_x) \otimes \Gamma_\Psi(\mathbf{T}),\ \Gamma(R_y) \otimes \Gamma_\Psi(\mathbf{T}),\ \Gamma(R_z) \otimes \Gamma_\Psi(\mathbf{T}) \tag{9.24}$$

Accordingly, the open shell states of rectangular CBD ($\mathbf{D}_{2h}$) have the irreps:

$$\Gamma_{\text{overall}}(\mathbf{S}) = B_{1g} \tag{9.25}$$
$$\Gamma_{\text{overall}}(\mathbf{T}_z) = B_{1g} \otimes B_{1g} = A_g \tag{9.26}$$
$$\Gamma_{\text{overall}}(\mathbf{T}_y) = B_{2g} \otimes B_{1g} = B_{3g} \tag{9.27}$$
$$\Gamma_{\text{overall}}(\mathbf{T}_x) = B_{3g} \otimes B_{1g} = B_{2g} \tag{9.28}$$

$\mathbf{T}_z$, the component of the triplet in which the orbital- and spin-angular momenta are both aligned along z, is totally symmetric. The overall symmetry of $\mathbf{T}_x$ is the same as the space symmetry of $\mathbf{T}_y$, and vice versa.

In square-planar CBD ($\mathbf{D}_{4h}$), the irreps of the overall state functions of the two lowest open shell states are:

$$\Gamma_{\text{overall}}(\mathbf{S}) \quad = \quad B_{1g} \tag{9.29}$$
$$\Gamma_{\text{overall}}(\mathbf{T}_z) = A_{2g} \otimes A_{2g} = A_{1g} \tag{9.30}$$
$$\Gamma_{\text{overall}}(\mathbf{T}_x, \mathbf{T}_y) = E_g \otimes A_{2g} = E_g \tag{9.31}$$

Here too, $\mathbf{T}_z$ is totally symmetric, but $\mathbf{T}_x$ and $\mathbf{T}_y$ are doubly degenerate.

9.1.2.4 Geminals

Since CBD, like all molecules beyond H_2, contains more than just two electrons, a word of explanation is called for to justify our having characterized its space-, spin- and overall symmetry on the basis of the two electrons occupying ϕ_2 and ϕ_3 alone. For this purpose it is convenient to regard the electrons in a molecule as coming in pairs, each pair occupying a *geminal* [9], i.e. a two-electron function that is individually either a singlet or a triplet. The $2n$ electrons in the n doubly-occupied MOs of a closed shell singlet can be regarded as occupying n geminals, each having a totally symmetric spin and space function. The overall wave function is the antisymmetrized product[7] of the n totally symmetric geminals and is itself totally symmetric. In an open shell state, $(2n - 2)$ electrons are paired in $(n-1)$ totally symmetric geminals, so its space symmetry is determined fully by the n-th geminal, *viz.* the antisymmetrized product of the two singly-occupied MOs (Equation 9.11). It follows quite generally that Equation 9.23 holds for any singlet and Equation 9.24 for any triplet.

9.2 Intersystem Crossing

The symmetry requirements for crossing from a singlet to a triplet potential energy surface have been dealt with in a variety of ways, with particular emphasis on *intersystem crossing* (IC), most commonly observed when an excited open

[7] See footnote 8 of Chapter 3.

shell singlet decays to the more stable triplet with the same electron configuration, but also in reference to *reactive intersystem crossing* (RIC), i.e. the direct conversion of a photochemically excited reactant to a product molecule with a different multiplicity. [2, pp. 270ff.], [10, 11], [12, pp. 228ff.]

Photochemical reactions, with and without retention of spin, will be deferred to the following chapter. The remainder of this chapter will be devoted almost entirely to reactions in which a molecule in its singlet ground-state decomposes or isomerizes thermally to produce a triplet of the product that, though not necessarily its own ground-state, is sufficiently stable that the overall energy of the products is comparable to that of the reactant. The procedure applied in both chapters, is an extension of *OCAMS* to *spin-forbidden* processes that was first suggested by the present author [13] on heuristic grounds, and justified theoretically soon thereafter in collaboration with Trindle [14].[8]

The condition for a *symmetry-allowed* pathway for a reaction in which a singlet (**S**) and triplet (**T**) are interconverted is the retention of overall symmetry along the pathway [2, p. 206]:

$$\Gamma_{\text{overall}}(\mathbf{S}) = \Gamma_{\text{overall}}(\mathbf{T}) \tag{9.32}$$

Consequently, the process is *allowed* if the irrep of the space function of the singlet equals the overall irrep of any one of the triplet components, i.e. that of its space function multiplied by the irrep of R_x, R_y, or R_z:

$$\Gamma_\Psi(\mathbf{S}) = \Gamma_{R_x} \otimes \Gamma_\Psi(\mathbf{T}),\ \ \Gamma_{R_y} \otimes \Gamma_\Psi(\mathbf{T}) \ \text{ or } \ \Gamma_{R_z} \otimes \Gamma_\Psi(\mathbf{T}) \tag{9.33}$$

Equations 9.32 and 9.33 specify that the total angular momentum about each of the cartesian axes is retained. Therefore, since **T** has one unit ($\frac{h}{2\pi}$) of spin angular momentum aligned along one of the axes, it can only be formed from **S** with the simultaneous appearance of a unit of orbital-angular momentum antiparallel to the same axis, so that there is no change in net angular momentum. Conversely, a triplet can be converted to a singlet "allowedly" only if the unit of spin-angular momentum annihilated is accompanied by a compensating change in orbital-angular momentum.

The requirement of overall symmetry conservation is purely qualitative and gives no indication of the efficiency with which spin-orbit coupling promotes intersystem crossing. Evidently, in order for it to occur, the singlet and triplet potential energy surfaces should cross, or at least be close enough that thermal excitation of the species on the lower surface can bring it into range of the upper. The strength of the coupling increases with the nuclear charge of the atoms involved, so it is quite small in organic molecules made up entirely of first-row elements and hydrogen, and increases if heavy atoms are present – either as substituents or in the solvent. [2, pp. 185–186]

Returning to our example, let us ask whether rectangular CBD can be expected to cross easily to the triplet as it approaches square-planar geometry. Applying Equation 9.33 to the overall irreps of the singlet and triplet in $\mathbf{D}_{2h}$

[8] Lee [15] and Chiu [16] have proposed similar correlation methods independently.

(Equations 9.15–9.18), we see that it is not – and would not be even if the singlet and triplet surfaces were much closer in energy than they are – since:

$$B_{1g} \neq A_g, \; B_{3g}, \text{ or } B_{2g} \quad . \tag{9.34}$$

9.2.1 Intersystem Crossing of Carbenes

Methylene (CH_2) is one of the few small organic molecules with a triplet ground-state. [17], [18, pp. 128ff.] It was noted in Section 4.1.1 that both the triplet and the closed-shell singlet lying ≈10 kcal/mol above it have $\mathbf{C}_{2v}$ symmetry. As indicated schematically in Fig 9.1, the gap between the LUMO (b_1) and the HOMO (a_1) of the singlet narrows in the more nearly linear triplet, in which each is singly-occupied, and the latter becomes the ground state.

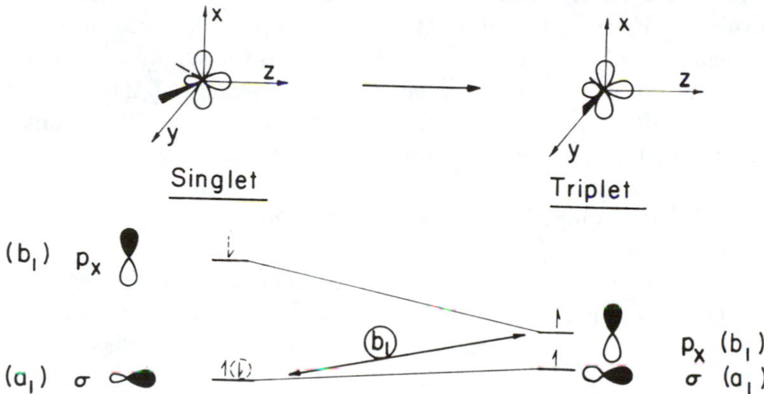

Figure 9.1. Intersystem crossing of a carbene ($\mathbf{C}_{2v}$). (From Fig. 1 of reference [14])

The relative energy of the two states can be increased or decreased by substitution, even to the extent of reversing their order. It was mentioned in Section 6.1 that replacing the H atoms of methylene with electronegative substituents pushes $^1A_1$ below $^3B_1$. The singlet of silylene, in which the central carbon atom by has been replaced by silicon, lies some 18. kcal/mol below the triplet [19, Table 1]. Metcalfe and the author [21] have shown computationally that the relative stability of the singlet and triplet of diphenylcarbene depends strongly on its momentary geometry, coplanarity of the phenyl rings and a large interbond angle at the carbenic C atom favoring the triplet. Accordingly, kinetic measurements place the triplet of diphenylcarbene 5.1 kcal/mol below the singlet, whereas in fluorenylidene, in which the rings are still coplanar but the central angle has decreased, the gap is reduced to 1.1 kcal mol [22]. Staab, Meier and their coworkers [23] investigated by electron spin resonance a series of carbenes with the structure of [1.n]paracyclophanes trapped in a matrix. They found that decreasing the number of C atoms binding the *para* atoms, and thus at the same time decreasing the carbenic angle and forcing the rings out of coplanarity, stabilizes the singlet and eventually brings it below the triplet.

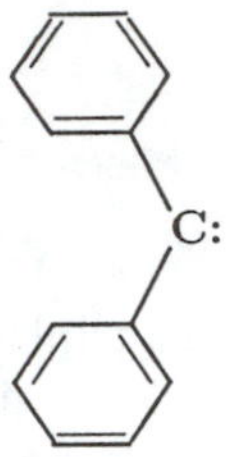

C: C: $(CH_2)_n$ C:

Diphenylcarbene 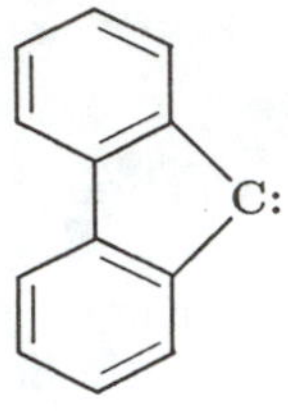 Fluorenylidene 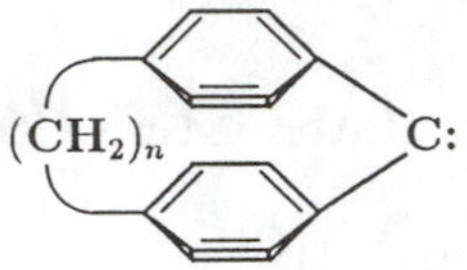 [1,n]Paracyclophane Carbene

9.2.1.1 Spin-Orbit Coupling

Whichever of the two states is the lower in any particular case, the other will be accessible by thermal excitation at moderate teperatures, so – provided that both of them have $\mathbf{C}_{2v}$ symmetry – Fig. 9.1 specifies the symmetry requirements of intersystem crossing for all of them: In the conversion of the singlet to the triplet, *spin-orbit coupling* allows the orbital- and spin-angular momenta to interact, and flips the spin of one electron. In order to satisfy Equation 9.33, the $\mathbf{T}_y$ component is generated: its space and spin functions both have b_1 symmetry, so their product is totally symmetric, like the closed shell singlet. For the reverse process to occur, one electron in the triplet geminal $[\sigma^1\, p_x^1]$ has to flip its spin and pair with an electron of opposite spin to generate the singlet geminal $[\sigma^2]$; the unit of orbital-angular momentum about y that is lost has to be annihilated together with a unit of spin-angular momentum oppositely aligned along the same axis.

Of the three components, only $\mathbf{T}_y$ has its spin oriented properly, but at ordinary temperatures, the interconversion of $\mathbf{T}_y$ with the two other components is more rapid than intersystem crossing. It can therefore be safely inferred that equilibrium between the components of the triplet is maintained throughout, whichever one of them is selected by symmetry to be formed from the singlet or to decay to it.

9.2.1.2 Spin-Vibronic Coupling

Although spin-orbit coupling is the principal mechanism of singlet–triplet interconversion, a secondary mechanism, *spin-vibronic coupling*, also plays a role, becoming important – as may occur – when spin-orbit coupling is prohibited by symmetry. It is illustrated for diphenylcarbene in Fig. 9.2.

For groups with real, one-dimensional, representations, i.e. $\mathbf{D}_{2h}$ and its subgroups, the formal requirement can be expressed in analogy to Equation 9.33:

$$\Gamma_{\mathrm{disp}} \otimes \Gamma_\Psi(\mathbf{S}) = \Gamma_{R_x} \otimes \Gamma_\Psi(\mathbf{T}),\ \ \Gamma_{R_y} \otimes \Gamma_\Psi(\mathbf{T})\ \text{ or }\ \Gamma_{R_z} \otimes \Gamma_\Psi(\mathbf{T}) \qquad (9.35)$$

in which Γ_{disp} is the irrep of the displacement of the molecule from its equilibrium geometry.

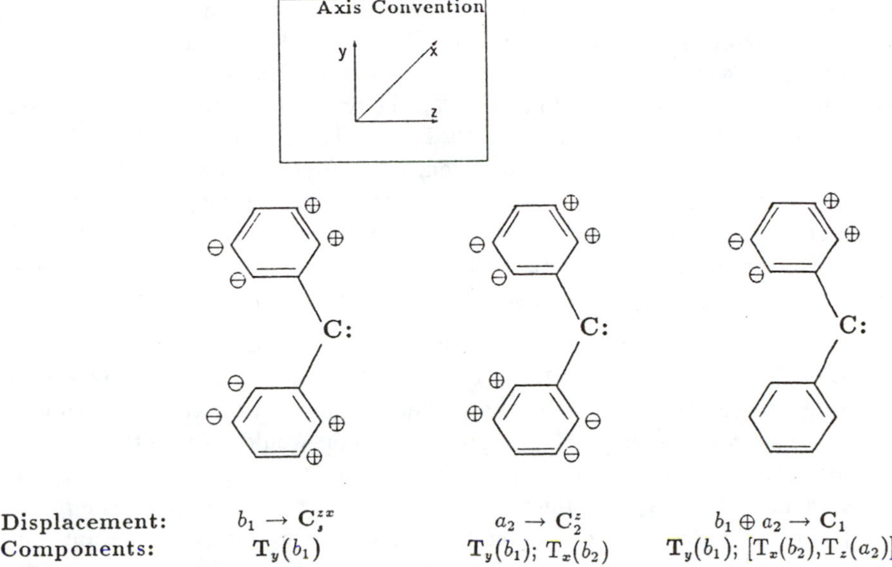

Figure 9.2. Spin-orbit and spin-vibronic coupling in diphenylcarbene. The triplet component induced by spin-orbit coupling is printed in bold letters. ($\oplus$: displacement towards $+x$; $\ominus$: displacement towards $-x$)

In our example, a totally symmetric vibration – such as the periodic increase and decrease of the central angle – merely augments the probability of generating $\mathbf{T}_y$, the component produced by spin-orbit coupling. The b_1 displacement shown in Figure 9.2, a disrotation of the phenyl groups, offers nothing new: the product $\Gamma_{\text{disp}} \otimes \Gamma_\Psi(\mathbf{S})$ has the same irrep (B_1) as the space function $\Psi(\mathbf{T})$, rather than its direct product with any of the three rotations. Conrotatory motion of the rings (a_2) induces intersystem crossing to $\mathbf{T}_x(b_2)$,[9] because $a_2 \otimes b_1 = b_2$. Out-of-plane rotation of one ring while the other stays in plane is a superposition of the disrotation and conrotation that produces $\mathbf{T}_z(a_2)$ as well, since $(b_1 \oplus a_2)$ introduces a second order b_2 term, and $b_2 \otimes b_1 = a_2$.

A somewhat different viewpoint can be adopted whenever a nuclear diplacement is not periodic, but is a major contributor to the reaction coordinate, as when the singlet and triplet have different geometries. In such a case, it is instructive to regard the displacement as desymmetrizing the molecule to the kernel of its irreducible representation. In our example, both the disrotation (b_1) and the conrotation (a_2) – superimposed on the totally symmetric closure of the central angle – lead to the geometry of the [1,n]paracyclophanes, in which the energy of the singlet has gone below the triplet. In the first case, the reaction path can be said to have been desymmetrized to C_s^{zx} and in the second to C_2^z:

[9] Triplet components are labeled by the irreps of their spin-functions; the labels are in lower case, to avoid confusion with the space symmetry of the triplet, which is the same for its three components.

1. In C_s^{zx}, R_x and R_z belong to a'', and R_y – rotation in the zx plane – to a'. When Equation 9.33 is applied in C_s^{zx}, the overall symmetry of the triplet component, like that of the singlet, is A'. Since its space symmetry is also A', the spin function has to transform like R_y, the only totally symmetric rotation.

2. In C_2^z, ^1A can interconvert with $\mathbf{T}_y$ and $\mathbf{T}_x$ of ^3B because both R_x and R_y belong to B, as does the space function, so the overall symmetry of both is A. These results are exactly the same as that obtained with Equation 9.35, but the distinction drawn in Fig. 9.2 between the components produced by spin-orbit and spin-vibronic coupling respectively has disappeared: spin-vibronic coupling in the parent group has become spin-orbit coupling in the subgroup.

In anticipation of the discussion of photochemical reactions in the next chapter, it might be noted that spin-orbit coupling alone cannot promote intersystem crossing between the first excited, open shell, singlet and the triplet. Both have the same space symmetry, so their interconversion would require the presence of a totally symmetric rotation – non-existent in $\mathbf{C}_{2v}$ – to produce a triplet component with an A_1 spin-function. Direct intersystem crossing in the parent methylene therefore has to rely on spin-vibronic coupling with the antisymmetric stretching vibration, $\xi_{as}(b_2)$[10] of Fig. 4.3, its only non-totally symmetric vibration, to generate $\mathbf{T}_x$ as the first-formed component. Alternatively, it may decay thermally to the closed-shell singlet, from which it can cross thermally to the triplet as described above. In diphenylcarbene, the disrotatory and conrotatory displacements of Fig. 9.2 can convert $^1B_1$ to $\mathbf{T}_y$ and $\mathbf{T}_z$ respectively, and the composite motion would produce $\mathbf{T}_x$ as well.

9.3 Reactive Intersystem Crossing

The possibility of crossing along the reaction path between a closed shell singlet reactant and a product in its triplet state must be considered whenever the singlet and triplet potential surfaces intersect. This will occur whenever one of the products of a thermal decomposition has a triplet ground state; familiar examples are the thermolysis of diazomethanes to produce carbenes and of methylenepyrazoline to form trimethylenemethane. It can also take place when the product is thermodynamically so much more stable than the reactant that an excited triplet of the latter is comparable in energy to the ground state of the former. Well known examples of this type of reaction, which has been given the evocative name "Photochemistry Without Light" [24], are the isomerization of Dewar benzene (DB) to the triplet of benzene and fragmentation of 1,2-dioxetanes to two molecules of ketone – one of which is in its triplet state. The remainder of this chapter will be devoted to these reactions; their analysis in the

---

[10] By happenstance, B_1 and B_2 are not interchanged despite the different axis conventions of Figs. 4.3 and 9.1.

following pages will be drawn largely from the paper by Trindle and the author that has already been cited [14]. First, however, a few words should be devoted to the kinetic behaviour of thermal reactions in which spin is not conserved.

9.3.1 The Arrhenius Parameters of Spin-Non-Conservative Reactions

In the standard treatment by transition state theory of a first-order reaction, such as a thermal fragmentation or isomerization [25, p. 97–101], the familiar Eyring equation is manipulated as follows:

$$k_1 = \kappa \frac{kT}{h} K^{\ddagger} \tag{9.36}$$

$$k_1 = \frac{kT}{h} e^{-\Delta G^{\ddagger}/RT} \tag{9.37}$$

$$k_1 = \frac{kT}{h} e^{\Delta S^{\ddagger}/R} \cdot e^{-\Delta H^{\ddagger}/RT} \tag{9.38}$$

The sign and magnitude of the entropy of activation, $\Delta S^{\ddagger}$, is taken as a gauge of the probability of forming the transition state. In unimolecular reactions, $\Delta S^{\ddagger}$ does not depend on concentration units and is assigned a "normal" value of zero, so that "normal" reactions have pre-exponential factors close to $\frac{kT}{h}$, or $\approx 10^{13}\,\mathrm{s}^{-1}$ at ordinary temperatures, and reactions with "improbable" transition states are expected to have negative entropies of activation and – as a result – pre-exponential factors significantly lower than $10^{13}\,\mathrm{s}^{-1}$.

Loosely speaking, one might say that reactive intersystem crossing – depending as it does on the incursion of a weak magnetic interaction as the reactant approaches the singlet–triplet intersection – is an inherently "improbable" process. Spin-non-conservative reactions should therefore have low values of the pre-exponential factor, which would be expected to become more "normal" when heavy atoms are present in the reactant or the reaction medium.

Note, however, that the *transmission coefficient* (κ) interpreted as "the fraction of crossings that are successful in leading to final products", disappears between Equation 36 and Equation 37 because "it is believed that κ ordinarily is rather close to unity" [25, p. 97]. Neglect of the transmission coefficient is no doubt justified for most thermal reactions, but hardly for those in which spin is not conserved, where there is a strong tendency for the reacting molecule to retain its original multiplicity, even when it is no longer on the surface of lowest potential energy. When κ is omitted, as it is in Equations 37 and 38, it – and any contribution to the rate from tunneling implicitly included with it – is lumped with the entropy of activation and interpreted as a negative contribution to $\Delta S^{\ddagger}$.[11] Be that as it may, the operative conclusion remains: Spin-non-conservative reactions are indeed characterized by abnormally low pre-exponential factors that are increased by the presence of heavy atoms in the reactant or its immediate surroundings.

[11] Different authors interpret κ more or less broadly [26, pp. 61ff.]; we are evidently adopting the broadest definition.

9.3.2 Thermolysis of Diazomethane

The fragmentation of diazomethane to methylene and dinitrogen has been dealt with computationally in considerable detail [27]. The computed potential curves displayed in Fig. 9.3 show that ($^3B_1$), which is an excited triplet at the equilibrium geometry of diazomethane, reaches a minimum near the geometry at which it crosses the closed shell singlet ($^1A_1$) surface. As a result, the intersection is quite flat, a circumstance that is conducive to intersystem crossing,[12] despite the weak spin-orbit coupling characteristic of molecules composed of light atoms.

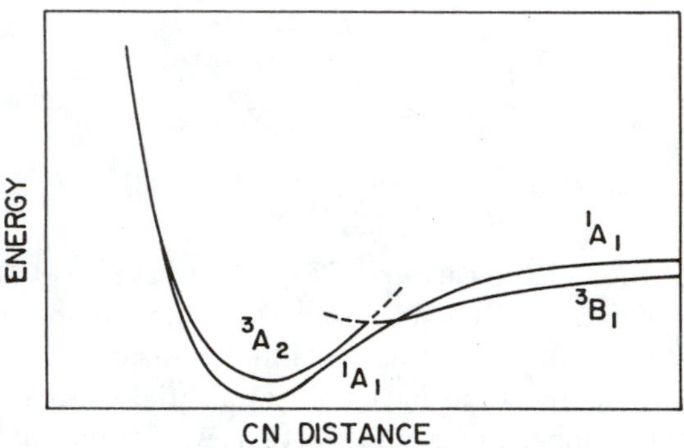

Figure 9.3. Variation of energy of Diazomethane with CN distance ($\mathbf{C}_{2v}$). (**a**) Lowest singlet; (**b**) lowest triplet. (From Fig. 2 of [27])

The correspondence diagram in Fig. 9.4 can be applied to either the selective production of $^1CH_2(A_1)$ or $^3CH_2(B_1)$ or to the simultaneous formation of both. The ground-state configuration of the valence electrons of N_2 (Section 3.2.2), desymmetrized from $\mathbf{D}_{\infty h}$ to $\mathbf{C}_{2v}^z$ with the aid of Table 3.2,[13] can be condensed to:

$$[\sigma_g^2\sigma_u^2\pi_u^4\sigma_g'^2] \Longrightarrow [a_1^6b_1^2b_2^2]$$

or are simply labeled by inspection; adding on the right the two MOs of methylene from Fig. 9.1 provides 7 MOs to accomodate 12 electrons. Neglecting the 4 electrons in the CH bonds of diazomethane, the remaining 12 are paired as shown on the left, yielding $[a_1^6b_1^4b_2^2]$, which would correlate with N_2 and CH_2 only if the latter had the doubly-excited singlet configuration $[p_x(b_1)^2]$.

In order to produce the closed shell singlet, the molecule has to bend in the zx plane, as shown in (a) of Fig. 9.5, so the N_2 fragment departs at an angle to the molecular axis. In order to form the triplet, one electron from the uppermost singlet geminal (b_1) of the reactant moves to an orbital of CH_2 with the same irrep; the other is obliged to go to the totally symmetric σ orbital of methylene,

[12] The flatter the curve the greater the density of vibrational states.

[13] σ_u is symmetric to reflection in all planes that include the z axis, so it is more properly referred to as σ_u^+.

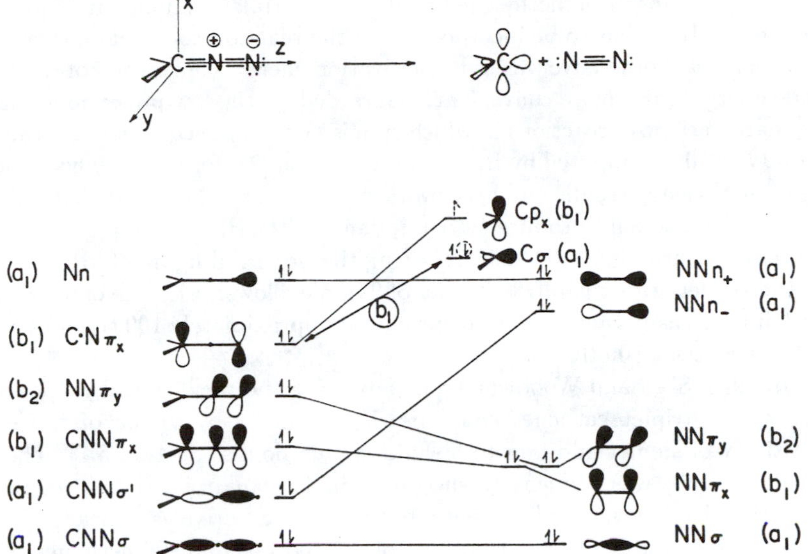

Figure 9.4. Correspondence diagram for thermolysis of Diazomethane (from Fig. 2 of reference [14])

losing a unit of orbital-angular momentum along the pathway. At the same time, a compensating unit of spin-angular momentum about y is generated; *viz.* the triplet component T_y, with space and spin symmetry B_1 and overall symmetry A_1, is produced. The process is represented pictorially in (**b**) of Fig. 9.5[14]. One electron is imagined to be moving from the p_x orbital of the terminal nitrogen atom to the σ orbital of the carbenic carbon atom. The momentary magnetic moment produced by this "rotation" of the electron in the xz plane induces it to flip its spin, producing a unit of spin-angular momentum oriented along y.

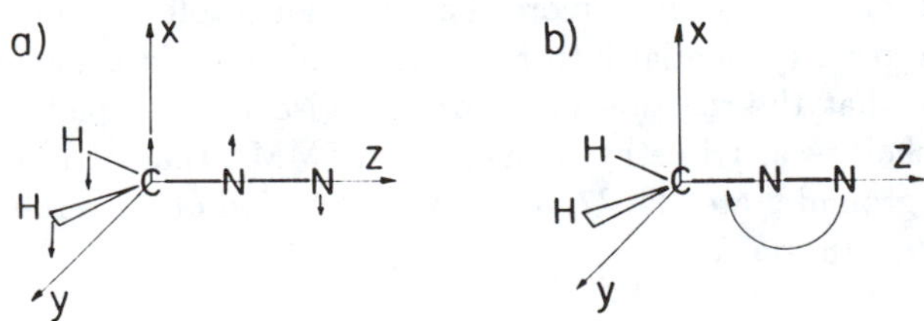

Figure 9.5. b_1 perturbations of the potential energy of CH_2N_2 during thermolysis ($\mathbf{C}_{2v}^z$). (**a**) Out-of-plane distortion; (**b**) Momentary magnetic moment about y. (From Fig. 3 of reference [14])

[14] The reader who deplores pictorializations of this kind can be reassured by the fact that the formal requirements of spin-orbit coupling are fulfilled.

Since the singlet of methylene lies above the triplet, and an out-of-plane distortion of CH_2N_2 has to be incorporated in the reaction coordinate, thermolysis to the singlet should have the higher activation energy. Generation of the triplet, while energetically more convenient, is retarded by the low pre-exponential factor characteristic of reactions in which spin is not conserved. Adopting the value of $\kappa \approx 5.\times 10^{-3}$ computed by Bader and Generosa [28] for the intersystem crossing of methylene, Trindle and his coworkers [29, 30] conclude that thermolysis of diazomethane should produce ${}^1CH_2(A_1)$ and ${}^3CH_2(B_1)$ at comparable rates in the temperature range 400 – 600°K. Using the preferred figure of ≈10. kcal/mol for the singlet-triplet gap [18] instead of Trindle's lower estimate of 6. kcal/mol, the triplet is easily calculated to be produced approximately 100 times as rapidly as the singlet at 500 °K.

In 1956, Skell and Woodworth [32] advanced the very fruitful proposal that singlet and triplet carbenes could be distinguished by virtue of the stereoselectivity of singlet addition to olefinic double bonds (Section 6.1). The *Skell hypothesis* [33] was immensely successful in the analysis of the reactions of photochemically generated carbenes, but was less effective as a means of identifying the spin state of the first product of thermolysis of diazomethanes. Triplet carbenes are very much less reactive than the corresponding singlets. If the lower-lying triplet is produced preferentially and intersystem crossing is much faster than its reaction with the olefin, an equilibrium will be set up between the two spin states, and the product of singlet addition will preponderate if its relative reactivity outweighs its lower concentration. This is certainly the case in the thermolysis of diphenyldiazomethane [31, p. 306] and the conclusion has been reasonably extended to methylene itself [29, 30].

9.3.3 Thermolysis of Methylenepyrazoline

Like other pyrazolines, 4-methylene-1-pyrazoline (MP) undergoes thermal extrusion of dinitrogen to form methylenecyclopropane (MCP) [35], but it does so much more rapidly: the energy of activation is about 9 kcal/mol lower than that of the parent 1-pyrazoline, more than enough to offset a hundredfold reduction in the pre-exponential factor [40, 41]. This kinetic behaviour is *prima facie* evidence that the reaction proceeds stepwise *via* a triplet intermediate. The obvious choice was trimethylenemethane (TMM), that had been shown to have a triplet ground-state [36, 37, 38], in confirmation of numerous theoretical predictions [39], [18, pp. 141ff.].

MP → TMM + N_2 → MCP

If the rate controlling step in the preferred pathway is spin-non-conservative formation of TMM, which then cyclizes – once more with spin inversion – to the more stable MCP, it is a reasonable deduction that the concerted extrusion of N_2 and closure to the product is forbidden by symmetry. The correspondence diagram for direct thermolysis of MP to MCP [14, Fig. 8], does not have to be reproduced here. The exocyclic π orbital is retained in the product and can be omitted, leaving just two MOs to be considered: In methylenepyrazoline these are the symmetric and antisymmetric σ_{CN} combinations; in the ultimate products they are the new σ_{CC} orbital that closes the cyclopropane ring and the in-plane π orbital of N_2, both of them totally symmetric. As a result, the diagram calls for an antisymmetric in-plane displacement, so thermolysis of MP on the singlet surface – like that of the parent pyrazoline – would be expected to proceed by stepwise cleavage of the two CN bonds. Thermolysis of 1-pyrazoline and its alkyl derivatives has the high energy of activation ($\geq$ 40. kcal/mol) and high pre-exponential factor ($\geq 10^{15}$) characteristic of reactions of this kind [43], [41, Table 1].

The much more elaborate correspondence diagram for generation of TMM [14, Fig. 9] is displayed in Fig. 9.6. As in previous examples in which CH bonds occupy different planes in the reactant and product, the occupied σ_{CH} orbitals are taken into account explicitly. They can be brought into correspondence by either a conrotation or a disrotation, respectively a_2 and b_2 in the axis convention adopted. The symmetry coordinates that have to be incorporated in the two reaction coordinates are illustrated schematically in Fig. 9.7: In one (a_2) the extruded N_2 molecule rotates about the symmetry axis as it departs and the methylene groups move into plane conrotatorily. In the other (b_2) it recedes above (or below) the molecular plane as the methylenes disrotate into plane.

The MOs on the left side of Fig. 9.6 are easily stacked in energetic order and labeled by irrep. Those on the right have to be considered more carefully, as follows. The degenerate π_{NN} orbitals split to a_1 and b_2; both are occupied, so their energetic order is immaterial. The two degenerate orbitals of trimethylenemethane are split by the receding dinitrogen molecule to a_2 and b_2: the former is stabilized by $\mathrm{NN}\pi_y^*(a_2)$ and the latter is destabilized by $\mathrm{NN}\pi_y(b_2)$. In the triplet, each is singly occupied, both electrons being derived from a pair in $\mathrm{CN}\sigma_-(b_1)$ of MP.

The correspondence diagram does not reject either pathway:

a) The conrotation takes the reacting system into $\mathbf{C}_2$, in which $\mathrm{CN}\sigma_-$ and ψ_3 have the same irrep (b), so one electron is transferred between these two MOs without change of spin; the other goes to ψ_2, producing $\mathbf{T}_x$ and $\mathbf{T}_y$ of TMM, both of which have spin-symmetry B in $\mathbf{C}_2^z$.

b) The disrotation leads to $\mathbf{C}_s^{yz}$, in which B_1 and A_2 map onto A''. Along this pathway, one electron of $\mathrm{CN}\sigma_-$ goes to ψ_2 with retention of its spin and the other goes to ψ_3 with a spin-flip, generating $\mathbf{T}_y$ and $\mathbf{T}_z$ since R_y and R_z transform as a'', the irrep onto which a_2, the direct product of of b_1 and b_2, maps in $\mathbf{C}_s^{yz}$.

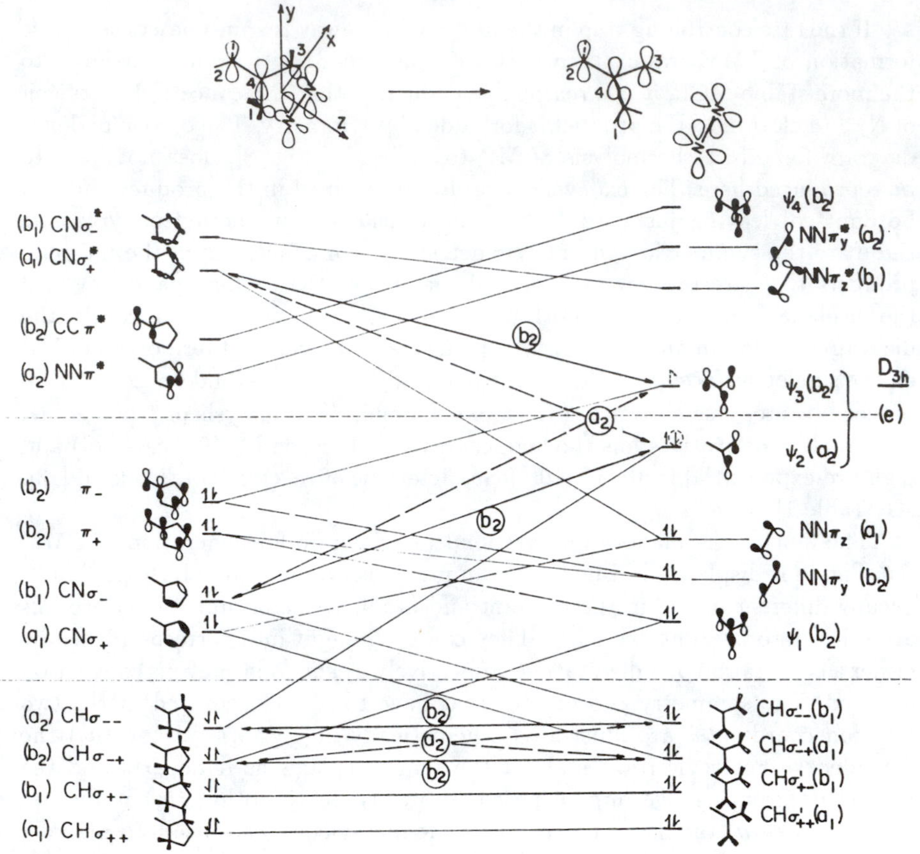

Figure 9.6. Correspondence diagram for thermolysis of Methylenepyrazoline (From Fig. 9 of reference [14])

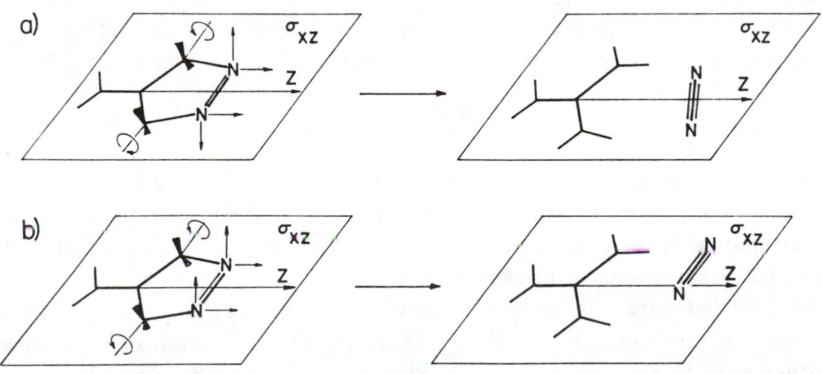

Figure 9.7a,b. Alternative pathways for thermolysis of methylenepyrazoline. (**a**) Disrotary Pathway (b_2) (**b**) Conrotary Pathway (a_2) (From Fig. 10 of reference [14])

Although both reaction paths are formally *allowed*, the conrotatory one is preferred: The crossing of the a_2 orbitals along this pathway indicates that the singlet and triplet surfaces cross as well, a necessary condition if the reaction is to take place without gratuitous vibrational excitation. A computational study [42] similar to that on the fragmentation of CH_2N_2 referred to above, finds formation of triplet TMM by the conrotatory pathway to be about five times as rapid as that of the singlet. The calculated ratio of pre-exponential factors, $3. \times 10^{-3}$, is similar to that experimentally observed, but the value adopted for the energy of activation, 7. kcal/mol, is too low. With the more realistic value of 9. kcal/mol, the triplet pathway is computed to preponderate by a factor of 50.

9.3.3.1 The Zwitterion Cascade Mechanism

Indirect evidence in support of the conrotatory mechanism for reactive intersystem crossing can be adduced from the fragmentation of 7-methylene-2,3-diaza[2.2.1]bicyclohept[2]ene (MDBH), that is constrained to follow the disrotatory pathway and does not produce the triplet directly. In MDBH, C_1 and C_3 of MP (numbered as in Fig. 9.6) are connected by a $-CH_2CH_2-$ bridge, so that it has $\mathbf{C}_s^{yz}$ symmetry, i.e. the disrotation is built into it substitutionally. Its thermolysis is much more rapid than that of MP; the Arrhenius parameters measured by Berson et al. [44] (E_a = 28 kcal/mol, A = 10^{15}) are consistent with a spin-conserving rate-limiting step that is much more facile than that for thermolysis of ordinary pyrazolines. The authors interpret their results as evidence that *singlet* TMM is formed first and then "cascades" rapidly via the triplet to the ultimate product, MCP.

There is no need to reproduce the correspondence diagram for thermolysis of MDBH to the bridged TMM in its closed shell singlet state [14, Fig. 11], since the relevant conclusions can be drawn as easily from Fig. 9.6: Under substitutional desymmetrization to $\mathbf{C}_s^{yz}$, both occupied π orbitals of the departing N_2 orbital interact unfavorably with ψ_3, increasing the gap between it and ψ_2 sufficiently for the latter to be doubly occupied. In the resulting closed shell singlet TMM, both electrons are on C_1 and C_3, leaving C_2 positively charged; the product is a zwitterion. As the dinitrogen molecule departs along the disrotatory pathway ((**b**) of Fig. 9.7) both electrons in $CN\sigma_-$ can pass directly to ψ_2 of TMM; the zwitterion then crosses to the two a'' components ($\mathbf{T}_y$, $\mathbf{T}_z$) of the triplet, that becomes increasingly stable as N_2 recedes, and – changing its spin once more – collapses to MCP.

9.3.3.2 Secondary Isotope Effects

Their extensive kinetic study of the thermolysis of deuterated methylenepyrazoline molecules at 170°C, and in particular the ratios of the differently labeled MCP molecules produced, led Crawford and Chang [40, 41] to reject reactive intersystem crossing of MP to TMM. They propose instead that, as in the case of 1-pyrazoline, initial rupture of one CN bond is followed by closure of the resulting diazenyl biradical to MCP with concomitant loss of N_2. The rates of the two successive processes are assumed to be comparable, so the isotope effect on the overall rate and the product ratio is determined by both.

It is difficult to reconcile the proposed mechanism with the observed low pre-exponential factor. Moreover, the isotope effects on the rate can be accomodated very simply on the assumption that the effect of dideuteration at the exocyclic and endocyclic C atoms on rate-limiting spin-non-conservative fragmentation of MP to MCP and N_2 is additive in free-energy of activation – i.e. multiplicative in $\frac{k_H}{k_D}$. The results of a least-squares fit of the experimental data [41, Table 3] are as follows:

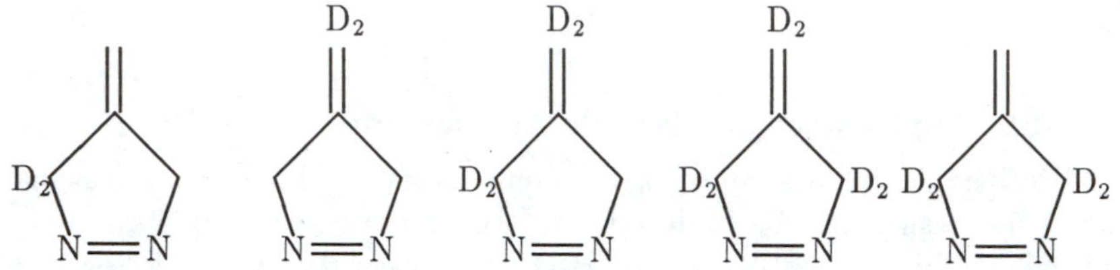

$(\frac{k_H}{k_D})_{exp}$	= 1.15	1.03	1.19	1.38	1.42
$(\frac{k_H}{k_D})_{calc}$	= 1.15	1.03	1.17	1.37	1.41

Cumulative isotopic retardation at each of the carbon atoms bonded to nitrogen indicates that the rupture of both bonds is synchronous, and the smaller effect at the exocyclic methylene group is consistent with loosening of the CC bond as the π orbital becomes delocalized. The moderate size of both effects, $\approx 17\%$ and 3% respectively, suggests that the transition state is an early one: crossing to the triplet surface occurs well before the CN bonds are broken.

At first sight, the isotopic product distribution displayed in Fig. 9.8 is disturbing. Product-determining closure of one of the three pairs of methylene groups of TMM, produced in a prior rate-limiting step, would be expected to yield the same ratio of isotopic isomers from a labeled TMM biradical, regardless of the substrate from which it originates. Instead, Fig. 9.8 shows preferential incorporation of deuterium in the exocyclic methylene group of MCP when deuterium is initially present on a ring C atom of MP.

The expectation that TMM would lose all "memory" of its precursor is based on the assumption that, once produced, it has time to reach thermal equilibrium before crossing once more to the singlet surface and closing to MCP. It should be recognized that the triplet can only be formed in a vibrationally excited state, specifically in the a_2 vibrational mode localized in the CH_2 and/or CD_2 groups that have just been released from bonding with nitrogen ((**a**) of Fig. 9.7). Its closure to MCP can be thought of as vibronically induced intersystem crossing,[15] that is known to be much less effective in deuterated molecules and to vary with the position of isotopic substitution [45, pp. 211–215]; as a result, deuteration can increase the lifetime of a given triplet very considerably [45, Fig. 4.9].

[15] TMM can be regarded as an excited triplet state of MCP, distorted from the equilibrium geometry of the latter.

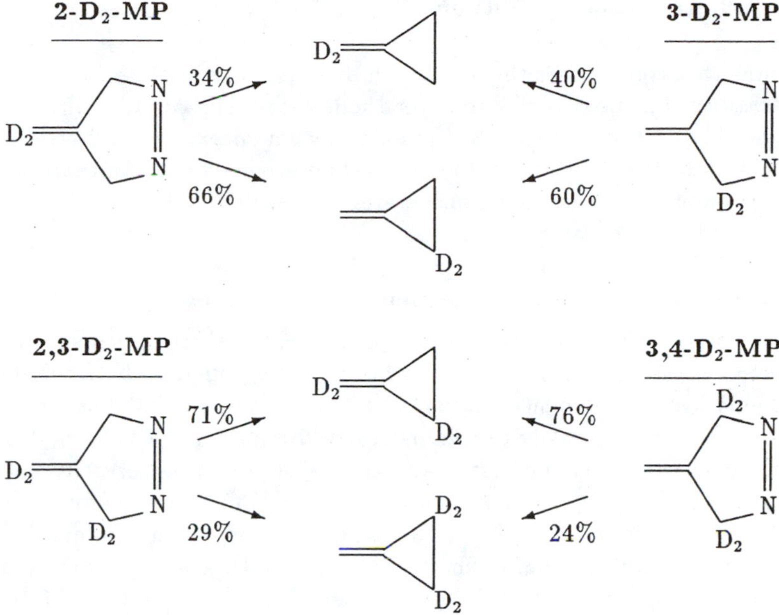

Figure 9.8. Secondary isotope effects on product distribution (MP → MCP). (From [40, Table 4])

A plausible extension of these ideas to the reaction under consideration suggests that tetradeuterated TMM lives longer than the dideuterated species and that – within each pair of similarly deuterated triplets – the one derived from ring-deuterated MP will tend less to cyclize to MCP before attaining thermal equilibrium than that derived from exocyclically deuterated MP. Accordingly, if intersystem crossing to the product is competitive with thermal relaxation, the biradical from 2-D_2-MP would be expected to close to MCP when it is the most vibrationally excited, whereas that from 3,4-D_4-MP will have come closest to thermal equilibrium before it reacts. Both tetradeuterated MP molecules show a marked preference for producing MCP with exocyclic CD_2, as might be expected for the reasons given in Section 6.4.1 in connection with the secondary isotope effect on allene-cycloaddition. The tendency decreases with departure from thermal equilibrium, until 2-D_2-MP, the most highly excited and therefore the least selective biradical, produces a nearly statistical (2:1) mixture of products.[16]

[16] The invocation of non-RRKM behavior in thermal reactions is ordinarily frowned upon by the chemical kinetic community. Here, however, we are dealing with rapid non-adiabatic isomerization of two constitutionally and structurally identical, but energetically different, triplets. It may well be a case in which "strong and sometimes peculiar effects (e.g. isotope effects) can be brought about by ... the non-constancy of the transmission coefficient as a function of kinetic energy" [47, 48].

9.3.4 "Photochemistry Without Light"

Reactions that come under this heading have been dealt with so extensively in the literature that no attempt to cover them within the confines of this book could posibly do them justice. In the following few pages, we will restrict our attention to the basic features of the two best known examples, concentrating on the question of how much light considerations of orbital and geminal symmetry can cast on their mechanism.

9.3.4.1 Isomerization of Dewar Benzene to Triplet Benzene

Like the corresponding isomerization of prismane, the thermal isomerization of DB to the closed shell ground-state of benzene was found to be forbidden by the schematic correspondence diagram in Fig 5.11. It was pointed out, however, that DB crosses quite easily to the first excited triplet $\mathbf{T}_1$ of benzene [49, 50]. This isomerization was the first – and only – spin-non-conservative reaction analysed in the popular presentation of OCAMS [13]; a more complete correspondence diagram, in which Dewar benzene was anasymmetrized and benzene desymmetrized to $\mathbf{D}_{2h}$, followed shortly thereafter [51]. The diagram in Fig. 9.9, set up in $\mathbf{C}_{2v}$, reproduces the third one published [14, Fig. 5], differing from it in

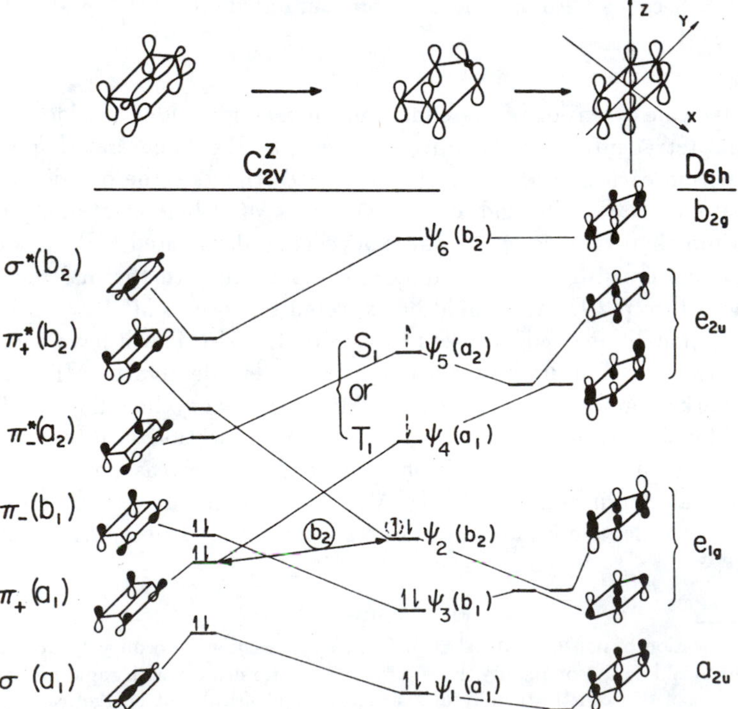

Figure 9.9. Correspondence diagram for isomerization of Dewar benzene to triplet benzene ($\mathbf{C}_{2v}$)

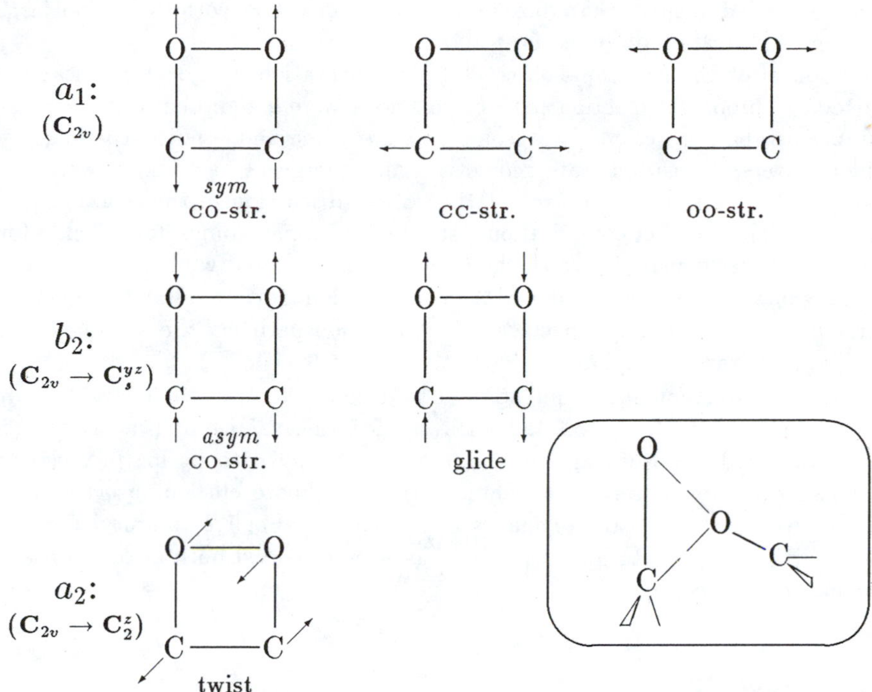

Figure 9.11. Skeletal symmetry coordinates of 1,2-dioxetane and (in frame) the TS (speculative) for its fragmentation

fragmentation of TMD to two vibrationally excited acetone molecules, both in the electronic ground-state, $\mathbf{S}_0$.

Recall, however, that TMD is much closer in energy to the excited surfaces corresponding to one molecule of acetone in $\mathbf{S}_0$ and one excited to either $\mathbf{S}_1$ or $\mathbf{T}_1$; crossing to either of them may well be competitive with reaction on the ground-state surface to form two (S_0) molecules. Consulting Fig. 9.10 once more, we see that the $\sigma_- \rightarrow \pi_+$ correspondence calls for the same in-plane displacement that is specified by the ground-state correspondence. In addition, however, only one electron of $\pi_x^*(a_2)$ passes smoothly into n_-^x; the other has to be induced to go into $\pi_-^*(a_1)$ by a perturbation of irrep a_2. In the case of the singlet, the perturbation can only be an additional, energetically costly, nuclear distortion: incorporation of the out-of-plane twist into the reaction coordinate. No nuclear displacement away from the TS in Fig. 9.11 is needed to "allow" crossing to the triplet surface. Rotation of electronic charge about the z axis, as one electron is transferred from an MO aligned along x in the reactant to a product orbital aligned along y, suppli es the spin-orbit coupling required to generate $\mathbf{T}_z$ of the product.[17] The TS for the spin-non-conservative reaction should be quite similar to that of the spin-retentive one. It is substantially

[17] As the reaction coordinate is desymmetrized to $\mathbf{C}_s^{yz}$, $\mathbf{T}_y$ comes in as well.

more extended in space than dioxetane, so it is consistent with the rather large volume of activation observed for the reaction [61].

Soon after the spin-non-conservative fragmentation of dioxetanes was reported, a number of reaction path calculations were attempted; diffferent assumptions about reaction path geometry were made and – not surprisingly – widely diverse transition state geometries and energies of activation were obtained [60, Table 1]. A more recent theoretical discussion of the reaction [62, pp. 183–188] cites later computational studies and comes comes down firmly for a biradical mechanism, in which the OO bond is ruptured while the CC bond remains intact. However, in none of the computations reviewed was the reaction surface mapped out fully; specifically, the reaction path *via* the TS suggested by Fig. 9.11 was not explored.

As an afterthought, it might be noted that the transition structure in Fig. 9.11 appears to be on the pathway for thermal isomerization of 1,2-dioxetane to 1,3-dioxetane; this reaction is not known to occur, presumably because spin-non-conservative fragmentation is a more efficient process. It is suggestive, however, that the analogous isomerization of 1,2-dimesityl-1,2-di-t-butyl-disiladioxetane to its 1,3 isomer, has been observed both in solution and in the solid state [63].

9.4 References

[1] R. McWeeny: *Spins in Chemistry.* Academic Press, New York 1970. An eminently readable introduction to spin.

[2] For a systematic presentation that covers all of the aspects of electron spin considered in this book, see: S.P. McGlynn, T. Azumi and M. Kinoshita: *Molecular Spectroscopy of the Triplet State.* Prentice-Hall, Englewood Cliffs 1969.

[3] T. Bally and S. Masamune: Tetrahedron *36*, 343 (1980).

[4] B.E. Douglas and C.A. Hollingsworth: *Symmetry in Bonding and Spectra.* Academic Press, London 1985.

[5] H. Kollmar and V. Staemmler: J. Amer. Chem. Soc. *99*, 3583 (1977).

[6] W.T. Borden, E.R. Davidson and P. Hart: J. Amer. Chem. Soc. *100*, 388 (1978).

[7] J.H Jafri and M.D. Newton: J. Amer. Chem. Soc. *100*, 7088 (1978).

[8] V. Bonačić-Koutecký, J. Koutecký and J. Michl: Angew. Chem. *99*, 216 (1987); Angew. Chem. Int. Ed. (English) *26*, 170 (1987).

[9] A.C. Hurley, J. Lennard-Jones and J.A. Pople: Proc. Roy. Soc. A-*220*, 446 (1953).

[10] L. Salem and C. Rowland: Angew. Chem. *84* , 86 (1972); Angew. Chem. Int. Ed. (English) *11*, 92 (1972).

[11] S. Shaik and N.D. Epiotis: J. Amer. Chem. Soc. *100*, 18 (1978).

[12] N.D. Epiotis: *Theory of Organic Reactions.* Springer, Berlin Heidelberg 1978.

[13] E.A. Halevi: Angew. Chem. *88*, 664 (1976); Angew. Chem. Int. Ed. (English) *15*, 593 (1976).

[14] E.A. Halevi and C. Trindle: Israel J. Chem, *16*, 283 (1977).

[15] T. Lee: J. Amer. Chem. Soc. *99*, 3909 (1977).

[16] Y.-N. Chiu: J. Amer. Chem. Soc. *104*, 6937 (1982).

[17] G.Herzberg and J. Shoosmith: Nature *183*, 1801 (1951).
[18] W.T. Borden and E.R. Davidson: Ann. Rev. Phys. Chem. *30*, 125 (1979). A critical survey of small molecules that have triplet ground states, with emphasis on theory.
[19] K.K. Baldridge, J.A. Boatz, S. Koseki and M.S. Gordon: Ann. Rev. Phys. Chem. *38*, 211 (1987).
[20] R.A. Moss: Accts. Chem. Res. *13*, 58 (1980).
[21] J. Metcalfe and E.A. Halevi: J. Chem. Soc. Perkin II *1977*, 634.
[22] E.V. Sitzman, J. Langan and K.B. Eisenthal: J. Amer. Chem. Soc. *106*, 1868 (1984).
[23] R. Alt, H.A. Staab, H.P. Reisenauer and G. Maier: Tetrahedron Letters *25*, 633 (1984).
[24] E.A. White, P.D. Wildes, J.Wiecke, H. Doshan and C.C. Wei: J. Amer. Chem. Soc. *95*, 7050 (1973).
[25] A.A. Frost and R.G. Pearson: *Kinetics and Mechanism, Second edition*, Wiley, New York 1961.
[26] W. Forst: *Theory of Unimolecular Reactions.* Academic Press, New York 1973.
[27] E.A. Halevi, R. Pauncz, I. Schek and H. Weinstein: Jerusalem Symp. on Quantum Chem. and Biochem. *4*, 167 (1974).
[28] R.F.W. Bader and J.I. Generosa: Canad. J. Chem. *43*, 1631 (1965).
[29] S.N. Datta, C.D. Duncan, H.O. Pamuk and C. Trindle: J. Phys. Chem. *81*, 923 (1977).
[30] C. Trindle and C.D. Duncan: Tetrahedron Letters *26*, 2251 (1977).
[31] P.P. Gaspar and G.S. Hammond: Chapter 6 in R.A. Moss and M. Jones (eds.) *Carbenes, vol. 2*. Wiley, New York 1972.
[32] P.S. Skell and R.C. Woodworth: J. Amer. Chem. Soc. *78*, 4496 (1956).
[33] For critical reviews of the Skell hypothesis and its applications, see [31, pp. 293–308], [34].
[34] G. Closs: Topics in Stereochemistry *3*, 193 (1968).
[35] P.S. Engel: Chem. Revs., *80*, 99 (1980).
[36] P. Dowd: J. Amer. Chem. Soc. *88*, 2587 (1966).
[37] M.S. Platz, J.M. McBride, R.D. Little, J.J. Harrison, A. Shaw, S.E. Potter and J.A. Berson: J. Amer. Chem. Soc. *98*, 5725 (1976).
[38] R.J. Baseman, D.W. Pratt, M. Chow and P. Dowd: J. Amer. Chem. Soc. *98*, 5726 (1976).
[39] P. Dowd: Accts. Chem. Res. *5*, 242 (1972).
[40] M.H. Chang and R.J. Crawford: Canad. J. Chem., *59*, 2556 1981.
[41] R.J. Crawford and M.H. Chang: Tetrahedron , *38*, 837 1982.
[42] C.D. Duncan, E.A. Halevi and C. Trindle: J. Amer. Chem. Soc. *101*, 2269 (1979).
[43] R.J. Crawford and A. Mishra: J. Amer. Chem. Soc. *88*, 3963 (1966).
[44] J.A. Berson, C.D. Duncan, G.C. O'Connell and M.S. Platz: J. Amer. Chem. Soc. *98*, 2358 (1976).
[45] B.R. Henry and W. Siebrand: Radiationless transitions, Chapter 4 of reference [46].
[46] J.B. Birks: *Organic Molecular Photophysics, vol. 1.* Wiley, London 1973.
[47] J.C. Lorquet and B. Leigh-Nihant: J. Phys. Chem. *92*, 4778 (1988).
[48] F. Remacle, D. Dehareng and J.C. Lorquet: J. Phys. Chem. *92*, 4778 (1988).

[49] P. Lechtken, R. Breslow, A.H. Schmidt and N.J. Turro: J. Amer. Chem. Soc. *95*, 3035 (1973).

[50] N.J. Turro and A. Devaquet: J. Amer. Chem. Soc. *97*, 3859 (1975).

[51] E.A. Halevi: Nouv. J. Chim. *1*, 229 (1977).

[52] G. Bieri, E. Heilbronner, T. Kobayashi, A. Schmelzer, M.J. Goldstein, R.S. Leight and M.S. Lipton: Helvet. Chim. Acta *59*, 2657 (1976).

[53] J.B. Birks: The spectroscopy of π-electronic states of aromatic hydrocarbons, Chapter 1 of reference [46].

[54] N.J. Turro and P. Lechtken: J. Amer. Chem. Soc. *94*, 2886 (1972).

[55] N.J. Turro, P. Lechtken, N.E. Schore, G. Schuster, H.-C. Steinmetzer and A. Yekta: Accts. Chem. Res. *7*, 97 (1974).

[56] N.J. Turro, P. Lechtken, G. Schuster, J.. Orell, H.-C. Steinmetzer and W. Adam: J. Amer. Chem. Soc. *96*, 1627 (1974).

[57] T. Wilson, D.E. Golan, M.S. Harris and A.C. Baumstark: J. Amer. Chem. Soc. *98*, 1086 (1976).

[58] E.J.H. Bechara, A.L. Baumstark and T. Wilson: J. Amer. Chem. Soc. *98*, 4648 (1976).

[59] J.Y. Koo and G.B. Schuster: J. Amer. Chem. Soc. *99*, 5403 (1977).

[60] E.A. Halevi: Internat. J. Quantum Chem. *12, Suppl. 1*, 289 (1977).

[61] R.Schmidt, H.-C. Steinmetzer, H.-D. Brauer and H. Kelm: J. Amer. Chem. Soc. *98*, 8181 (1976).

[62] K. Yamaguchi: Theoretical Calculations of Singlet Oxygen Reactions. In: A.A. Frimer (ed.) *Singlet* O_2*, vol. 3.* CRC Press, Boca Raton 1985.

[63] K.L. McKillop, G.R. Gillette, D.R. Powell and R. West: J. Amer. Chem. Soc. *112*, (1990) (in press). The author is grateful to Prof. West for communicating these results prior to their publication.

Chapter 10

Excited State Reactions

The analysis of the fragmentation of tetramethyl-1,2-dioxetane at the end of the preceding chapter, in which three competing processes had to be considered, foreshadows the difficulty of applying the criteria of symmetry conservation to the much more complex reactions that originate in an excited state of the reactant. An attempt will nevertheless be made in the following sections to show how the approach developed in the preceding chapters can be extended to deal with them. Each of the photochemical reactions chosen for discussion was selected in order to illustrate as convincingly as possible a particular point that has to be kept in mind when applying symmetry criteria to excited state reactions. In no case is it claimed that the mechanistic analysis is conclusive.

10.1 The Basic Photophysical Processes

Excited states are ordinarily – but not invariably – produced by photoexcitation. Before the photochemical reaction proper can be taken up, the photophysical processes preceding it [1], each with its own symmetry requirements, have to be listed. These are summarized in Fig. 10.1.

When, as is nearly always the case, the initial state is $\mathbf{S}_0$, photoexcitation to a triplet is assumed to be so highly *forbidden* as to be negligible. The selection rules for excitation to the various higher singlets will not be restated [3]; let us merely note three qualitative points:

1. Photon absorption and its inverse, fluorescence, are allowed only if the electronic transition changes the molecular dipole moment, however fleetingly. They are particularly effective when the molecule is polar in its ground-state and the electronic excitation is polarized longitudinally: on absorption of a photon, electronic charge shifts along the direction of the permanent dipole, increasing, decreasing, or perhaps even reversing it.

2. If an excited state higher than $\mathbf{S}_1$ is produced, it will ordinarily obey *Kasha's Rule* [7] and relax rapidly to $\mathbf{S}_1$. From there it can either fluoresce to $\mathbf{S}_0$ or relax non-radiatively by internal conversion (IC) to vibrationally excited $\mathbf{S}_0$; the excess vibrational energy is lost rapidly in solution – less so in the dilute gas phase. Kasha's Rule finds expression in Fig. 10.1 by the absence of all processes originating in $\mathbf{S}_2$ except internal conversion to $\mathbf{S}_1$.

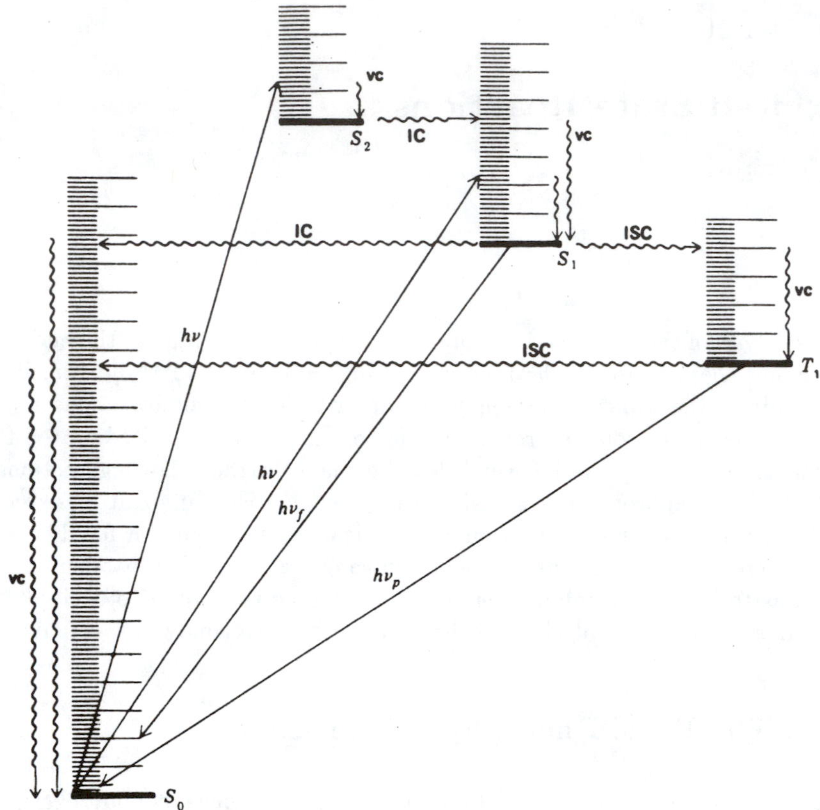

Figure 10.1. Modified Jablonski diagram showing transitions between excited states and the ground-state (Reproduced with permission from Reference [2]). Radiative processeses are shown by straight lines, radiationless processes by wavy lines. IC = internal conversion; ISC = intersystem crossing, VR = vibrational relaxation; $h\nu_f$ = fluorescence; $h\nu_p$ = phosphorescence.

3. The selection rules for intersystem crossing (ISC) were given in Section 9.2. If a higher triplet is produced, it too will obey Kasha's Rule and relax to $\mathbf{T}_1$. Since its two modes of relaxation to $\mathbf{S}_0$, phosporescence and ISC, both involve spin-inversion, $\mathbf{T}_1$ will be comparatively long-lived. Except when trapped at very low temperatures,[1] there is ample time for the three components of the triplet: $\mathbf{T}_x$, $\mathbf{T}_y$ and $\mathbf{T}_z$, to reach equilibrium before it phosphoresces, relaxes by ISC or reacts chemically.

[1] Interconversion of the triplet components can be frozen out when the triplet is produced photochemically at very low temperatures in a matrix or host crystal. In these conditions, photolysis of diphenyldiazomethane produces the $\mathbf{T}_y(b_1)$ component of diphenylcarbene selectively [8]. This finding is consistent with Fig. 9.1 if the carbene is produced in its closed shell singlet state and then crosses to the more stable triplet. It is also consistent with direct reactive intersystem crossing RISC from the photoexcited reactant molecule [9, footnote on p. 285].

10.1.1 Fluorescence: The Azulene Anomaly

The molecule often cited as "the exception that proves Kasha's Rule" is azulene, that fluoresces preferentially to $\mathbf{S}_0$ from its second excited singlet, [4, pp. 8,22–23], [6, 147–148]. The anomaly has been ascribed to a rather large $\mathbf{S}_1$–$\mathbf{S}_2$ energy gap and to a remarkably weak fluorescence from $\mathbf{S}_1$, that cannot compete with vibronically induced internal conversion to $\mathbf{S}_0$ and subsequent relaxation to its vibrational ground-state. It is clear, however, that orbital symmetry cannot be an insignificant factor.

Unlike its isomer naphthalene, azulene is a polar molecule, negatively charged on its five-membered ring and positively charged on its seven-membered ring [2, p. 47]. The two highest occupied and two lowest unoccupied MOs show the characteristic alternation between π orbitals that are symmetric to rotation about the symmetry axis and those that are symmetric to reflection in the plane perpendicular to the σ-frame.[2] Assigning z to the symmetry axis and placing the σ frame in the yz plane, leads to the ground-state orbital occupancy [10, p. 143]: $[\ldots b_1^2\, a_2^2 \,|\, b_1^0\, a_2^0]$. The HOMO $\to$ LUMO excitation to $\mathbf{S}_1(^1B_2)$ is y-polarized,

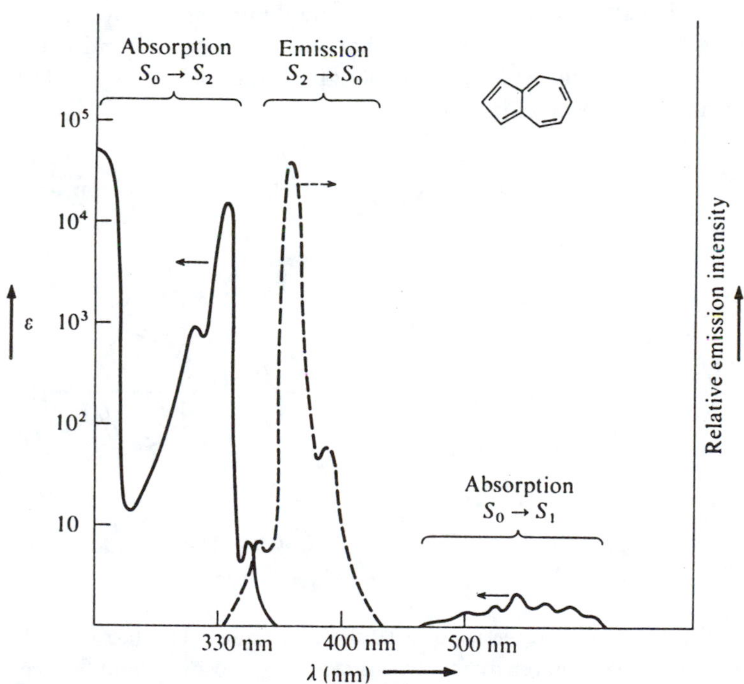

Figure 10.2. The anomalous $\mathbf{S}_2 \to \mathbf{S}_0$ flourescence of Azulene. (Reproduced with permission from Reference [6].)

[2] See, for example, Figs 1.5, 5.3, 5.8, 5.9, 7.2, 7.3 and 9.9. The symmetry of the molecule has to be low enough that the π orbitals do not come in degenerate pairs.

i.e. the shift of charge during the excitation is at right angles to the direction of the permanent dipole. In contrast, $\mathbf{S}_2(2^1A_1)$ is totally symmetric;[3] whether the transition leading to it is (HOMO−1) → LUMO or HOMO → (LUMO+1), or – since the transitions are of the same symmetry species – a superpositon of the two, it implies a shift of charge parallel to the symmetry axis and a consequent change in the dipole moment. As a result, it can be seen in Fig. 10.2 that the intensity of absorption to $\mathbf{S}_2$ is $\approx 10^4$ times greater than to $\mathbf{S}_1$. Fluorescence, which has the same symmetry requirements as absorption, has to compete with internal conversion. This is extremely rapid for large molecules in fluid solution, so only ≈ 2% of the molecules in $\mathbf{S}_2$ survive non-radiative decay to $\mathbf{S}_1$ and fluoresce to $\mathbf{S}_0$ instead [6, p. 148]. The very much weaker $\mathbf{S}_1 \rightarrow \mathbf{S}_0$ fluorescence cannot compete with internal conversion and is not observed.

While Kasha's Rule only states specifically that non-radiative relaxation to $\mathbf{S}_1$ is generally more rapid than fluorescence from $\mathbf{S}_n$ (n>1), it carries the additional implication that all photochemical reactions that occur on the singlet potential energy surface will originate in $\mathbf{S}_1$, whether it is produced directly or by internal conversion of a higher singlet. The occasional violation of Kasha's Rule, however rare an occurrence, suggests that there may be cases in which photochemical reactions originate in higher singlets; we will see that this is indeed so, and that – as in the case of azulene fluorescence – symmetry is the determining factor. Another possibility that must be kept in mind is that after IC from $\mathbf{S}_1$ to $\mathbf{S}_0$ the molecule retains enough vibrational energy to fall apart; here too symmetry plays an important rôle.

10.1.2 Stereoselectivity of Photophysical Processes: Bimanes

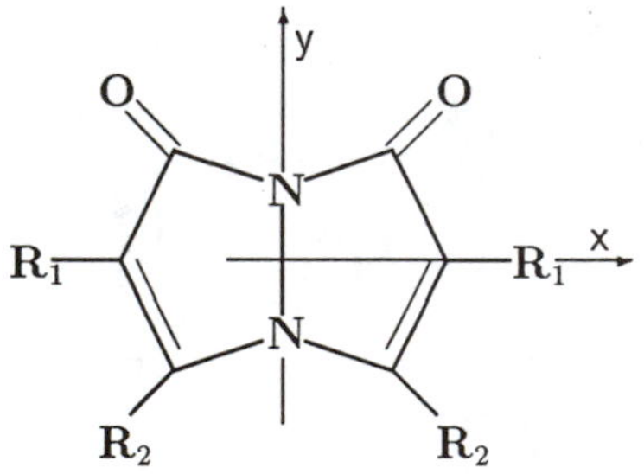

syn-Dioxabimanes $\mathbf{C}_{2v}^{y}$ ($\mathbf{C}_{s}^{yz}$)

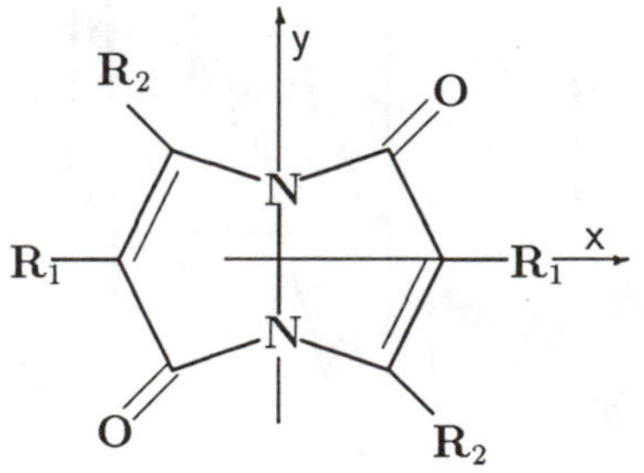

anti-Dioxabimanes $\mathbf{C}_{2h}^{z}$ ($\mathbf{C}_{2}^{z}$)

The photophysical properties of 9,10-dioxabimanes (1,5-diazabicyclo[3.3.0]-octadienediones) were extensively investigated by Kosower and his coworkers [11]. The most striking feature of their results is the stereoselectivity of the relaxation processes from $\mathbf{S}_1$. The *syn*-dioxabimanes fluoresce so well that they can

[3] In order to distinguish it from the closed shell ground state, the two totally symmetric states are relabelled 1^1A_1 and 2^1A_1 in order of increasing energy.

be used as fluorescent markers. Their phosphorescence is ordinarily a hundredfold weaker; heavy atom substitution increases the phosphorescence quantum yield, but even an iodo-substituent raises it to no more than half that of fluorescence. In stark contrast, the fluorescence of *anti*-dioxabimanes is completely swamped by phosphorescence. Evidently, as the authors point out, intersystem crossing in the *anti*-bimanes is remarkably efficient. We will see that the much greater efficiency of ISC in the *anti*-dioxabimanes than in their *syn* isomers emerges directly from their different symmetry properties.

The molecules of both series are non-planar: *anti*-dioxabimanes relax from $\mathbf{C}_{2h}$ to $\mathbf{C}_2$ and *syn*-dioxabimanes from $\mathbf{C}_{2v}$ to $\mathbf{C}_s$. The parent molecules of the two series ($R_1 = R_2 = H$) were investigated computationally [12]; the principal results are summarized in Table 10.1. The most stable conformations of both stereoisomers depart from planarity by some 20–30°, but the gain in energy is no more than 5–6 kcal/mol in either case. The HOMO and LUMO retain their identity in the subgroups, remaining essentially π and π^* orbitals. In both cases, the energetic separation from the adjacent orbitals is sufficiently large that $\mathbf{S}_1$ and $\mathbf{T}_1$ are derived from the HOMO $\rightarrow$ LUMO transition. As these MOs are affected to only a minor extent by π-σ interaction, it is instructive to carry out the symmetry analysis in the point groups of the planar molecules, in which the formal distinction between π and σ orbitals is maintained.

Table 10.1. Computed properties of *syn*- and *anti*-dioxabimanes (AM1-SCF) [12]

Stereoisomer	**Symmetry point group**	$\Delta \mathrm{H}_f$ (kcal/mol)	**Configuration** [...HOMO^2:LUMO^0]
syn-dioxa-	$\mathbf{C}_{2v}^{y}$	59.82	$[...b_1^2 : a_2^0]$
bimane	$\mathbf{C}_{s}^{yz}$	53.59	$[...a'^2 : a''^0]$
anti-dioxa-	$\mathbf{C}_{2h}^{z}$	55.37	$[...b_g^2 : a_u^0]$
bimane	$\mathbf{C}_{2}^{z}$	50.62	$[...b^2 : a^0]$

The *syn*-bimane has a substantial permanent dipole moment, $\mu = 7.45D$ [12], but $\mathbf{S}_1$ has B_2 symmetry: it is polarized along x, at right angles to the dipole axis, so fluorescence – though *allowed* – is not an overwhelmigly favored process. Intersystem crossing from $\mathbf{S}_1$ to $\mathbf{T}_1$ is strictly forbidden in the planar *syn*-bimane, because both have the same space symmetry, and conservation of overall symmetry would require crossing to a triplet component with a totally symmetric spin function that cannot exist in $\mathbf{C}_{2v}$. R_x is totally symmetric in $\mathbf{C}_s^{yz}$, so production of $\mathbf{T}_x(a'')$ is weakly *allowed* in the non-planar *syn*-bimane, but would hardly be expected to compete with fluorescence and internal conversion to $\mathbf{S}_0$.

The *anti*-bimane is, of course, non-polar in the planar conformation, and acquires a weak dipole along z ($\mu = 1.77D$) when it bends out of plane to $\mathbf{C}_2^z$. $\mathbf{S}_1$ has the same irrep in $\mathbf{C}_{2h}^z$, as x and y (B_u), so fluorescence is *allowed* in this isomer as well. Here, however, ISC is also *allowed*: R_z is totally symmetric in $\mathbf{C}_{2h}^z$, so $\mathbf{S}_1$ can cross directly to the $\mathbf{T}_z$ component of the triplet. The reader who is not inclined to reject out of hand the fanciful portrayal in Fig. 9.5b of an electron "rotating" about an axis as it moves from one MO to another, will recognize that R_z implies a momentary *ring-current* about the symmetry axis. As in other examples where a similar pictorialization is possible (*vide infra*), intersystem crossing turns out to be particularly efficient. It is strong enough in the present example to become the dominant photophysical process.

10.1.3 Chemical Sensitization: Singlet Dioxygen

Singlet dioxygen has generated intense interest in recent years [13, 14], not least as a result of its anti-tumor activity when generated in its lowest singlet state ($^1\Delta$) by *in situ* photosensitization [15]. It lies only 22.5 kcal/mol above the triplet ground-state, so it is thermally accessible at ordinary temperatures. Therefore, although it is in an excited state, $^1O_2$ is a closed shell molecule and its reactions can be analysed by the methods applied in Part II to closed shell molecules in their ground-states.

In Section 3.2, O_2 and other homonuclear diatomic molecules were desymmetrized artificially from cylindrical symmetry ($\mathbf{D}_{\infty h}$) to $\mathbf{D}_{2h}$. A hypothetical quadrupolar field was invoked that splits the degeneracy of ($^1\Delta$): its ^1A_g component is stabilized whereas the $^1B_{1g}$ component, with the same configurational symmetry as the triplet ground-state ($^3B_{1g}$), remains unaffected. We now note that an approaching reactant molecule necessarily reduces the symmetry of O_2 below $\mathbf{D}_{\infty h}$, almost invariably[4] to $\mathbf{D}_{2h}$ or one of its subgroups. It is therefore legitimate to identify the reactive component of singlet dioxygen with its ^1A_g component,[5] that has the same form and symmetry properties as the π orbital of ethylene.

We limit ourselves here to a single example, due to Turro et al. [16]. In it, ground-state $^3O_2$ is apparently converted to $^1O_2$ by interaction with a strained acetylene, with which it then reacts to produce an α-diketone in a chemiluminescent excited state:

$^{1,3}O_2$ + Acetylene ⟶ Dioxetene ⟶ Diketone*

[4] Exceptions would be the approach of an atom or axially symmetric molecule along the internuclear axis of O_2 ($\rightarrow \mathbf{C}_{\infty v}$) or the orthogonal approach of two O_2 molecules to one another ($\rightarrow \mathbf{D}_{2d}$).

[5] Kasha [14, Vol. 1, p. 4] has cautioned against the "multiple error" of identifying $^1\Delta$ with ^1A_g. As long as it is kept in mind that the latter is one component of the former and – lying lower than its partner – is the reactive species in most of its chemical reactions, there is little danger of falling into serious error.

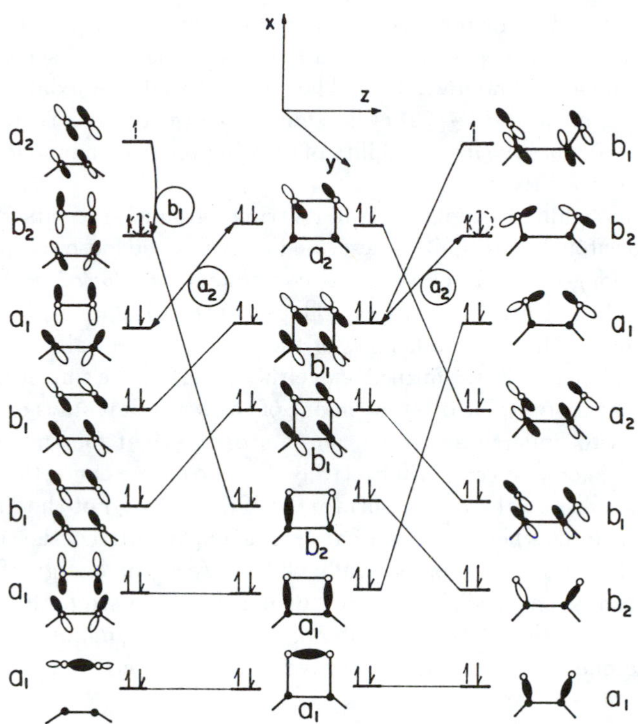

Figure 10.3. Correspondence diagram in $\mathbf{C}_{2v}^{x}$ for: $^1O_2$ + acetylene → dioxetene → α-diketone. (The C_2H_2 molecule is slightly bent in-plane)

The detailed symmetry-analysis by Trindle and the author [17] of the parent reaction is summarized concisely in Fig. 10.3. Two O atoms, two C atoms and two H atoms provide twenty-two valence electrons altogether. Ten of these: six in the CC and CH bonds of ethylene and four oxygen lone-pair electrons – need not be considered explicitly, because the five orbitals housing them retain their symmetry properties across the diagram. Fig. 10.3 makes do with six MOs for the closed shell dioxetene intermediate; these are supplemented by one more in the reactants and product, where open shell states have to be taken into account. The ordering of the orbitals in the three species is intuitive, but – apart from the upper two, that determine the nature of the lowest excited states – it is immaterial.

Reading Fig. 10.3 from left to right, we consider first the cycloaddition of $^1O_2$ to the strained acetylene. For the purpose of the analysis, the reactants are assumed to approach one another in the coplanar $[_{\pi}2_s + {_{\pi}2_s}]$ orientation, in which the degeneracy of the π^* orbitals of O_2 is split by the approaching acetylene. Favorable interaction with the vacant π^* MOs of the alkyne should stabilize both, but better overlap with the in-plane component is expected to push the one labeled b_2 below its partner and select it as the doubly-occupied HOMO. The only orbital mismatch with the dioxetene intermediate is between an a_1

and an a_2 orbital. This merely means that instead of approaching the acetylene along a strictly coplanar pathway, it can approach axially and screw itself into the multiple bond "allowedly". It will be recalled that the axially symmetric pathway is *forbidden* in $[_\pi 2 + {}_\pi 2]$-cycloaddition of ethylene (cf. Section 1.4.2.2), illustrating once more the inadvisability of drawing too close an analogy between superficially related reactions.

If the approaching O_2 molecule is in its triplet ground-state, its π^* orbitals – each singly occupied – are split as described above. If the loose complex formed is sufficiently long-lived, ISC of its $\mathbf{T}_z(b_1)$ component[6] is *allowed*, and the singlet complex can then either dissociate or collapse to the dioxetene.

Depending on the relative heights of the various potential energy barriers, the dioxetene, however it was formed, can either revert to the singlet reactants or isomerize to the more stable diketone in one of two ways. If it stays on the singlet surface, it has to undergo an a_2 deformation of the tight four-membered ring at substantial energetic cost. Alternatively, the dioxetane can retain its planar geometry but cross to the triplet surface via ISC to $\mathbf{T}_x(a_2)$ of the diketone. As with other reactions where there is an accessible triplet surface, the choice of the dominant pathway depends on a "trade-off" between the energy of activation and the transmission coefficient, which depends on the strength of the spin-orbit coupling (Section 9.3). The observation of chemiluminescence from the product diketone is evidence that reactive intersystem crossing (RISC) to the triplet diketone is preferred.

10.2 Photofragmentation

10.2.1 Photolysis of Cyclobutadiene

It is clear from Fig. 10.4 that thermal $[_\pi 2 + {}_\pi 2]$-cycloaddition of acetylene, like that of ethylene (Fig. 6.2), is *forbidden* in $\mathbf{D}_{2h}$ by a mismatch between a b_{1u} and a b_{3u} orbital. The same conclusion applies to the reverse reaction, thermal fragmentation of cyclobutadiene (CBD) to two acetylene molecules [18]. Here too the b_{2g} correspondence between $\sigma_-(b_{3u})$ and $\pi_+^z(b_{1u})$ that formally "allows" the thermal reaction is geometrically unsuitable. The HOMOs and LUMOs of the reactant and product-pair are in cross-correspondence under an a_u displacement, that also induces a correspondence between $\sigma_-(b_{3u})$ and $\pi_+^y(b_{3g})$. The two C_2H_2 fragments then only have to twist about the C_2-axis as they recede to form two acetylene molecules, one in the closed shell ground-state and the other as an excited singlet.

Does the possible formation of the acetylene pair in $\mathbf{T}_1$ have to be taken into account? There are two reasons why it need not be. First, at the orbital level of approximation, the electrons in the crossing orbitals have to be considered separately. Each is transferred from a u to a g MO; a spin-flip would require

[6] See footnotes 9 in Chapter 9 and 20 in Chapter 3.

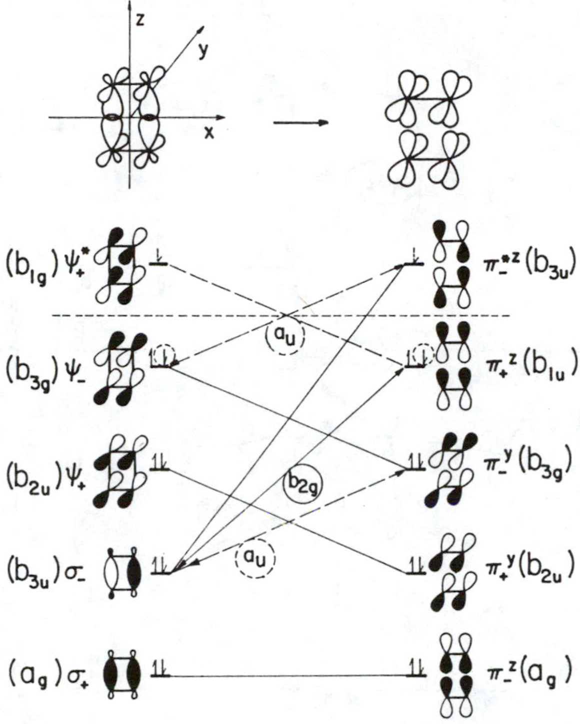

Figure 10.4. Correspondence diagram for fragmentation of Cyclobutadiene ($\mathbf{D}_{2h}$). Solid lines: thermolysis; broken lines: photolysis. (From Fig. 6 of reference [18])

formation of a triplet component with a space function that is antisymmetric to inversion – and none exists. Secondly, at the more rigorous state level such a process would be described as: $\mathbf{S}_1(B_{2g}) \rightarrow \mathbf{T}_1(B_{2g})$, requiring the generation of a totally symmetric (A_g) triplet component, whereas, like the rotations, these transform in $\mathbf{D}_{2h}$ as B_{1g}, B_{2g} and B_{3g}.

The symmetry analysis is consistent with the experimental results reported by Chapman et al. [19]. Photolysis of CBD in an argon matrix produces a pair of caged acetylene molecules that diffuse apart on warming to 35°K. If the matrix is thawed without being irradiated, CBD does not fall apart but dimerizes to *syn*-tricyclo[4.2.0.0]^{2,5}octa-3,7-diene, as described in Section 7.3.1.

10.2.2 Photochemical Decomposition of Formaldehyde

Gas phase irradiation of formaldehyde raises it to its $\mathbf{S}_1$ state, which – depending on the frequency of the exciting radiation – decomposes along two competing pathways [20, 21]:

(I) : $H_2CO^* \rightarrow H_2 + CO$
(II): $HCO^* \rightarrow H + HCO$

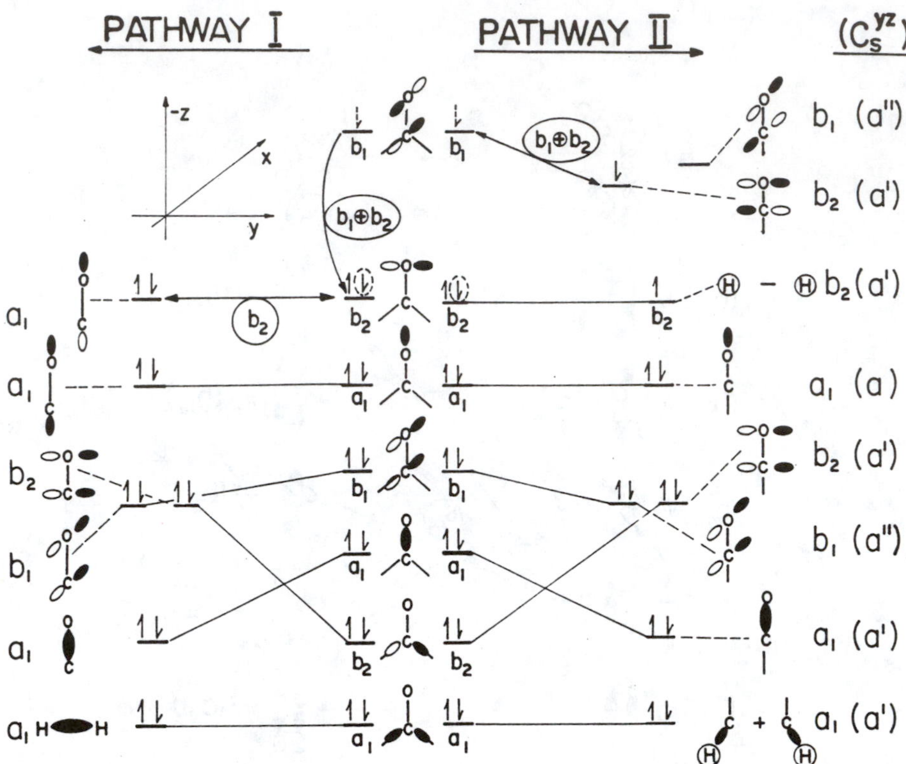

Figure 10.5. Correspondence diagrams for photolysis of formaldehyde

The radical pathway (**II**) is dominant at higher frequencies; at lower frequencies, when $\mathbf{S_1}$ is not formed with enough vibrational energy to break into two radical fragments, it undergoes ISC to vibrationally excited $\mathbf{S_0}$ and falls apart along the molecular fragmentation pathway (**I**) [22]. A detailed symmetry analysis of the relevant processes, bolstered by *ab initio* computations, was published by Bachler and the author [23]. Its principal conclusions can be deduced directly from the schematic corespondence diagrams in Fig. 10.5.

10.2.2.1 Pathway I: $S_1(H_2CO) \rightarrow S_0(H_2CO) \rightarrow H_2 + CO$

$\mathbf{S_1}$ is the HOMO→LUMO state with the configuration $[...b_2^1 b_1^1]$ and consequent state symmetry $^1A_2$. A vibrational perturbation of irrep a_2 is required in order to induce internal conversion to the totally symmetric $\mathbf{S_0}$. Fig. 10.6 depicts the six symmetry coordinates of H_2CO, none of which has the proper irrep to promote IC. As has been demonstrated experimentally [24] the process occurs *via* coupling of antisymmetric in-plane (b_2) and out-of-plane (b_1) vibrations, that together provide a second-order term of the proper a_2 symmetry (cf. Section 6.1.2.1).

Only one non-correspondence between ground-state H_2CO and its molecular fragmentation products appears on the left side of Fig. 10.5: $a_1 \not\leftrightarrow b_2$. The

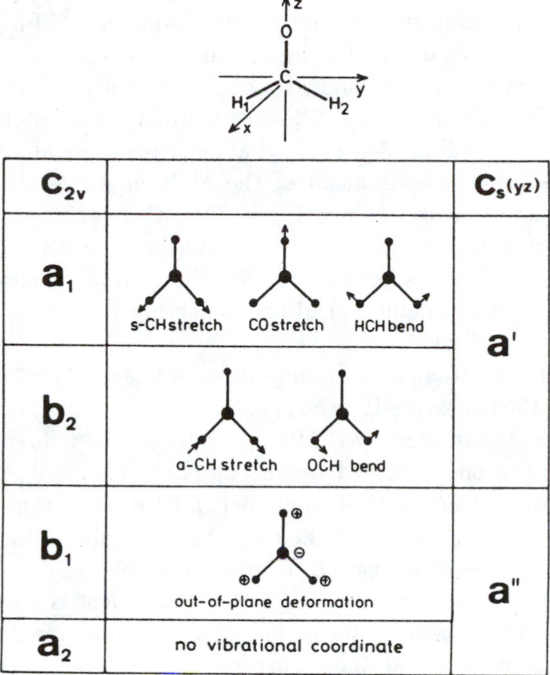

Figure 10.6. Symmetry coordinates of formaldehyde. (From Fig. 2 of reference [23])

reaction coordinate therefore has to include at least one b_2 component. Consulting Fig. 10.6, we see that it comprises all five of the in-plane components. Of the three totally symmetric coordinates, *s*-CH stretch extends the CH bonds, HCH bend – in opposite phase to that depicted – brings the two departing H atoms into bonding distance of one another, and CO stretch makes a minor adjustment to the CO bond length. The correspondence diagram in Fig. 10.5 then specifies that the two b_2 coordinates should be included as well: *a*-CH stretch allowing one bond to be stretched more than the other, and OCH bend allowing in-plane bond-angle optimization.

In full agreement with the symmetry analysis, the reaction path obtained from refined high-level calculations of the ground-state fragmentation pathway [25, 26, 27], is indeed in-plane, retaining $\mathbf{C}_s^{yz}$ but not full $\mathbf{C}_{2v}^{z}$ symmetry.

10.2.2.2 Pathway II: $\mathbf{S}_1(H_2CO) \rightarrow H + HCO$

First, we recognize that although formaldehyde itself is non-planar in $\mathbf{S}_1$, the inversion barrier is very low (1–2 kcal/mol) [28, 29], so formal imposition of planarity cannot introduce any error. The radical pathway, involving rupture of a single CH bond, does not retain $C_2(z)$ or $\sigma(zx)$; its highest possible symmetry is $\mathbf{C}_s^{yz}$. Formal $\mathbf{C}_{2v}^{z}$ symmetry can be retained for the purpose of the analysis by *anasymmetrization* (Section 7.1.1.1) – a particularly simple procedure in the present instance.

The HCO radical is portrayed on the right-hand side of Fig. 10.5 as nearly fully-formed and the departing H atom as nearly but not quite detached. The carbon and oxygen atoms are aligned along z, so they are transformed into themselves by whatever sym-op of $\mathbf{C}_{2v}^{z}$ – $\sigma(zx)$ or $C_2(z)$ – is chosen to be the *anasymmetrizer*. According to Rule 1, that requires their atomic orbitals not to vanish on anasymmetrization, all of the MOs in which they are involved are simply characterized by their irreps in $\mathbf{C}_{2v}^{z}$. Only the two non-equivalent AOs of the H atoms are transformed by the anasymmetrizer into one another. Rule 2 then specifies that, since the CH-bonding orbital is necessarily totally symmetric, the $1s(H)$ combination of the departing H atoms has to be taken with negative sign and labeled b_2.

In the fully formed radical, assuming it to be linear, the π^* orbitals are degenerate. Before the departing H atom is completely detached, however, residual bonding stabilizes the in-plane (b_2) MO and puts it below its out-of-plane (b_1) partner, so it is the one occupied by the unpaired electron. As a result, the diagram indicates that one pair of singly-occupied orbitals remains out of correlation and an a_2 perturbation is called for. As noted above, formaldehyde has no a_2 vibration, so – as in the case of IC from $\mathbf{S}_1$ to $\mathbf{S}_0$ – a composite ($b_2 \oplus b_1$) motion is invoked instead. The in-plane (b_2) displacement is inherent to the reaction pathway, so the necessity for including a b_1 displacement in the reaction coordinate merely means that the transition state for the radical pathway is non-planar, retaining no symmetry elements at all. As noted above, $\mathbf{S}_1(H_2CO)$ is itself pyramidal, but – since the inversion barrier is low – the reaction path would be expected to take it considerably farther from planarity. As in the case of the molecular pathway, the qualitative results are in complete agreeement with the results of detailed computations [30].

10.2.2.3 Sidelight: Coping with the Limitations

One of the professed objects of *OCAMS* is to provide a qualitative guide to reaction path computations [31] by selecting the symmetry coordinates that have to be included in the reaction coordinate and postponing the inclusion of others for subsequent optimization steps. It is gratifying that the symmetry analysis of formaldehyde fragmentation is fully consistent with previously published reaction-path computations, but it cannot be denied that it would have been only marginally useful as a prior guide to how these calculations should be carried out. This is because *OCAMS*, like other purely qualitative procedures, has no way of estimating the weight with which different symmetry coordinates of the same irrep contribute to the reaction coordinate, a weight – moreover – that varies along the reaction path.

For example, it is intuitively obvious that the two principal a_1 components of the reaction coordinate for fragmentation to H_2 and CO are s-CH stretch and HCH bend, whereas CO stretch contributes very little. However, all that the analysis based on Figs. 10.5 and 10.6 can state firmly is that the out-of-plane bend (b_1) can be omitted from the reaction coodinate for the molecular fragmentation (Pathway I) but must be included with the other five in calculation of the homolytic dissociation (Pathway II). This information is not negligible,[7] but it certainly leaves much to be desired.

Instead of going directly to an detailed computational mapping of the potential energy surface, which can be a time-consuming exercise for large molecules, Bachler [34] has proposed a semi-quantitative procedure[8] for selecting the energetically most favorable symmetry coordinates from among all of those that are symmetry allowed. The procedure has been tested successfully on the thermal fragmentation of formyl fluoride (CHFCO) [35], in which all five of the in-plane symmetry coordinates have the same irrep, a', in $\mathbf{C}_s$. This approach has to be applied to photochemical reactions in two steps:

1. Using the symmetry properties of vibrational coordinates to predict the changes in molecular geometry that occur on photoexcitation from the electronic ground-state to a given excited state; and
2. correlating that excited state with its possible reaction products.

Bachler [36], has shown how to accomplish the first step in his extension and amplification of the Bader-Pearson-Salem approach (See footnote 15 in Chapter 1), according to which the symmetry properties of vibrational modes activated in an electronically excited state are determined by the irreps of the orbitals between which the transition takes place. The way is open to following with the second step.

10.3 Photoisomerization of Benzene

Let us return to the three thermodynamically unstable polycyclic C_6H_6 isomers: Prismane, Dewar benzene (DB) and benzvalene (BV), all of which can be produced photochemically from benzene. According to Bryce-Smith and Gilbert's exhaustive survey of the photochemistry of benzene [37], prismane is the product of a secondary step, so it will not be considered. Excitation to $\mathbf{S}_1$ produces benzvalene; so does excitation to $\mathbf{S}_2$, but – in the liquid state – DB is formed as well.

As was noted in Section 9.3.4 in connection with spin-non-conservative isomerization of Dewar benzene, the HOMO→LUMO excitation of benzene gives rise to two configurations, one for $\mathbf{S_1}$ – labeled $^1B_1$ in $\mathbf{C}_{2v}^z$, and one for $\mathbf{T_1}$ – that has the same B_2 configuration as the second excited singlet, $\mathbf{S_2}$. The assignment of each state to a transition between one pair of orbitals is legitimate in Fig. 9.9, in which benzene is depicted as not yet having become plane-hexagonal. In the $\mathbf{D}_{6h}$ geometry of ground-state benzene, its two HOMOs and two LUMOs come in degenerate pairs. As a result, each of the two lowest excited singlets involves

---

[7] Fragmentation of formaldehyde along Pathway I has been used as a model in theoretical investigations of mode selectivity [32, 33]. In them the reaction coordinate is assumed to be in rapid equilibrium with the five in-plane modes but not with the out-of-plane bending mode, which interacts with it more slowly.

[8] The method is based on the form of the overlap density function of non-correlating orbitals.

two configurations, both constructed from all four of the frontier orbitals in different combinations:

$$\mathbf{S}_1(^1B_{2u}) \in [...\psi_2^1\psi_5^1] \text{ and } [...\psi_3^1\psi_4^1] \tag{10.1}$$
$$\mathbf{S}_2(^1B_{1u}) \in [...\psi_2^1\psi_4^1] \text{ and } [...\psi_3^1\psi_5^1] \tag{10.2}$$

They have the same energy in the primitive MO approximation as the corresponding triplets, so it might be thought that a symmetry analysis based on that simple model would not be applicable; perhaps suprisingly, it is.

The configurations that contribute to the two singlets differ in an essential respect. The transition leading to $\mathbf{S}_1$ is commonly referred to as $\mathbf{L}_b$, or *transverse*. An electron being transferred from ψ_2 to ψ_5 or from ψ_3 to ψ_4 changes the electron density on the various AOs drastically: it is decreased on some of the carbon atoms and increased on others. The $\mathbf{L}_a$ transition leading to $\mathbf{S}_2$ and – as noted in Section 9.3.4.1 – to $\mathbf{T}_1$ as well, is *longitudinal*: excitation of an electron from ψ_2 to ψ_4 or ψ_3 to ψ_5 leaves it occupying the same atomic orbitals, merely changing their relative phases. The resulting difference in electron interaction breaks the degeneracy of $\mathbf{S}_1$ and $\mathbf{S}_2$, separating them quite widely. It is therefore not surprising that photoexcitation to the two singlets leads the benzene molecule along different reaction pathways. The symmetry of the isomerization products is lower than hexagonal, so the natural procedure is to carry out each analysis in the subgroup of $\mathbf{D}_{6h}$ characterizing the presumed product. This is done for the isomerizations to benzvalene and Dewar benzene in Fig. 10.7.

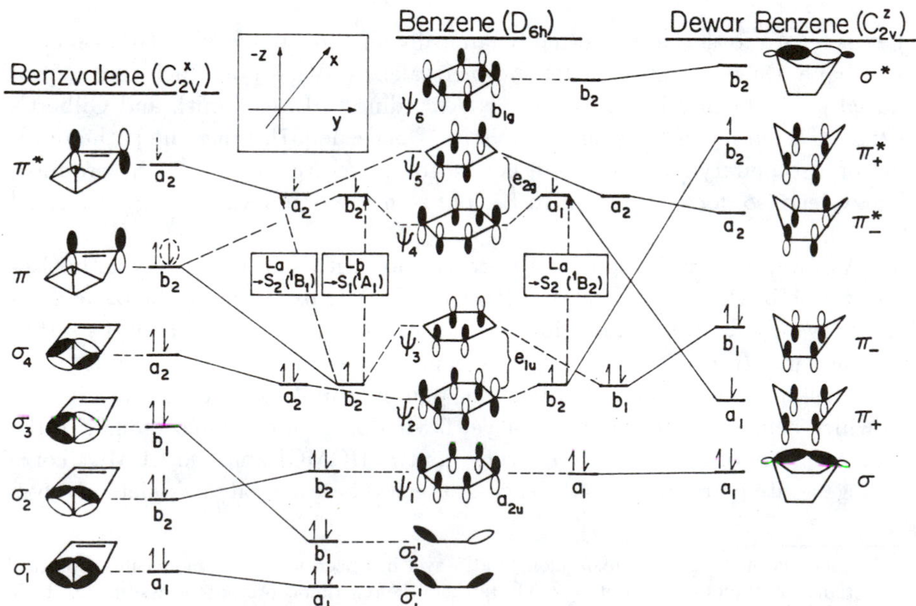

Figure 10.7. Correspondence diagrams for photoisomerization of benzene ($\mathbf{D}_{6h}$) to benzvalene ($\mathbf{C}_{2v}^x$) and Dewar benzene ($\mathbf{C}_{2v}^z$)

10.3.1 Photoisomerization to Benzvalene

The MOs of benzvalene on the left of Fig. 10.7 are the five CC-bonding MOs in Fig. 5.9, labeled in $\mathbf{C}_{2v}^{x}$ as before, to which is added $\pi^*(a_2)$, that is singly-occupied in the first excited singlet. The σ_{CH} orbitals are not included in Fig. 10.7, but it will be recalled from Section 5.3.2 that they are the factor that forces the reaction-path of the ground-state isomerization to retain at most $\mathbf{C}_2^x$ symmetry. The two combinations of the benzene σ bonds - derived from those of benzvalene that are not broken - are carried over from Fig. 5.9 as well.

The π MOs are drawn for the symmmetrical benzene molecule, numbered as in Fig 9.9 but labeled on their left by their irreps in $\mathbf{C}_{2v}^{x}$, the subgroup of $\mathbf{D}_{2h}$ appropriate to benzvalene. It is noteworthy that ψ_3 and ψ_4 have the irrep b_2 in $\mathbf{C}_{2v}^{x}$ while ψ_2 and ψ_5 are a_2. Both components of $\mathbf{S}_1$ are therefore totally symmetric, whereas those of $\mathbf{S}_2$ have $^1B_1$ symmetry like $\mathbf{S}_1$ of BV. Only one component of each transition appears in the correspondence diagram, because - at the orbital level of approximation - excitation from $\psi_2(a_2)$ would leave it singly-occupied, and destroy its direct correspondence with $\sigma_4(a_2)$. However, one orbital correspondence is sufficient to " allow" the reaction.

Isomerization from $\mathbf{S}_1$ is clearly *allowed* by orbital symmetry; two elctrons occupying different b_2 MOs of benzene can be thought of as moving to $\pi(b_2)$, the HOMO of benzvalene. The process does violate the *non-crossing rule* between states (Section 3.3.1), since it implies the correlation of $\mathbf{S}_0$(BV) with $\mathbf{S}_1$, that is totally symmetric in $\mathbf{C}_{2v}^{x}$, instead of with the closed shell ground-state, $\mathbf{S}_0$, that is of course also totally symmetric. However these two states, distinguished as 1^1A_1 and 2^1A_1, are coupled by totally symmetric vibrations that allow the isomerizing molecule to *funnel* rapidly into the lower of the two.

Formally, photoexcitation to $\mathbf{S}_2$ can also lead directly to $\mathbf{S}_1$ of benzvalene, but the absorbed photon would have to be highly energetic to reach it. Under normal photolytic conditions, it will obey Kasha's Rule and undergo internal conversion to vibrationally excited $\mathbf{S}_1$, from which it can proceed to BV as above. This process would produce $\mathbf{S}_0$ of BV with sufficient vibrational excitation to isomerize to other C_6H_6 isomers, as the experimental evidence suggests [37, Fig. 11].

Bryce-Smith and Gilbert [37] postulate the presence along the singlet pathway of a biradical intermediate, prefulvene, with fused three-membered and five-membered rings. A triplet biradical of that structure was found by Oikawa et al. [38] in their computational investigation of the lowest triplet pathway. The $\mathbf{T}_1$ states of benzene and benzvalene have the same space-symmetry (B_1) in $\mathbf{C}_{2v}^{x}$, and $\mathbf{S}_1(A_1)$ (but not $\mathbf{S}_2(B_1)$!) can indeed cross to the $\mathbf{T}_z(b_1)$ triplet component at the surface crossing. Spin-non-conservative formation of the triplet prefulvene biradical from $\mathbf{S}_1$ can therefore not be excluded, particularly since fulvene is produced along with benzvalene [37].

10.3.2 Photoisomerization to Dewar Benzene

The symmetry analysis that follows is similar to that of Haller [39] and reaches identical conclusions. In Fig. 10.7, the MOs of benzene are relabeled to their right by their irreps in $\mathbf{C}_{2v}^z$; those of DB are carried over from Fig. 9.9. The non-correspondence that blocks thermal isomerization of DB on the ground-state potential energy surface was shown in Section 9.3.4.1 to be between $\psi_2(b_2)$ and $\pi_+(a_1)$. Therefore, when an electron is excited out of $\psi_2(b_2)$ to $\psi_4(a_1)$, a cross-correlation is established between the $^1B_2$ states of benzene and Dewar benzene. The required transition is the longitudinal one, $\mathbf{L}_a$, leading to $\mathbf{S}_2$ of benzene; it correlates nominally with a higher excited state of DB, but one that has the same B_2 state-symmetry as $\mathbf{S}_1$, its HOMO–LUMO state. As a result, the isomerizing molecule – obeying Kasha's Rule – can undergo rapid IC into the latter under the influence of totally symmetric vibrations.

In order for isomerization to occur from $\mathbf{S}_2$, the excited benzene molecule has to withstand internal conversion to $\mathbf{S}_1$ long enough to react as such. IC from $\mathbf{S}_2(B_{1u})$ to $\mathbf{S}_1(B_{2u})$ of benzene is activated most effectively by its two out-of-plane b_{2g} vibrations [40, Chapter 10]. Vibronic coupling between the two electronic states is apparently good enough in the gas phase to convert all of the molecules in $\mathbf{S}_2$ to $\mathbf{S}_1$, whence those with enough vibrational energy can isomerize to benzvalene – and perhaps beyond. Vibrational excitation is quenched so rapidly in the liquid phase that an appreciable fraction of $\mathbf{S}_2$ molecules can disobey Kasha's rule and survive long enough to isomerize to DB.

Bachler has applied his extension of the Bader-Pearson-Salem approach [36] to both isomerization processes. His preliminary results indicate that vibrational modes that would be expected to contribute to the reaction coordinates for isomerization to benzvalene are indeed excited in $\mathbf{S}_1$ and those leading to Dewar benzene are excited in $\mathbf{S}_2$ [41].

10.4 Spin-Non-Conservative Photoisomerization: Naphthvalene

Like its analog benzvalene, naphthvalene shows remarkable kinetic stability. Its behavior on photoexcitation to its lowest excited singlet is reported by Turro and his coworkers to be "unexpected". Spin-non-conservative isomerization predominates over all of the other photophysical and photochemical processes combined. Of the excited naphthvalene (NV) molecules, 70% undergo reactive intersystem crossing (RISC) to $\mathbf{T}_1$ of naphthalene (N), as compared to 10% that fluoresce to their own ground state. Since the quantum yields of these two processes account for 80% of the photons absorbed , no more than 20% of the excited NV molecules undergo either or both of the two non-radiative relaxation processes: internal conversion to the ground-state of naphthvalene and reactive internal conversion (RIC) to the ground-state of naphthalene. Most surprisingly, fluorescence from the first excited singlet of naphthalene was not observed, indicating that the *a priori* most probable chemical reaction, isomerization on the $\mathbf{S}_1$ potential energy surface, does not take place.

$\mathbf{S}_0$(NV)
hν
$\phi_f = 0.1$
$\mathbf{S}_1$(NV)
$\mathbf{S}_1$(N)
$\phi_{ric} \leq 0.2$
$\phi_{risc} = 0.7$
$\mathbf{T}_1$(N)
$\mathbf{S}_0$(N)

Thiel and the author [43] adopted a dual approach in investigating this reaction: qualitative theory (OCAMS) and computation (MNDOC)[9]. The computation places $\mathbf{S}_1$(NV) well above the first few excited singlets of naphthalene, so the failure to observe fluorescence from $\mathbf{S}_1$(N) cannot be waved away with the argument that it is energetically inaccesssible. The reason for the unusual behavior of NV on photoexcitation evidently has to be sought in the symmetry properties of the excited states involved.

When the analysis was first carried out, the nature of $\mathbf{S}_1$(NV) was uncertain, so Fig. 10.8 comprises two correspondence diagrams based on different assignments. It can be assumed that the excitations to $\mathbf{S}_1$ and $\mathbf{S}_2$ are essentially localized in the benzene ring, but – as Gleiter et al. [46] established by photoelectron spectroscopy – the splitting of the π orbitals by the bicyclobutane moiety is quite large. It is therefore a not unreasonable working hypothesis that $\mathbf{S}_1$ is the HOMO-LUMO state, labeled $^1B_1$ in $\mathbf{C}_{2v}^{x}$. On the other hand, a comparison of the orbitals of NV at the left of Fig 10.8 with those of benzene in Fig. 10.7 makes it clear that $^1B_1$ is derived from the $\mathbf{L}_a$ transition, and corresponds to $^1B_{1u}$, the second excited singlet of benzene. As shown at the right of Fig 10.8, the alternative transition $\mathbf{L}_b$ produces a singlet that is stabilized by interaction between the two totally symmetric open shell configurations: $[\phi_3(b_2)^1\phi_2^*(b_2)^1]$ and $[\phi_2(a_2)^1\phi_1^*(a_2)^1]$. Therefore, the possibility must be kept in mind that the perturbation of the π system may not be strong enough to invert the benzene order: $^1B_{2u}[(2^1A_1(\mathbf{C}_{2v}^{x})]$ below $^1B_{1u}[(^1B_1(\mathbf{C}_{2v}^{x})]$. The isomerization was therefore analysed separately on the basis of both assumptions, in the hope that consistency with the experimental results would point to the correct choice between them.

[9] MNDOC [44, 45], correlated version of MNDO, is particularly well suited for dealing with excited states of organic molecules.

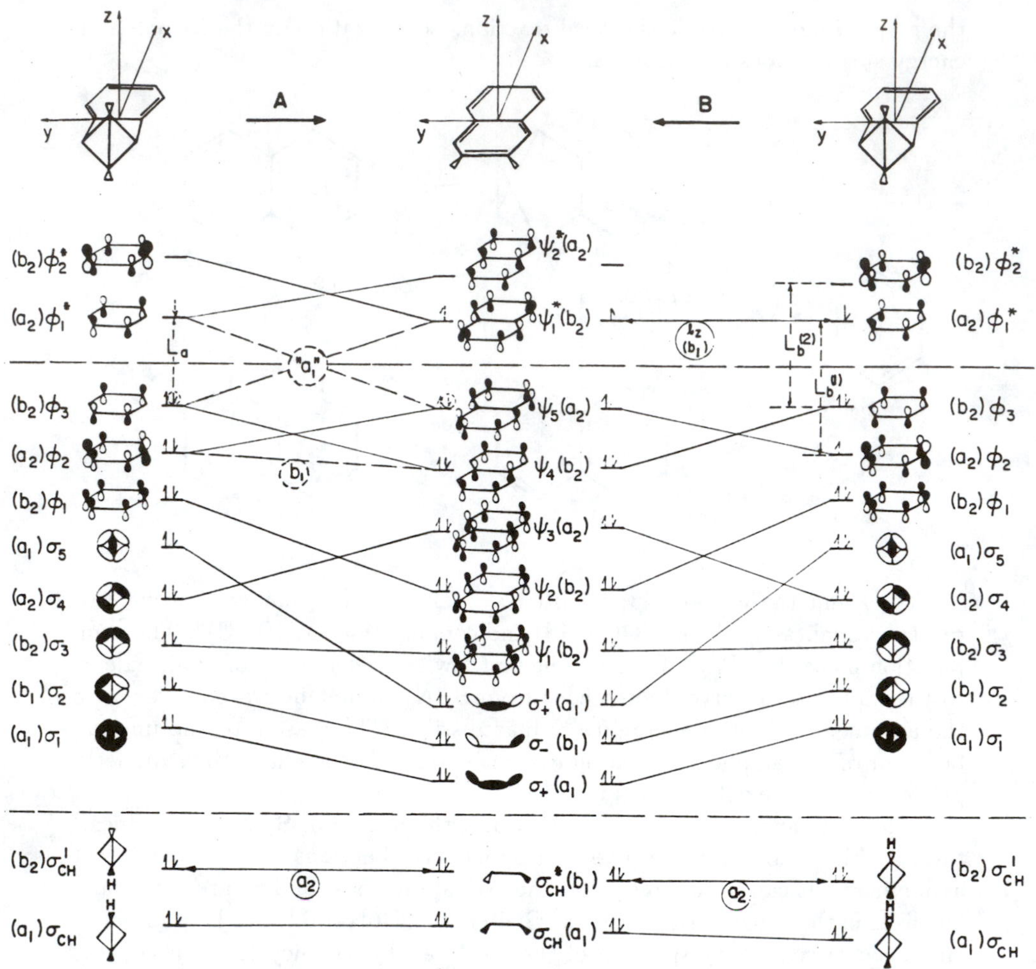

Figure 10.8 . Correspondence diagrams for photoisomerization of naphthvalene. ($\mathbf{C}_{2v}^{x}$): **(A)** $\mathbf{S}_1$(NV) is assumed to be the HOMO–LUMO state ($^1B_1$); **(B)** $\mathbf{S}_1$(NV) is assumed to be 2^1A_1 (From Fig. 2 of reference [43])

The orbital order is the same on both sides of Fig. 10.8, they differ only in the identity of the excitation associated with $\mathbf{S}_1$. All of the CC-bonding MOs that are doubly-occupied in the closed shell ground-state correlate across the diagram, but correlation of the CH-bonding orbitals calls for an a_2 twist. Isomerization on the ground-state surface is thus no more *forbidden* than that of benzvalene, but has a substantial activation energy for the same reason: the need to initiate rupture of two σ_{CC} bonds before any energetic gain can accrue from the formation of two new π bonds (Section 5.3.2).

Turning to the photochemical isomerization, the consequences of the alternative assumptions as to the nature of $\mathbf{S}_1$ are explored separately:

(**A**) $\mathbf{S}_1 = {}^1B_1$. Excitation from $\phi_3(b_2)$ to $\phi_1^*(a_2)$, derived from a component of $\mathbf{L}_a$ in benzene ($\psi_3 \rightarrow \psi_5$ in Fig. 10.9), leaves $\phi_2(a_2)$ of NV and $\psi_4(b_2)$ of N out of correlation with doubly occupied partners; they are therefore induced to correspond with one another under an in-plane, b_1 displacement, which – together with the necessary a_2 twist – desymmetrizes the reaction path all the way down to $\mathbf{C}_1$ and adds to the activation energy for isomerization. If enough energy is available for vibrational excitation, the electronically excited molecule can go over to $\mathbf{S}_1$(N), because of the HOMO↔LUMO cross-correlation. RISC to $\mathbf{T}_1$(N) is *forbidden*: it has the same space-symmetry as is assumed for $\mathbf{S}_1$(NV), so that – in the absence of spin-vibronic coupling – the triplet component formed would have to be totally symmetric, an impossibility in $\mathbf{C}_{2v}$. A reasonable ranking of the relaxation processes would be: Fluorescence first, formation of $\mathbf{S}_1$(N) a rather poor second, and crossing to $\mathbf{T}_1$(N) a particularly bad third.

(**B**) $\mathbf{S}_1 = 2^1A_1$. One of the two $\mathbf{L}_b$ components, excitation from $\phi_2(a_2)$ to $\phi_1^*(a_2)$, permits correlation between ψ_4(N) and a doubly-occupied b_2 orbital of NV; no distortion beyond the essential a_2 twist is called for by the doubly-occupied orbitals. The singly-occupied bonding MOs, both of them a_2, also correlate; the only correspondence remaining to be induced is between the two HOMOs: $\phi_1^*(a_2)$ and $\psi_1^*(b_2)$. As in (**A**) above, formation of $\mathbf{S}_1$(N) would require excitation of an in-plane distortion that destroys all symmetry and raises the activation energy gratuitously. Instead, a spin-flip of the electron being transferred between the two orbitals generates the $\mathbf{T}_z$ component of $\mathbf{T}_1$(N). The observed preponderance of reactive ISC to $\mathbf{T}_1$(N) over reactive IC to $\mathbf{S}_1$(N) is therefore reasonable. Fluorescence to $\mathbf{S}_0$(NV) is *allowed*. Purely qualitative considerations can hardly be expected to predict which of two *allowed* processes will predominate. As in the case of *trans*-dioxabimane (Section 10.1), the b_1 perturbation that accompanies transfer of an electron from an a_2 to a b_2 MO can be regarded as a momentary "ring-current", producing a transient magnetic moment perpendicular to the molecular plane that is particularly effective in flipping the spin of a π electron.[10] With this in mind, the sevenfold greater efficiency of reactive ISC over fluorescence, while it could not have been predicted, is not surprising.

Having concluded on qualitative grounds that $\mathbf{S}_1$ of naphthvalene can only be 2^1A_1, the assignment was confirmed by computation.[43] Before the calculations were complete, a paper appeared by Gleiter et al. [47] in which the same assignment was established spectroscopically and confirmed using another computational method (CNDO-CI).

10.5 Rydberg Photochemistry: Photolysis of Methane

An important family of chemical reactions occur after photoexcitation to molecular *Rydberg states* [48, 49, 50], a term borrowed from atomic spectroscopy. A concise definition is [4, p. 132]: "A Rydberg state is a state in which the spatial extent of the excited molecular orbital is large relative to the size of the molecular skeleton. Because of this the details of the molecular skeleton cease to be important and the energy of the excited orbital is given by an atomic-like term."

For the purposes of orbital symmetry analysis, let us note several features that distinguish Rydberg states from the *valence states* to which the discussion has been limited so far in this chapter. Rydberg orbitals are treated as if they were diffuse atomic orbitals of higher principal quantum number, in organic molecules usually $3s$ or $3p$, that are not necessarily centered on a particular atom. Excitation to them ordinarily requires irradiation with short wavelengths, so their energy content is high. Their disregard of the details of the molecular skeleton implies poor interaction with nuclear motion, so they tend to withstand vibrational relaxation to lower lying valence states. Therefore, when they lie above the antibonding valence orbitals, as is often though by no means invariably the case, Kasha's rule can be violated with relative impunity.

Rydberg states have well-defined symmetry properties. Excitation to an s-like orbital ($\mathcal{R}$) produces a state of the same symmetry as that of the MO from which the electron was excited – generally the HOMO. When the Rydberg orbital has the symmetry properties of a p AO, the state symmetry is the direct product of the irrep of the HOMO with that of x, y or z. Though coupling of the Rydberg orbital with antibonding valence orbitals is poor, it is not negligible and is no less symmetry-dependent than other molecular interactions. As a result, on *derydbergization* the highly excited molecule finds itself in a repulsive state and falls apart.

Photolysis of methane is chosen as the illustrative example, from which it can be seen that orbital symmetry conservation influences the mode of fragmentation even in the high energy regime characteristic of Rydberg photochemistry. Our discussion of this thoroughly investigated reaction follows Lee and Janoschek [51], who preceded their computations of various fragmentation pathways by qualitative symmetry analyses of state- and orbital symmetry conservation.

The two primary fragmentation modes of the excited methane molecule are:

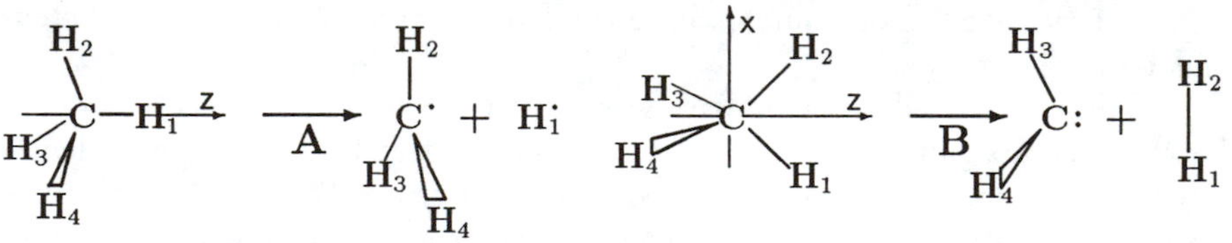

In (**A**), the least-motion pathway has $\mathbf{C}_{3v}$ symmetry, whereas (**B**) can retain no more than $\mathbf{C}_{2v}$. It can be confirmed in the Table of Kernels and Co-kernels (Appendix B) that both of these subgroups of $\mathbf{T}_d$ are co-kernels of T_2, differing

[10] Formally, the spin-orbit coupling operator for one electron is proportional to $l_x s_x + l_y s_y + l_z s_z$,[4, p. 188] each term of which couples the orbital- and spin-angular momenta about one of the cartesian axes. The implication is that when z is the axis normal to an aromatic ring, l_z can be thought of as a momentary ring-current, and coupling with the corresponding spin factor s_z will make production of the $\mathbf{T}_z$ component particularly efficient.

in the orientation of the rotational axis. In the former, the C_3 axis lies along one of the CH bonds; in the latter the C_2 axis bisects two opposing HCH angles. The first few excited states of methane are of Rydberg type, distorted from the tetrahedral symmetry of the ground-state to $\mathbf{C}_{2v}$ [52]. This in itself predisposes the molecule to decompose to methylene and H_2 rather than to methyl radical and a hydrogen atom. The preference for molecular fragmentation (**B**) is borne out both by experiment and computation, so we will restrict ourselves to it. A correlation diagram for mode (**A**) can be found in Fig. 1 of reference [51].

The four CH bonds of CH_4 combine to one totally symmetric MO and a triply-degenerate set of the irrep T_2. The corresponding CH-antibonding orbitals similarly combine to three of T_2 and – above them – one of A_1, but a Rydberg orbital ($\mathcal{R}$) that can be regarded loosely as a $3s(a_1)$ orbital centered on the C atom, lies just below the triply degenerate set. When the z axis is singled out as the direction in which the departing pair of H atoms is expected to move, the system is formally desymmetrized to $\mathbf{C}_{2v}^z$, in which the Rydberg orbital and the valence MOs aligned along z are labeled a_1 and the other two of each set split to b_1 and b_2. The symmetry properties of the MOs occupied in the ground- and first excited state of CH_4 are summarized:

$$\begin{aligned} \mathcal{R} = \text{“}3s\text{”} &\ :\ a_1(\mathbf{T}_d) \rightarrow a_1(\mathbf{C}_{2v}^z) \\ \sigma_x = (\sigma_1 - \sigma_2)/\sqrt{2} &\ :\ t_2(\mathbf{T}_d) \rightarrow b_1(\mathbf{C}_{2v}^z) \\ \sigma_y = (\sigma_3 - \sigma_4)/\sqrt{2} &\ :\ t_2(\mathbf{T}_d) \rightarrow b_2(\mathbf{C}_{2v}^z) \\ \sigma_z = (\sigma_1 + \sigma_2 - \sigma_3 - \sigma_4)/2 &\ :\ t_2(\mathbf{T}_d) \rightarrow a_1(\mathbf{C}_{2v}^z) \\ \sigma_+ = (\sigma_1 + \sigma_2 + \sigma_3 + \sigma_4)/2 &\ :\ a_1(\mathbf{T}_d) \rightarrow a_1(\mathbf{C}_{2v}^z) \end{aligned}$$

Excitation from any one of the t_2 orbitals leads to a ${}^1T_2$ Rydberg state, in which two CH bonds are elongated equally; we label the axis bisecting the angle between them z and recognize that the excited molecule has genuinely gone into $\mathbf{C}_{2v}^z$.

Labeling the MOs of the products by irrep is self-evident. $\sigma(H_1H_2)$ is totally symmetric and the irreps of the methylene orbitals can be read from Fig. 6.1A: σ_{CH}^+ is a_1 and σ_{CH}^- is b_2. These three MOs will be doubly-occupied in the ground- and lower excited product states, so the only one of the three isoenergetic components of the ${}^1T_2$ Rydberg state of methane can correlate with any of them is ${}^1B_1(\mathbf{C}_{2v}^z)$, in which three MOs with the same irreps: $\sigma_+(a_1)$, $\sigma_z(a_1)$ and $\sigma_y(b_2)$, are doubly-occupied, and excitation has occurred from the third t_2 orbital, $\sigma_x(b_1)$, to $3s(a_1)$. The ${}^1B_1$ component of the Rydberg state correlates directly with ground-state H_2 and CH_2 in its excited open shell ${}^1B_1$ state:
$\sigma_+(a_1)^2\sigma_z(a_1)^2\sigma_y(b_2)^2\sigma_x(b_1)^1\mathcal{R}(a_1)^1 \Leftrightarrow \sigma_{HH}(a_1)^2\sigma_{CH}^+(a_1)^2\sigma_{CH}^-(b_2)^2n(a_1)^1p_x(b_1)^1$
In order to cross to the lower-lying closed shell singlet, in which the lone-pair orbital, $n(a_1)$, is doubly occupied, a b_1 displacement is called for. Computations with a variety of methods [51, 52, 53] concur that this is precisely what occurs, the resultant ${}^1A_1$ methylene eventually undergoing intersystem crossing to its ground-state, ${}^3B_1$.

10.6 References

[1] The standard work on the subject is J.B. Birks: *Organic Molecular Photophysics.*, Wiley, London: *vol. 1*, 1973; *vol. 2*, 1975.

[2] J. March: *Advanced Organic Chemistry.* Third edition. Wiley, New York 1985.

[3] For a concise summary of the photochemical terms and principles employed in this chapter, see Chapter 1 of reference [4]. For fuller details see any standard photochemistry text, e.g. References [5], [6].

[4] S.P. McGlynn, T. Azumi and M. Kinoshita: *Molecular Spectrroscopy of the Triplet State.* Prentice-Hall, Englewood Cliffs 1969.

[5] D.O. Cowan and R.L. Drisko, *Elements of Organic Photochemistry.* Plenum, New York 1976.

[6] N.J. Turro, *Modern Molecular Photochemistry.* Benjamin/Cummings, Menlo Park 1978.

[7] M. Kasha: Disc. Faraday Soc. *9*, 14 (1950).

[8] D.C. Doetschman: J. Phys. Chem. *80*, 2167 (1976).

[9] E.A. Halevi and C. Trindle: Israel J. Chem. *16*, 283 (1977).

[10] E. Heilbronner and H. Bock: *The HMO-Model, vol. 3*, Verlag Chemie, Weinheim 1976.

[11] E.M. Kosower, H. Kanety and H. Dodiuk: J. Photochem. *21*, 171 (1983).

[12] J. Maxka and E.A. Halevi: *Unpublished results.*

[13] H.H. Wasserman and R.W. Murray: *Singlet Oxygen.* Academic Press, New York 1979.

[14] A.A. Frimer (ed.) *Singlet* O_2, *vols. 1-4.* CRC Press, Boca Raton 1985.

[15] See e.g. J. Moan: Photochem. Photobiol. *43*, 681 1986.

[16] N.J. Turro, V. Ramamurthy, K.-C.Liu, A. Krebs and R. Kemper: J. Amer. Chem. Soc. *98*, 6758 (1976).

[17] C. Trindle and E.A. Halevi: Int. J. Quantum Chem: Quantum Biol. Symp. *5*, 281 (1978).

[18] E.A. Halevi: Angew. Chem. *88*, 664 (1976); Angew. Chem. Int. Ed. (English) *15*, 593 (1976).

[19] O.L. Chapman, C.L. McIntosh and J. Pacansky: J. Amer. Chem. Soc. *95*, 793 (1973).

[20] R.D. McQuigg and J.G. Calvert: J. Amer. Chem. Soc. *91*, 1590 (1969).

[21] A. Horowitz and J.G. Calvert: Int. J. Chem. Kinetics *10*, 805 (1978).

[22] E.S. Yeung and C.B. Moore: J. Chem. Phys. *58*, 3988, (1973).

[23] V. Bachler and E.A. Halevi: Theoret. Chim Acta *59*, 595 (1981).

[24] R.G. Miller and E.K.C. Lee: J. Chem. Phys. *68*, 4448, (1978).

[25] R.L. Jaffe, D.M. Hayes and K. Morokuma: J. Chem. Phys. *60*, 5108, (1974).

[26] R.L. Jaffe and K. Morokuma: J. Chem. Phys. *64*, 4881, (1976).

[27] J.D. Goddard and H.F. Schaefer III: J. Chem. Phys. *70*, 5117, (1979).

[28] V.T. Jones and J.B. Coon: J. Mol Spectr. *31*, 137 (1969).

[29] R.J. Buenker and S.D. Peyerimhoff: J. Chem. Phys. *64*, 4881 (1976).

[30] D.M. Hayes and K. Morokuma: Chem. Phys. Lett. *12*, 539 (1972).

[31] E.A. Halevi: Internat. J. Quantum Chem. *12, Suppl. 1*, 289 (1977).

[32] W.H. Miller, J. Amer. Chem. Soc. *105*, 216 (1983).

[33] R. Schatzberger, E.A. Halevi and N. Moiseyev: J. Phys. Chem. *89*, 4691 (1985).

[34] V. Bachler and E.A. Halevi: Theoret. Chim Acta *63*, 83 (1983).

[35] V. Bachler, E.A. Halevi and O.E. Polansky: Theoret. Chim Acta *65*, 81 (1984).
[36] V. Bachler: J. Amer. Chem. Soc. *110*, 5972, 5977 (1988).
[37] D.Bryce-Smith and A. Gilbert: Tetrahedron *32*, 1309 (1976).
[38] S. Oikawa, M. Tsuda, Y. Okamura and T. Urabe: J. Amer. Chem. Soc. *106*, 6751 (1984).
[39] I. Haller: J. Chem. Phys. *47*, 1117 (1967).
[40] E.B. Wilson: J.C. Decius and P.C. Cross: *Molecular Vibrations.* McGraw-Hill, New York 1955.
[41] V. Bachler: *Results personally communicated.*
[42] N.J. Turro, P. Lechtken, A. Lyons, R.R. Hautala, E. Carnahan; and T.J. Katz J. Amer. Chem. Soc. *95*, 2035 (1973).
[43] E.A. Halevi and W. Thiel: J. Photochem. *28*, 373 (1985).
[44] W. Thiel: J. Amer. Chem. Soc. *103*, 1413,1420 (1981).
[45] W. Thiel and A. Schweig: J. Amer. Chem. Soc. *103*, 1425 (1981).
[46] R. Gleiter, K. Gubernator, M. Eckert-Maksić, J. Spanget-Larsen, B. Bianco, G. Gandillon and U. Berger: Helvet. Chim. Acta *64*, 1312 1981.
[47] J. Spanget-Larsen, K. Gubernator, R. Gleiter, E.W. Thulstrup, B. Bianco, G. Gandillon and U. Berger: Helvet. Chim. Acta *66*, 676 1983.
[48] E.M. Evleth and E. Kassab: Theoretical Analysis of the Role of Rydberg States in the Photochemistry of Some Small Molecules. In: R. Daudel, A. Pullmann, L. Salem and A. Veillard (eds.) *Quantum Theory of Chemical Reactions, vol. 2* Reidel, Paris 1980.
[49] E.M. Evleth, H.Z. Cao and E. Kassab: Stud. Phys. Theor. Chem. *35*, 497 (1985).
[50] E.M. Evleth and E. Kassab: Mol. Struct. Energ. *6*, 353 (1988).
[51] H.U. Lee and R. Janoschek: Chem. Phys. *39*, 271 (1979) and references therein.
[52] M. Gordon: Chem. Phys. Lett. *52*, 161 (1977).
[53] M. Gordon: Chem. Phys. Lett. *44*, 507 (1976).

Chapter 11

Into Inorganic Chemistry

Even minimally adequate coverage of the application of orbital symmetry criteria to inorganic reactions would increase the scope of this book inordinately and – in any case – is outside the author's competence. The modest attempt to address them in the final pages of this book does not merit *Part* status, so the chapter is included as a matter of necessity in Part IV: Spin and Photochemistry. A somewhat labored justification might run as follows: The principal new element that distinguishes inorganic from organic reactions is the ever-present possibility that d orbitals have to be taken into account. As we will see, the need to consider them when dealing with reactions of the main-group elements arises in connection with their photochemistry. They achieve crucial importance in the reactions of transition metal complexes, where they determine one of the essential properties of the reacting molecule or ion: its spin state.

11.1 Main-Group Elements

In the preceding chapters (Sections 5.2.1.1, 7.3.3.2, 8.1.1), silicon was treated analogously to carbon. The valence orbitals comprising its atomic basis set were taken to be: $3s$, $3p_x$, $3p_y$ and $3p_z$; its $3d$ orbitals were ignored. The role of d orbitals in bonding to silicon has been the subject of lively debate [1, p. 71]. While molecular dimensions, energies and physical properties are reproduced better when the basis set is augmented by the inclusion of d orbitals, this is also true of the analogous carbon compounds. In the latter case they are effective when their spatial extent is much more contracted than a normal $3d$ orbital would be; their function is to allow polarization of the electronic charge under the influence of the electrostatic field of the other nucleii in the molecule. Similarly, expanding the atomic basis set of silicon and other second-row elements by including *d-type polarization functions* [1, p. 64] does not imply that the occupied MOs are better described as combinations of d rather than of s and p orbitals, but merely that the MOs constructed from s and p orbitals are improved as a result of the added flexibility afforded by the larger basis set.

It might appear that d orbitals have to be included in the case of hypervalent compounds, such as SF_6, PCl_5 or $SiF_6^{=}$, because the $3s$, and $3p$ AOs can be used to form no more than four covalent bonds. Thus, the textbook description of SF_6 goes back to Pauling's classical paper [2] and postulates six equivalent SF bonds formed after prior sp^3d^2 hybridization of the sulphur AOs. Doubts cast by Rundle [3] on the need to expand the valence shell in this way have been repeatedly confirmed [4]. From their their modern computational study, Reed and Weinhold [5] were led to conclude that while the sulphur d orbitals contribute greatly to the binding energy, they are so thinly populated that the description in terms of sp^3d^2 hybridization should be discarded.

It should be recognized, however, that the validity or otherwise of the Pauling model of hypervalent molecules is irrelevant to the symmetry of their ground-state electron configurations. Consider the hypothetical formation of sulfur hexafluoride as six F^- ions, equidistantly situated in pairs along positive and negative x, y and z, approach a central S^{6+} ion, i.e. a sulfur atom stripped of its valence electrons. At long distances, the fluoride ions can be regarded as point charges that impose an octahedral electrostatic field on the central ion, in which its nine AOs – all vacant – are: $3s(a_1), 3\times 3p(t_{1u}), 5\times 3d(e_g(2) \oplus t_{2g}(3))$. The six combinations of F^- closed shells split as follows:

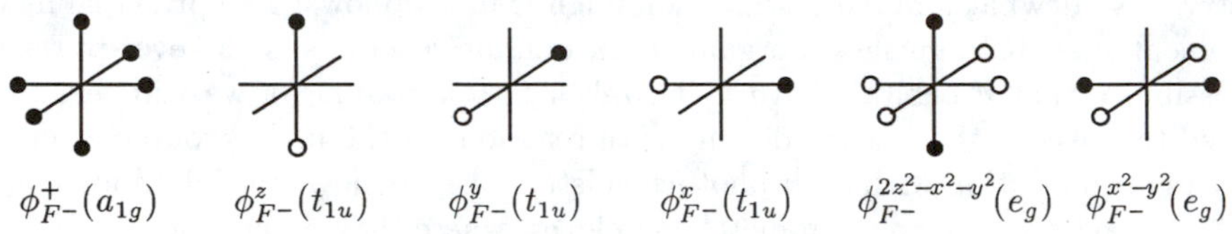

As the flouride ions approach, retaining $\mathbf{O}_h$ symmetry, two electrons[1] of each octet can be utilized for SF bonding. Initially, all twelve bonding electrons are localized in the six MOs of the F^- combinations; the configuration is $[a_1^2\, t_{1u}^6\, e_g^4]$. As the ions approach, each of these orbitals interacts favorably and unfavorably with a sulphur orbital of the same irrep, producing a bonding and antibonding combination; the six lowest of these are doubly-occupied in the closed shell ground-state.

In the limit of pure covalent bonding, the six occupied MOs utilize six equivalent sp^3d^2 hybrids, in which the two e_g orbitals participate fully. According to the currently prevalent view, covalent bonding is confined almost entirely to the s and p orbitals, leaving nearly three full electronic units of positive charge on the central S atom. The sulphur $3d_{2z^2-x^2-y^2}$ and $3d_{x^2-y^2}$ – like the other three $3d$ orbitals – remain virtually unoccupied, the residual electronic charge being distributed equally among the six F^- combinations. Along the entire approach, however, from the purely ionic extreme to the stable molecule with minimal d-orbital involvement, the electron configuration remains $[a_1^2\, t_{1u}^6\, e_g^4]$, as it would be in Pauling's purely covalent picture. Thus, while the energy and charge distribution depend strongly on the extent to which the d orbitals are involved in bonding, the symmetry of the electron configuration is indifferent to it.

[1] It may be helpful to draw an analogy with six hydride ions, where the entire two-electron closed shell of each can enter into an SiH bond.

11.1.1 Ground-State Isomerization: "Berry Pseudorotation"

Berry pseudorotation is the process whereby the two apical and three equatorial CX bonds of trigonal bipyramidal phosphorus pentahalides are scrambled [6, p. 1320]. The widely accepted mechanism proposed by Berry [7] for this ligand-reorganization reaction is analyzed very simply by OCAMS. The PX_5 molecule has $\mathbf{D}_{3h}$ symmetry; we align z along the threefold axis and x along the bond from the central P atom to one of the equatorial ligand atoms. The bond to this atom, numbered X_1, is chosen as the *pivot*, i.e. the one equatorial bond that remains in place as the other two (X_2 and X_3) become apical and the two apical bonds (X_4 and X_5) become equatorial. At the end of the process, the C_3 axis has "rotated" by 90° and now lies along y, whereas the bonds to X_1, X_4 and X_5 are in the new equatorial zx plane.

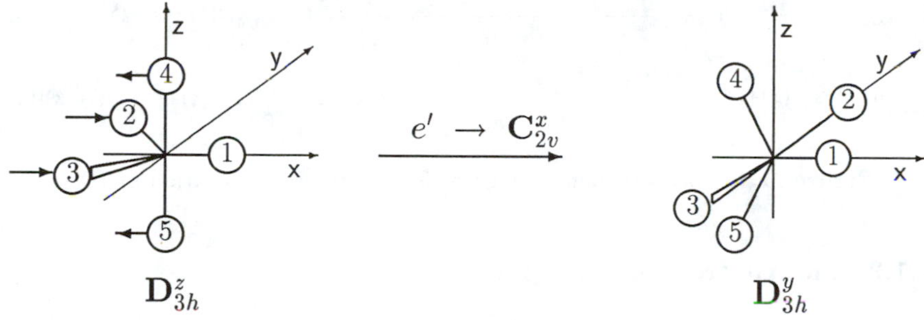

The hybridization traditionally assumed for the trigonal bipyramid is sp^3d, the configuration in the initial geometry being that of the constituent AOs: $3s(a')^2, 3p_z(a'')^2, (3p_x, 3p_y)(e')^4, 3d_{2z^2-x^2-y^2}(a')^2)$. It can be seen on the left side of Fig. 11.1 that the configuration based on the five combinations of halide-ion closed shells is identical. From the point of view of configurational symmetry, inclusion of a d orbital in the hybridization scheme in order to produce five equivalent covalent bonds is harmless but unnecessary.

The reaction begins in $\mathbf{D}_{3h}^z$ and ends in $\mathbf{D}_{3h}^y$, so the analysis is carried out in the largest subgroup common to both, $\mathbf{C}_{2v}^x$, which can be recognized as being the co-kernel of E'. $\mathbf{D}_{3h}^y$, the symmetry point group of the "pseudorotated" molecule, differs from $\mathbf{D}_{3h}^z$ merely by an interchange of z and y. All of the doubly occupied orbitals correlate in $\mathbf{C}_{2v}^x$, so there is no symmetry-imposed barrier to the Berry mechanism for pseudorotation and no need to invoke more elaborate alternative pathways, that may be conceptually distinct but operationally indistinguishable, such as "turnstile rotation" [6, Fig 29.11].

This conclusion is in accord with detailed *ab initio* calculations [8], according to which $\mathbf{C}_{2v}$ symmetry is retained along the pseudorotation pathway of PF_5. The calculated transition state, in which the lengths of the bonds to F_2, F_3, F_4 and F_5 are equal and the symmetry has risen to $\mathbf{C}_{4v}$, lies ≈ 4 kcal/mol above the ground state, in good agreement with experiment.

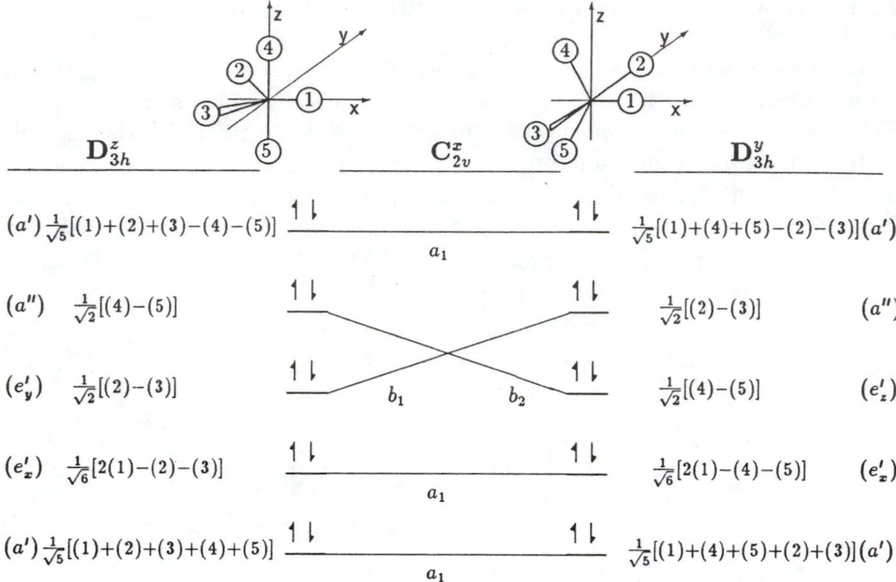

Figure 11.1. Correspondence diagram for Berry pseudorotation of PX_5

11.1.2 The Allotropy of Phosphorus

Elemental phosphorus exists in a wide variety of allotropic forms, most of them polymeric [9, Chap. 2]. Its monomeric forms, in addition to atomic P, are diatomic P_2 and the more stable tetrahedral P_4, whereas P_6 and P_8 species are conspicuously absent. Bock and Müller [10] investigated the $P_4 \rightleftharpoons 2\,P_2$ equilibrium by photoelectron spectroscopy, and detected no other species up to 1470°K. The experimental enthalpy of reaction ($\Delta H^o_{\text{exp}} = 55$ kcal/mol) was reproduced reasonably well by the value computed with MNDO ($\Delta H^o_{\text{calc}} = 43$ kcal/mol).

Consequent to this gratifying agreement between theory and experiment, the same computational method was applied to the equilibrium between P_4 and the unknown P_8. The latter was found to have a stable cubic structure and – surprisingly, in view of its non-existence – a calculated enthalpy of formation 68 kcal/mol lower than that of two P_4 molecules.[2] If the MNDO computation was to be believed, dimerization of P_4 to P_8 is strongly exothermic, so the fact that the latter has never been observed could only be ascribed to an insuperable potential barrier along the reaction path, resulting from incompatibility between the configurational symmetry of the approaching pair of P_4 tetrahedra and that of cubic P_8 [11].

[2] The computation was repeated very recently [12] using AM1 with limited CI; the results are virtually identical.

Of the two computations that had already been published, both using pseudopotential methods[3], one [13] found $\Delta H^o_{\text{exp}} = -47$ kcal/mol for the dissociation: $P_8 \rightleftharpoons 2\,P_4$, whereas the other [14] obtained $\Delta H^o_{\text{exp}} = +10$ kcal/mol, in qualitative, but hardly quantitative, agreement with the results of the MNDO computation. Subsequent more sophisticated calculations using a variety of methods [15, 16, 17, 18] reverse the energetic order: two P_4 molecules are more stable than P_8 by some 30 kcal/mol. The energy difference and even the relative energetic order depend crucially on the basis set used: "3d contributions definitely tip the balance in favor of P_4" [17].

Although the most reliable calculations agree that the dimerization of P_4 to P_8 is endothermic, the computed energy difference is not large enough to explain the complete non-observability of P_8. It seems that three necessary conditions must be fulfilled:

1. There is a symmetry-imposed barrier to the dimerization that is not easily circumvented.
2. No such barrier inhibits polymerization to amorphous red phosporus, a process that takes place readily in the liquid phase [6, p. 86].
3. Under conditions, if any, in which P_8 can be formed, it is kinetically unstable with respect to some form of phosphorus other than P_4.

Condition 1 has been demonstrated [11] as will be shown below; there are preliminary indications [12] that condition 2 holds as well. The validity of condition 3 has yet to be explored.

The least motion pathway for dimerization of P_4 is depicted in Fig. 11.2. The two tetrahedral molecules are set up face-to-face, rotated relative to each other by 60°. The z axis is passed through atoms 1 and 8, atoms 4 and 7 are in the yz

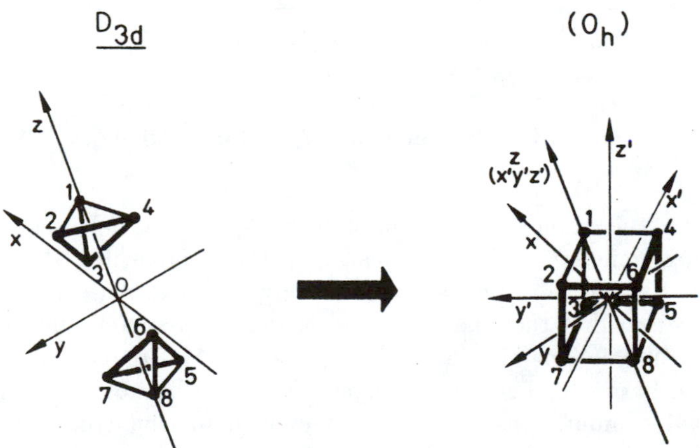

Figure 11.2. $2 \times P_4(\mathbf{D}_{3d}) \rightarrow P_8(\mathbf{O}_h)$. The six bonds broken and six bonds made in the reaction are emphasized. (From Fig. 1 of reference [11])

[3] Only the valence electrons are treated explicitly, the effect of the core electrons being simulated by an empirical function.

plane, which bisects bonds 23 and 56. When the two tetrahedral P_4 molecules are pushed together to form cubic P_8, three bonds of each are ruptured and six new bonds are formed in a cyclic array with the conformation of *chair*-cyclohexane. The z axis now lies along one diagonal of the cube, whereas the y axis bisects two of the new bonds (27 and 45) and the x axis bisects two of the six bonds that remain intact (12 and 58).

The symmetry analysis is carried out in Table 1.1. Proceeding as in Section 4.4.3, we take the six PP bonds broken in the reaction as the basis vector for the approaching pair of P_4 molecules and construct its MOs as *linear combinations of bond orbitals (LCBOs)* in $\mathbf{D}_{3d}$. We do the same for the newly formed bonds of the P_4 molecule, after conceptually stretching it slightly along the appropriate diagonal to desymmetrize it from $\mathbf{O}_h$ to $\mathbf{D}_{3d}$. The detailed form of the LCBOs, which can be found in the original publication [11], is not necessary for the purposes of the analysis; the two direct sums suffice. There is a single orbital mismatch: between an a_{2u} and an a_{1u} MO, that can be induced to correspond provided that a suitable a_{2g} displacement can be found. When the representation of the cartesian coordinates of the eight phosphorus atoms is reduced to the irreps of $\mathbf{D}_{3d}$, the only symmetry coordinate that transforms as a_{2g} is found to be R_z, which does not affect the potential energy.

Table 11.1. Symmetry analysis of 2 $P_4 \rightarrow P_8$ in $\mathbf{D}_{3d}$

$\mathbf{D}_{3d}$	E	$2C_3$	$3C_2$	i	$2S_6$	$3\sigma_d$	Direct sum
2 P_4	6	0	0	0	0	2	$a_{1g} \oplus e_g(2) \oplus a_{2u} \oplus e_u(2)$
P_8	6	0	2	0	0	0	$a_{1g} \oplus e_g(2) \oplus a_{1u} \oplus e_u(2)$
Symmetry coordinates	24	0	0	0	0	4	$3\times a_{1g} \oplus a_{2g} \oplus 4\times e_g(2)$ $\oplus a_{1u} \oplus 3\times a_{2u} \oplus 4\times e_u(2)$

of which: $(T_x, T_y) \in E_u$, $T_z \in A_{2u}$, $(R_x, R_y) \in E_g$, ; $R_z \in A_{2g}$
Formally, correspondence can be induced by: $A_{2u} \otimes A_{1u} = A_{2g}$
but
No symmetry coordinate other than R_z belongs to A_{2g}.

Evidently, there is no displacement of the phosphorus nuclei – short of nearly complete desymmetrization[4] – that can help the approaching P_4 tetrahedra circumvent the symmetry-imposed barrier between them and the P_8 cube. It must be stressed that the qualitative conclusion is completely independent of the relative thermodynamic stablity of P_4 and P_8. MNDO computations [11] estimate the barrier to be ≈ 100 kcal/mol high; it is doubtful whether more sophisticated computations will reduce it enough to lend credence to direct dimerization to cubic P_8 as a viable pathway.

[4] Desymmetrization to $\mathbf{S}_6$ by a superposition of a_{1u} and a_{2u} displacements or – more drastically – all the way down to $\mathbf{C}_i$, would formally "allow" the dimerization. The necessary distortions are opposed by large restoring forces, and would not be expected to reduce the activation energy significantly.

This is not to say that P_4 does not dimerize. Edge-to-edge dimerization, like paradigmatic [2+2]-cycloaddition (Section 6.2), is easily shown to be *allowed* under an in-plane displacement, naturally leading to a 1,4-biradical, or perhaps a zwitterion – as a result of *sudden polarization* [19].

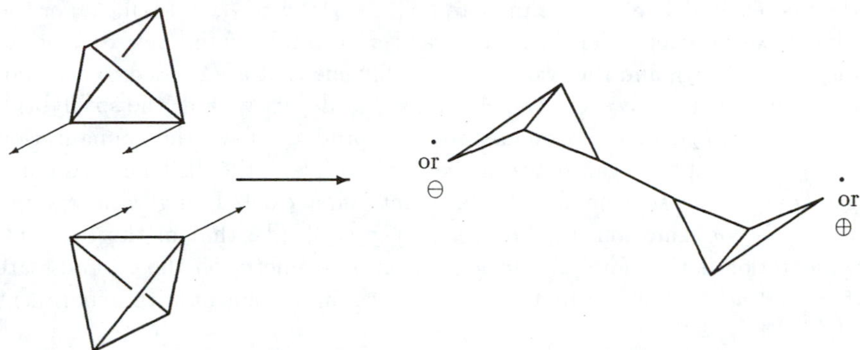

Exploratory computations [12] indicate that both the triplet biradical and the singlet zwitterion are energetically accessible. Either of these two species could serve as the intermediate for polymerization to red phosphorus, which comprises chains of just such open tetrahedra [9, Chap. 2].

11.1.3 An Excited-State Reaction: Photoextrusion of Silylene

An interestng attempt was made by Ramsey [20] to determine the nature of the reactive excited state of a cyclic trisilane by means of an orbital symmetry analysis of its photofragmentation. The question addressed was whether the relevant excitation might be to a low-lying $3d$ orbital rather than to an antibonding valence orbital, as is usually assumed.

The photolysis is known to occur readily [21] and can therefore be presumed to be *allowed* by orbital symmetry conservation.

Ramsey's argument can be recast as follows. Electrons localized on silicon atoms or involved in SiSi bonding are held more loosely than others in the molecule, so the only orbitals of the trisilane that need be considered in the analysis are the bonding and antibonding combinations of σ_{SiSi} bonds:

$$\begin{aligned} \sigma_+(a_1) &= (\sigma_{12} + \sigma_{23})/\sqrt{2} \\ \sigma_-(b_2) &= (\sigma_{12} - \sigma_{23})/\sqrt{2} \end{aligned}$$

$$\sigma_+^*(a_1) = (\sigma_{12}^* + \sigma_{23}^*)/\sqrt{2}$$
$$\sigma_-^*(b_2) = (\sigma_{12}^* - \sigma_{23}^*)/\sqrt{2}$$

The configuration of the ground-state is therefore $[\sigma_+(a_1)^2, \sigma_-(b_2)^2]$ and that of the first excited singlet is assumed to be $[\sigma_+(a_1)^2, \sigma_-(b_2), \sigma_+^*(a_1)]$. As for the products, we need only consider the σ and σ^* orbitals of the new SiSi bond, $\sigma_{13}(a_1)$ and $\sigma_{13}^*(b_2)$, and the two orbitals of silylene that are needed to accomodate – pairwise or singly – the two electrons that depart with it: the sp^2 hybrid, $h_z(a_1)$, and $p_y(b_2)$. In the ground-state, the products have the configuration $[\sigma_{13}(a_1)^2, h_z(a_1)^2]$. The gap between the two orbitals of the disilane is so much larger than in the silylene that the products' first excited singlet necessarily has the $^1B_2$ configuration: $[\sigma_{13}(a_1)^2, h_z(a_1), p_y(b_2)]$, like that of the reactant. Fragmentation is therefore *forbidden* by orbital symmetry on the ground-state surface and *allowed* on the first excited surface, in agreement with experiment [21].

Ramsey went on to consider the possibility that empty *d* orbitals of silicon are sufficiently low in energy for one of their combinations to be singly occupied in the first excited state of the trisilane. The lowest such combination is the mutually bonding $\pi_d(a_2)$ orbital[5]: $\frac{1}{\sqrt{3}}\{d_{xy}(1) + d_{xy}(2) + d_{xy}(3)\}$. If this orbital were indeed occupied in the first excited singlet, the latter would be $^1B_1$ and fail to correlate with the products on the lowest excited singlet surface. Ramsay's conclusion: the lowest combinations of silicon *d* orbitals, even in a situation as favorable as this, lie sufficiently far above the antibonding valence orbitals that they can be ignored in photochemical and – a fortiori – in thermal reactions.

A careful computational study by Janoschek and his coworkers [22] discloses flaws in Ramsey's argument, but confirms his conclusions. The excited singlets of a model trisilane with $\mathbf{C}_{2v}$ geometry, $(H_3Si)_2SiR_2$ (R = H), were computed with increasingly flexible basis sets. At the lowest level of computation, in which only the minimal basis set of valence AOs were included, a SiSi-bonding but SiH-antibonding $\pi(b_1)$ orbital appears unexpectedly below the antibonding $\sigma^*(a_1)$ orbital. The anomaly disappears at all higher levels of computation; the upper singly-occupied orbital invariably has a_1 symmetry, and the lowest excited state is $^1B_2$. It is stabilized by the incorporation of diffuse $4s$ and $4p$ orbitals in the basis set, endowing it with considerable Rydberg character, whereas the inclusion of $3d$ orbitals hardly affects it at all. The bonding $\pi_d(a_2)$ orbital is indeed the lowest *d*-orbital combination, but it lies very high above the Rydberg-valence orbitals that are singly-occupied in the lower excited states.

The $^1B_2 \Leftrightarrow {}^1B_2$ state correspondence postulated by Ramsey was confirmed with the more realistic model molecule, $(H_3Si)_2SiR_2$ ($R = CH_3$). Here too the importance of diffuse $4s$ and $4p$ orbitals for stabilizing the lower excited states and establishing their energetic order is manifest, as is the unimportance of silicon $3d$ orbitals.

[5] It can be pictured as three positively overlapping *d* orbitals, one on either side of the *d* orbital in Fig. 1.3.

11.2 Transition Metals: Isomerization of $NiX_4^=$

It is in transition metal chemistry that d orbitals acquire their overriding importance; as is well known, their symmetry properties determine the structure of high-symmetry complex molecules and ions [6, Chap. 17]. The author is convinced that these properties can be used to advantage in the analysis of their reaction mechanisms as well. A single example will be cited, illustrating the interplay between orbital and spin symmetry that has to be taken into account.

Divalent nickel ($3d^8$) forms tetrahedral and square-planar complexes of comparable energy, their relative thermodynamic stability depending on the identity of the ligands [6, p. 751]. The discussion of their interconversion that follows is abstracted from the symmetry analysis published by Knorr and the author [23].

Unless retarded by repulsion between bulky substituents in the more crowded planar isomer, the tetrahedral-to-planar isomerization has a low enthalpy of activation, $\Delta H^\ddagger \approx 10 \pm 4$ kcal/mol [24]. Its entropy of activation is ordinarily quite negative for a unimolecular isomerization ($\Delta S^\ddagger \leq -10$ kcal/mol°K) but is substantially less so – occasionally approaching zero – in nickel(II) complexes with halogen atoms as coordinating ligands [25]. This pattern of Arrhenius parameters, characteristic of reactions that occur with spin inversion (see Chapter 9), is hardly surprising in view of the fact that the tetrahedral complex is high-spin ($S = 1$) whereas the square-planar complex is low-spin ($S = 0$).

Starting in the tetrahedral geometry ($\mathbf{T}_d$), isomerization of the complex ion $NiX_4^=$ can proceed intermolecularly along either of two pathways, each incorporating one of the degenerate displacements of irrep E illustrated in Fig. 11.3. The $\mathbf{S}_a$ mode is a compression of the molecule parallel to the z axis that pushes the NiX bonds into the xy plane. Along this pathway, the symmetry is lowered from $\mathbf{T}_d$ to $\mathbf{D}_{2d}^z$, the co-kernel of E, and then rises to $\mathbf{D}_{4h}^z$ as the ion becomes planar. The alternative $\mathbf{S}_b$ mode is a twist about the z axis that reduces the symmetry directly to $\mathbf{D}_2$, the kernel of E; when square-planar geometry is eventually attained, the NiX bonds find themseves in either the zx or yz plane, depending on the sense of the twisting motion. Both of these pathways are spin-non-conservative, so neither can be entertained seriously unless it can be shown

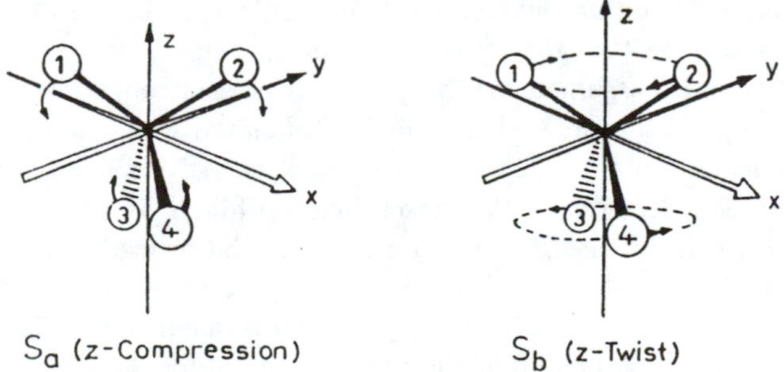

Figure 11.3. The degenerate (e) displacements in Tetrahedral $NiX_4^=$ ($\mathbf{T}_d$). (From Fig. 1 of reference [23])

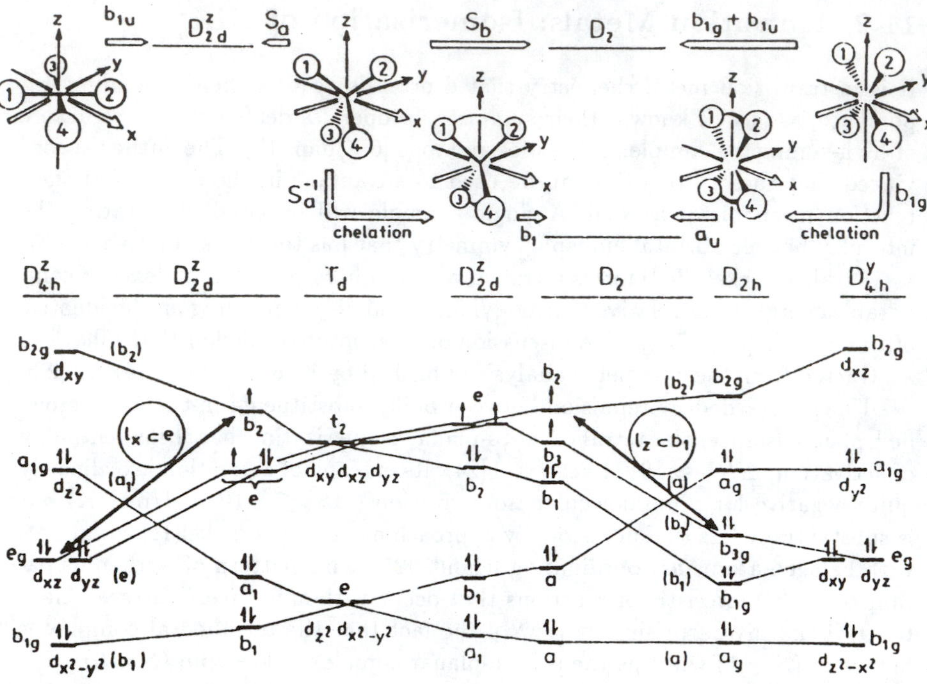

Figure 11.4. Correspondence diagram for tetrahedral → planar isomerization of $NiX_4^=$. *Left side:* Along the compresion coordinate ($\mathbf{S}_a$); *right side:* Along the twist coordinate($\mathbf{S}_b$), with or without chelation. (From Fig. 2 of reference [23])

that the "spin-forbiddenness" is mitigated by an appropriate compensatory interchange of spin- and orbital-angular momenta.

This is done in Fig. 11.4, that comprises two correspondence diagrams: on the left for path $\mathbf{S}_a$ in $\mathbf{D}_{2d}^z$, and on the right for $\mathbf{S}_b$ in $\mathbf{D}_2$. The tetrahedral complex in the center is flanked by its square-planar isomer in the two planes appropriate to the different pathways. Only the *d* orbitals are included in the diagram, so we should convince ourselves that the neglect of all the other orbitals is justified. The Ni^{++} ion has sixteen electrons in its valence shell and the four ligand X^- ions can be thought of as contributing eight[6] more, adding up to twenty-four. It can be assumed for the purposes of electron bookkeeping that the electrons supplied by the X^- ions remain associated with them, as they are at large internuclear distances. The valence basis set of Ni is $[3s,\ 3\times3p,\ 5\times3d]$. Eight of the sixteen electrons in the nickel dication fill its $3s$ and $3p$ orbitals in pairs, leaving eight to be distributed among the five $3d$ orbitals. Six are obliged to double up in three of them, and the remaining two have the choice of either being paired in one of the two still unoccupied $3d$ orbitals if they differ in energy, or occupying both of them singly if they are degenerate.

[6] See footnote 2.

In the square-planar complex, the in-plane orbital with its lobes directed towards the negatively charged ligands (d_{xy} on the left and d_{xz} on the right) is higher in energy than any of the others, so it remains unoccupied and the complex has a closed shell singlet ($S = 0$) ground-state. In the tetrahedral complex the representation of the four ligand ions can be reduced easily (see Section 4.4.3) to the direct sum $a_1 \oplus t_2(3)$. The t_2 combinations interact repulsively – and equally – with the $3d$ orbitals that have the same irrep. Four of the eight d electrons occupy the lower energy set of e orbitals and the remaining four are distributed as equally as possible among the components of the destabilized triply degenerate set: Two are paired in one orbital and the remaining two, in accordance with Hund's rule, occupy the other two orbitals with parallel spin, producing a complex with a triplet ($S = 1$) ground-state.

Consider first the $\mathbf{S}_a$ pathway: The analysis is carried out in $\mathbf{D}_{2d}^z$, the highest subgroup common to $\mathbf{T}_d$ and $\mathbf{D}_{4h}^z$, in which the degeneracy of the t_2 orbitals is split to $e(2)$ and b_2. The latter is the $3d_{xy}$ orbital, that goes up in energy as the NiX bonds move into coplanarity, and is therefore the upper singly-occupied MO. The lower set of e orbitals, $3d_{x^2-y^2}$ and $3d_{z^2}$ split to b_1 and a_1 respectively, and correlate directly with the corresponding orbitals of the planar isomer. If the lower singly-occupied orbital is $3d_{xz}$ – as shown in the diagram, its electron can cross to a similarly labeled orbital, as do both electrons of its doubly-occupied partner, $3d_{yz}$. The only correspondence that has to be induced is that of a single electron between $3d_{xy}$ and $3d_{xz}$; it calls for a perturbation that transforms like e, the irrep of yz. This is also the irrep of R_x, so the $\mathbf{T}_x$ component of the tetrahedral triplet undergoes reactive intersystem crossing to the square-planar singlet in its closed shell ground-state. Alternatively, if the lower singly-occupied orbital is $3d_{yz}$, a correspondence is induced between $3d_{xy}$ and $3d_{yz}$, and the $l_y s_y$ component of the spin-orbit coupling operator converts the T_y component to the singlet. In either case, the *spin-forbidden* isomerization becomes *allowed*, thanks to the interconversion of spin- and orbital-angular momenta.

Let us now suppose that the complex is chelated: ligand **1** is linked to **2** and **3** to **4** by a short chain of atoms. Chelation desymmetrizes the tetrahedral complex substitutionally to $\mathbf{D}_{2d}$ but blocks the $\mathbf{S}_a$ mode, leaving the twist mode ($\mathbf{S}_b$) as the only isomerization pathway open to it. At the same time, it reduces the symmetry of the square-planar complex to $\mathbf{D}_{2h}$. The highest common subgroup is now $\mathbf{D}_2$, which is the kernel of (E) in $\mathbf{T}_d$ and the point group of highest symmetry that can be maintained along the $\mathbf{S}_b$ pathway for either the chelated or unchelated complex. In this subgroup, the orbitals that are singly-occupied in the ground-state are $3d_{yz}(b_3)$ and $3d_{xz}(b_2)$, so the state label of the tetrahedral complex is $^3B_1$, and T_z is the component that crosses to the planar singlet.

Finally, let us assume that the ligands are distinguishable; say that **1** and **2** are Cl whereas **3** and **4** are Br. There is only one isomer of the tetrahedral complex but the square-planar complex has two: *cis* and *trans*. Both $\mathbf{S}_a$ and $\mathbf{S}_b$ thus offer geometrically convenient intramolecular pathways for *cis*↔*trans* isomerization, but imply the occurrence of two successive spin-flips along the pathway, as the low-spin complex is converted to the high-spin tetrahedral complex and back again to the other planar isomer. Since both pathways permit interconversion of spin- and orbital-angular momenta, the mechanism would appear to be a viable one.

11.3 Afterword

The few reactions discussed in this mini-chapter hardly scratch the surface of inorganic chemistry. The illustrative symmetry analyses, drawn for the most part from work in which the author was directly involved, were presented in the hope that a few bona fide inorganic chemists may be persuaded that the ideas developed in the preceding chapters are not limited to the organic domain. In having thus ventured gingerly into an area that is not his own, the author draws sustenance from the precept enunciated by an ancient sage [26]: "The task is not yours to complete, but neither are you at liberty to shirk it."

11.4 References

[1] Y. Apeloig: "Theoretical aspects of organosilicon compounds" in S. Patai and Z. Rappoport (eds.) *The Chemistry of Organic Silicon Compounds.* Wiley, New York 1989.
[2] L. Pauling: J Amer. Chem. Soc. *53*, 1369 (1931).
[3] R.E. Rundle: J Amer. Chem. Soc. *85*, 112 (1963).
[4] J.Musher: Angew. Chem. *81*, 68 (1969); Angew. Chem. Internat. Ed. (English) *8*, 54 (1969).
[5] A.E. Reed and F. Weinhold: J Amer. Chem. Soc. *108* , 3586 (1986).
[6] F.A. Cotton and G. Wilkinson: *Advanced Inorganic Chemistry, Fifth edition.* Wiley, New York 1988.
[7] R.S. Berry: J. Chem Phys. *32*, 633 (1960).
[8] C.J. Marsden: J. Chem. Soc. Chem. Commun. *1984*, 401, and references therein.
[9] D.E.C. Corbridge: *The Structural Chemistry of Phosphorus.* Elsevier, Amsterdam 1974.
[10] H. Bock and H. Müller: Inorg. Chem. *23*, 437? (1984).
[11] E.A. Halevi, H. Bock and B. Roth: Inorg. Chem. *23*, 4376 (1984).
[12] E.A. Halevi: *unpublished computations.*
[13] E. Fluck, C.M.E. Pavlidou and R. Janoschek: Phosphorus and Sulfur *6*, 469 (1979).
[14] G. Trinquer, J.-P. Malrieu, J.-P. Daudey: Chem. Phys. Lett. *80*, 522 (1981).
[15] G. Trinquer, J.-P. Daudey and N. Komiha: J. Amer. Chem. Soc. *107*, 7210 (1985).
[16] M.W. Schmidt and M.S. Gordon: Inorg. Chem. *24*, 4503 (1985).
[17] L.A. Ahlrichs, S. Brode and C. Ehrhardt: J. Amer. Chem. Soc. *107*, 7260 (1985).
[18] K. Raghavachari, R.C. Haddon and J.S. Binkley: Chem. Phys. Lett. *122*, 219 (1985).
[19] V. Bonačić-Koutecký, P. Bruckmann, P. Hiberty, J. Koutecký, C. Leforestier and L. Salem: Angew. Chem. *87*, 599 (1975); Angew. Chem. Internat. Ed. (English) *14*, 575 (1975).

[20] B.G. Ramsey: J. Organomet. Chem. *67*, C67 (1974).
[21] H. Sakurai, Y. Kobayashi and N. Nakadaira: J. Amer. Chem. Soc. *93*, 5252 (1971).
[22] E.A. Halevi, G. Winkelhofer, M. Meisel and R. Janoschek: J. Organomet. Chem. *294*, 151 (1985).
[23] E.A. Halevi and R. Knorr: Angew. Chem. Suppl. *1982*, 622.
[24] R. Knorr, A. Weiss, H. Polzer and A. Räpple: J. Amer. Chem. Soc. *99*, 550 (1977) and references therein.
[25] R. Knorr and F. Ruf: J. Amer. Chem. Soc. *101*, 5424 (1979).
[26] Attributed to Rabbi Tarfon (ca. 100 C. E.): *Pirke Avot (Ethics of the Fathers)* 2,17.

Appendices

Appendix A

Character Tables* of the More Common Symmetry Point Groups

1. Cyclic Groups

C_1	E			
A	1	x, y, z	R_x, R_y, R_z	$x^2, y^2, z^2, xy, yz, zx$

C_2	E	C_2			
A	1	1	z	R_z	x^2, y^2, z^2, xy
B	1	−1	x, y	R_x, R_y	yz, zx

C_3	E	C_3	C_3^2	$\omega = \exp(2\pi i/3)$		
A	1	1	1	z	R_z	$x^2 + y^2, z^2$
E {	1	ω	ω^2	} x, y	} R_x, R_y	} $x^2 - y^2, xy,$
	1	ω^*	ω^{*2}			yz, zx

C_4	E	C_4	C_2	C_4^3			
A	1	1	1	1	z	R_z	$x^2 + y^2, z^2$
B	1	−1	1	−1			$x^2 - y^2, xy$
E {	1	i	−1	$-i$	} x, y	} R_x, R_y	} yz, zx
	1	$-i$	−1	i			

* Adapted from I. Gutman and O.E. Polansky: Mathematical Concepts in Organic Chemistry. Springer, Berlin Heidelberg New York London Paris Tokyo 1986. Appendix 5, pp. 184–199

$\mathbf{C}_5$	E	C_5	C_5^2	C_5^3	C_5^4	$\omega = \exp(2\pi i/5)$		
A	1	1	1	1	1	z	R_z	$x^2 + y^2, z^2$
E_1	1	ω	ω^2	ω^3	ω^4	x, y	R_x, R_y	yz, zx
	1	ω^*	ω^{*2}	ω^{*3}	ω^{*4}			
E_2	1	ω^2	ω^4	ω	ω^3			$x^2 - y^2, xy$
	1	ω^{*2}	ω^{*4}	ω^*	ω^{*3}			

$\mathbf{C}_6$	E	C_6	C_3	C_2	C_3^2	C_6^5	$\omega = \exp(2\pi i/6)$		
A	1	1	1	1	1	1	z	R_z	$x^2 + y^2, z^2$
B	1	-1	1	-1	1	-1			
E_1	1	ω	ω^2	-1	$-\omega$	$-\omega^2$	x, y	R_x, R_y	yz, zx
	1	ω^*	ω^{*2}	-1	$-\omega^*$	$-\omega^{*2}$			
E_2	1	ω^2	ω^4	1	ω^2	ω^4			$x^2 - y^2, xy$
	1	ω^{*2}	ω^{*4}	1	ω^{*2}	ω^{*4}			

2. C_{nv} Groups

$\mathbf{C}_{2v}$	E	C_2	σ_v zx	σ_d yz			
A_1	1	1	1	1	z		x^2, y^2, z^2
A_2	1	1	-1	-1		R_z	xy
B_1	1	-1	1	-1	x	R_y	zx
B_2	1	-1	-1	1	y	R_x	yz

$\mathbf{C}_{3v}$	E	$2C_3$	$3\sigma_v$			
A_1	1	1	1	z		$x^2 + y^2, z^2$
A_2	1	1	-1		R_z	
E	2	-1	0	x, y	R_x, R_y	$x^2 - y^2, xy$; yz, zx

$\mathbf{C}_{4v}$	E	$2C_4$	C_2	$2\sigma_v$	$2\sigma_d$			
A_1	1	1	1	1	1	z		x^2+y^2, z^2
A_2	1	1	1	−1	−1		R_z	
B_1	1	−1	1	1	−1			x^2-y^2
B_2	1	−1	1	−1	1			xy
E	2	0	−2	0	0	x, y	R_x, R_y	yz, zx

$\mathbf{C}_{5v}$	E	$2C_5$	$2C_5^2$	$5\sigma_v$	$\varphi = 2\pi/5$		
A_1	1	1	1	1	z		x^2+y^2, z^2
A_2	1	1	1	−1		R_z	
E_1	2	a	b	0	x, y	R_x, R_y	yz, zx
E_2	2	b	a	0			x^2-y^2, xy

$a = 2\cos\varphi = 2\cos 4\varphi = (\sqrt{5}-1)/2 = 0.618034,$
$b = 2\cos 2\varphi = -(\sqrt{5}+1)/2 = -1.618034.$

$\mathbf{C}_{6v}$	E	$2C_6$	$2C_3$	C_2	$3\sigma_v$	$3\sigma_d$			
A_1	1	1	1	1	1	1	z		x^2+y^2, z^2
A_2	1	1	1	1	−1	−1		R_z	
B_1	1	−1	1	−1	1	−1			
B_2	1	−1	1	−1	−1	1			
E_1	2	1	−1	−2	0	0	x, y	R_x, R_y	yz, zx
E_2	2	−1	−1	2	0	0			x^2-y^2, xy

3. $\mathbf{C}_{nh}$ Groups

$\mathbf{C}_{1h}$	E	σ_h xy	$\mathbf{C}_{1h} = \mathbf{C}_s$		
A'	1	1	x, y	R_z	x^2, y^2, z^2, xy
A''	1	−1	z	R_x, R_y	yz, zx

$\mathbf{C}_{2h}$	E	C_2	i	σ_h			
A_g	1	1	1	1		R_z	x^2, y^2, z^2, xy
A_u	1	1	−1	−1	z		
B_g	1	−1	1	−1		R_x, R_y	yz, zx
B_u	1	−1	−1	1	x, y		

$\mathbf{C}_{3h}$	E	C_3	C_3^2	σ_h	S_3	S_3^2			
A'	1	1	1	1	1	1		R_z	$x^2 + y^2, z^2$
A''	1	1	1	−1	−1	−1	z		
E'	1	ω	ω^2	1	ω	ω^2	x, y		$x^2 - y^2, xy$
	1	ω^*	ω^{*2}	1	ω^*	ω^{*2}			
E''	1	ω	ω^2	−1	$-\omega$	$-\omega^2$		R_x, R_y	yz, zx
	1	ω^*	ω^{*2}	−1	$-\omega^*$	$-\omega^{*2}$			

$\mathbf{C}_{4h}$	E	C_4	C_2	C_4^3	i	S_4^3	σ_h	S_4			
A_g	1	1	1	1	1	1	1	1		R_z	$x^2 + y^2, z^2$
A_u	1	1	1	1	−1	−1	−1	−1	z		
B_g	1	−1	1	−1	1	−1	1	−1			$x^2 - y^2, xy$
B_u	1	−1	1	−1	−1	1	−1	1			
E_g	1	i	−1	$-i$	1	i	−1	$-i$		R_x, R_y	yz, zx
	1	$-i$	−1	i	1	$-i$	−1	i			
E_u	1	i	−1	$-i$	−1	$-i$	1	i	x, y		
	1	$-i$	−1	i	−1	i	1	$-i$			

C_{5h}	E	C_5	C_5^2	C_5^3	C_5^4	σ_h	S_5	S_5^2	S_5^3	S_5^4	$\omega = \exp(2\pi i/5)$		
A'	1	1	1	1	1	1	1	1	1	1		R_z	$x^2 + y^2, z^2$
A''	1	1	1	1	1	−1	−1	−1	−1	−1	z		
E_1'	1	ω	ω^2	ω^3	ω^4	1	ω	ω^2	ω^3	ω^4	x, y		
	1	ω^*	ω^{*2}	ω^{*3}	ω^{*4}	1	ω^*	ω^{*2}	ω^{*3}	ω^{*4}			
E_1''	1	ω	ω^2	ω^3	ω^4	−1	$-\omega$	$-\omega^2$	$-\omega^3$	$-\omega^4$		R_x, R_y	yz, zx
	1	ω^*	ω^{*2}	ω^{*3}	ω^{*4}	−1	$-\omega^*$	$-\omega^{*2}$	$-\omega^{*3}$	$-\omega^{*4}$			
E_2'	1	ω^2	ω^4	ω	ω^3	1	ω^2	ω^4	ω	ω^3			$x^2 - y^2, xy$
	1	ω^{*2}	ω^{*4}	ω^*	ω^{*3}	1	ω^{*2}	ω^{*4}	ω^*	ω^{*3}			
E_2''	1	ω^2	ω^4	ω	ω^3	−1	$-\omega^2$	$-\omega^4$	$-\omega$	$-\omega^3$			
	1	ω^{*2}	ω^{*4}	ω^*	ω^{*3}	−1	$-\omega^{*2}$	$-\omega^{*4}$	$-\omega^*$	$-\omega^{*3}$			

C_{6h}	E	C_6	C_3	C_2	C_3^2	C_6^5	i	S_3^2	S_6^5	σ_h	S_6	S_3	$\omega = \exp(2\pi i/6)$		
A_g	1	1	1	1	1	1	1	1	1	1	1	1		R_z	$x^2 + y^2, z^2$
A_u	1	1	1	1	1	1	−1	−1	−1	−1	−1	−1	z		
B_g	1	−1	1	−1	1	−1	1	−1	1	−1	1	−1			
B_u	1	−1	1	−1	1	−1	−1	1	−1	1	−1	1			
E_{1g}	1	ω	ω^2	−1	$-\omega$	$-\omega^2$	1	ω	ω^2	−1	$-\omega$	$-\omega^2$		R_x, R_y	yz, zx
	1	ω^*	ω^{*2}	−1	$-\omega^*$	$-\omega^{*2}$	1	ω^*	ω^{*2}	−1	$-\omega^*$	$-\omega^{*2}$			
E_{1u}	1	ω	ω^2	−1	$-\omega$	$-\omega^2$	−1	$-\omega$	$-\omega^2$	1	ω	ω^2	x, y		
	1	ω^*	ω^{*2}	−1	$-\omega^*$	$-\omega^{*2}$	−1	$-\omega^*$	$-\omega^{*2}$	1	ω^*	ω^{*2}			
E_{2g}	1	ω^2	ω^4	1	ω^2	ω^4	1	ω^2	ω^4	1	ω^2	ω^4			$x^2 - y^2, xy$
	1	ω^{*2}	ω^{*4}	1	ω^{*2}	ω^{*4}	1	ω^{*2}	ω^{*4}	1	ω^{*2}	ω^{*4}			
E_{2u}	1	ω^2	ω^4	1	ω^2	ω^4	−1	$-\omega^2$	$-\omega^4$	−1	$-\omega^2$	$-\omega^4$			
	1	ω^{*2}	ω^{*4}	1	ω^{*2}	ω^{*4}	−1	$-\omega^{*2}$	$-\omega^{*4}$	−1	$-\omega^{*2}$	$-\omega^{*4}$			

4. S_{2n} Groups

S_2	E	i	$S_2 = C_i$		
A_g	1	1		R_x, R_y, R_z	$x^2, y^2, z^2, xy, yz, zx$
A_u	1	−1	x, y, z		

S_4	E	S_4	C_2	S_4^3			
A	1	1	1	1		R_z	$x^2 + y^2, z^2$
B	1	−1	1	−1	z		$x^2 - y^2, xy$
E	1	i	−1	$-i$	$\}\ x, y$	$\}\ R_x, R_y$	$\}\ yz, zx$
	1	$-i$	−1	i			

S_6	E	C_3	C_3^2	i	S_6^5	S_6	$\omega = \exp(2\pi i/3)$		
A_g	1	1	1	1	1	1		R_z	$x^2 + y^2, z^2$
A_u	1	1	1	−1	−1	−1	z		
E_g	1	ω	ω^2	1	ω	ω^2		$\}\ R_x, R_y$	$\}\ x^2 - y^2, xy, yz, zx$
	1	ω^*	ω^{*2}	1	ω^*	ω^{*2}			
E_u	1	ω	ω^2	−1	$-\omega$	$-\omega^2$	$\}\ x, y$		
	1	ω^*	ω^{*2}	−1	$-\omega^*$	$-\omega^{*2}$			

5. D_n Groups

D_2	E	C_2 (z)	C_2 (y)	C_2 (x)	$D_2 = V$		
A	1	1	1	1			x^2, y^2, z^2
B_1	1	1	−1	−1	z	R_z	xy
B_2	1	−1	1	−1	y	R_y	zx
B_3	1	−1	−1	1	x	R_x	yz

$\mathbf{D}_3$	E	$2C_3$	$3C_2$			
A_1	1	1	1			x^2+y^2, z^2
A_2	1	1	−1	z	R_z	
E	2	−1	0	x, y	R_x, R_y	x^2-y^2, xy, yz, zx

$\mathbf{D}_4$	E	$2C_4$	C_2	$2C_2'$	$2C_2''$			
A_1	1	1	1	1	1			x^2+y^2, z^2
A_2	1	1	1	−1	−1	z	R_z	
B_1	1	−1	1	1	−1			
B_2	1	−1	1	−1	1			
E	2	0	−2	0	0	x, y	R_x, R_y	x^2-y^2, xy, yz, zx

$\mathbf{D}_5$	E	$2C_5$	$2C_5^2$	$5C_2'$	$\varphi = 2\pi/5$		
A_1	1	1	1	1			x^2+y^2, z^2
A_2	1	1	1	−1	z	R_z	
E_1	2	a	b	0	x, y	R_x, R_y	yz, zx
E_2	2	b	a	0			x^2-y^2, xy

$a = 2\cos\varphi = 2\cos 4\varphi = (\sqrt{5}-1)/2 = 0.618034,$
$b = 2\cos 2\varphi = -(\sqrt{5}+1)/2 = -1.618034.$

$\mathbf{D}_6$	E	$2C_6$	$2C_3$	C_2	$3C_2'$	$3C_2''$			
A_1	1	1	1	1	1	1			x^2+y^2, z^2
A_2	1	1	1	1	−1	−1		R_z	
B_1	1	−1	1	−1	1	−1			
B_2	1	−1	1	−1	−1	1	z		
E_1	2	1	−1	−2	0	0	x, y	R_x, R_y	yz, zx
E_2	2	−1	−1	2	0	0			x^2-y^2, xy

6. $\mathbf{D}_{nd}$ Groups

$\mathbf{D}_{2d}$	E	$2S_4$	C_2	$2C_2'$	$2C_2''$	$\mathbf{D}_{2d} = \mathbf{V}_d$		
A_1	1	1	1	1	1			x^2+y^2, z^2
A_2	1	1	1	−1	−1		R_z	
B_1	1	−1	1	1	−1			x^2-y^2
B_2	1	−1	1	−1	1	z		xy
E	2	0	−2	0	0	x, y	R_x, R_y	yz, zx

D_{3d}	E	$2C_3$	$3C_2'$	i	$2S_3$	$3\sigma_d$			
A_{1g}	1	1	1	1	1	1			x^2+y^2, z^2
A_{1u}	1	1	1	−1	−1	−1			
A_{2g}	1	1	−1	1	1	−1		R_z	
A_{2u}	1	1	−1	−1	−1	1	z		
E_g	2	−1	0	2	−1	0		R_x, R_y	x^2-y^2, xy, yz, zx
E_u	2	−1	0	−2	1	0	x, y		

D_{4d}	E	$2S_8$	$2C_4$	$2S_8^3$	C_2	$4C_2'$	$4\sigma_d$			
A_1	1	1	1	1	1	1	1			x^2+y^2, z^2
A_2	1	1	1	1	1	−1	−1		R_z	
B_1	1	−1	1	−1	1	1	−1			
B_2	1	−1	1	−1	1	−1	1	z		
E_1	2	$\sqrt{2}$	0	$-\sqrt{2}$	−2	0	0	x, y		
E_2	2	0	−2	0	2	0	0			x^2-y^2, xy
E_3	2	$-\sqrt{2}$	0	$\sqrt{2}$	−2	0	0		R_x, R_y	yz, zx

D_{5d}	E	$2C_5$	$2C_5^2$	$5C_2'$	i	$2S_{10}^3$	$2S_{10}$	$5\sigma_d$	$\varphi = 2\pi/5$		
A_{1g}	1	1	1	1	1	1	1	1			x^2+y^2, z^2
A_{1u}	1	1	1	1	−1	−1	−1	−1			
A_{2g}	1	1	1	−1	1	1	1	−1		R_z	
A_{2u}	1	1	1	−1	−1	−1	−1	1	z		
E_{1g}	2	a	b	0	2	a	b	0		R_x, R_y	yz, zx
E_{1u}	2	a	b	0	−2	$-a$	$-b$	0	x, y		
E_{2g}	2	b	a	0	2	b	a	0			x^2-y^2, xy
E_{2u}	2	b	a	0	−2	$-b$	$-a$	0			

$a = 2\cos\varphi = 2\cos 4\varphi = (\sqrt{5}-1)/2 = 0{,}618034,$
$b = 2\cos 2\varphi = -(\sqrt{5}+1)/2 = -1{,}618034\,.$

D_{6d}	E	$2S_{12}$	$2C_6$	$2S_4$	$2C_3$	$2S_{12}^5$	C_2	$6C_2'$	$6\sigma_d$			
A_1	1	1	1	1	1	1	1	1	1			x^2+y^2, z^2
A_2	1	1	1	1	1	1	1	−1	−1		R_z	
B_1	1	−1	1	−1	1	−1	1	1	−1			
B_2	1	−1	1	−1	1	−1	1	−1	1	z		
E_1	2	$\sqrt{3}$	1	0	−1	$-\sqrt{3}$	−2	0	0	x, y		
E_2	2	1	−1	−2	−1	1	2	0	0			x^2-y^2, xy
E_3	2	0	−2	0	2	0	−2	0	0			
E_4	2	−1	−1	2	−1	−1	2	0	0		R_x, R_y	yz, zx
E_5	2	$-\sqrt{3}$	1	0	−1	$\sqrt{3}$	−2	0	0			

7. D_{nh} Groups

D_{2h}	E	C_2	C_2	C_2	i	σ	σ	σ	$D_{2h} = V_h$		
		z	y	x		xy	zx	yz			
A_g	1	1	1	1	1	1	1	1			x^2, y^2, z^2
A_u	1	1	1	1	−1	−1	−1	−1			
B_{1g}	1	1	−1	−1	1	1	−1	−1		R_z	xy
B_{1u}	1	1	−1	−1	−1	−1	1	1	z		
B_{2g}	1	−1	1	−1	1	−1	1	−1		R_y	zx
B_{2u}	1	−1	1	−1	−1	1	−1	1	y		
B_{3g}	1	−1	−1	1	1	−1	−1	1		R_x	yz
B_{3u}	1	−1	−1	1	−1	1	1	−1	x		

D_{3h}	E	$2C_3$	$3C_2'$	σ_h	$2S_3$	$3\sigma_v$			
A_1'	1	1	1	1	1	1			$x^2 + y^2, z^2$
A_1''	1	1	1	−1	−1	−1			
A_2'	1	1	−1	1	1	−1		R_z	
A_2''	1	1	−1	−1	−1	1	z		
E'	2	−1	0	2	−1	0	x, y		$x^2 - y^2, xy$
E''	2	−1	0	−2	1	0		R_x, R_y	yz, zx

D_{4h}	E	$2C_4$	C_2	$2C_2'$	$2C_2''$	i	$2S_4$	σ_h	$2\sigma_v$	$2\sigma_d$			
A_{1g}	1	1	1	1	1	1	1	1	1	1			$x^2 + y^2, z^2$
A_{1u}	1	1	1	1	1	−1	−1	−1	−1	−1			
A_{2g}	1	1	1	−1	−1	1	1	1	−1	−1		R_z	
A_{2u}	1	1	1	−1	−1	−1	−1	−1	1	1	z		
B_{1g}	1	−1	1	1	−1	1	−1	1	1	−1			$x^2 - y^2$
B_{1u}	1	−1	1	1	−1	−1	1	−1	−1	1			
B_{2g}	1	−1	1	−1	1	1	−1	1	−1	1			xy
B_{2u}	1	−1	1	−1	1	−1	1	−1	1	−1			
E_g	2	0	−2	0	0	2	0	−2	0	0		R_x, R_y	yz, zx
E_u	2	0	−2	0	0	−2	0	2	0	0	x, y		

$\mathbf{D}_{5h}$	E	$2C_5$	$2C_5^2$	$5C_2'$	σ_h	$2S_5$	$2S_5^3$	$5\sigma_v$	$\varphi = 2\pi/5$		
A_1'	1	1	1	1	1	1	1	1			$x^2 + y^2, z^2$
A_1''	1	1	1	1	−1	−1	−1	−1			
A_2'	1	1	1	−1	1	1	1	−1		R_z	
A_2''	1	1	1	−1	−1	−1	−1	1	z		
E_1'	2	a	b	0	2	a	b	0	x, y		
E_1''	2	a	b	0	−2	$-a$	$-b$	0		R_x, R_y	yz, zx
E_2'	2	b	a	0	2	b	a	0			$x^2 - y^2, xy$
E_2''	2	b	a	0	−2	$-b$	$-a$	0			

$a = 2\cos\varphi = 2\cos 4\varphi = (\sqrt{5} - 1)/2 = 0{,}618\,034,$

$b = 2\cos 2\varphi = -(\sqrt{5} + 1)/2 = -1{,}618\,034\,.$

$\mathbf{D}_{6h}$	E	$2C_6$	$2C_3$	C_2	$3C_2'$	$3C_2''$	i	$2S_3$	$2S_6$	σ_h	$3\sigma_d$	$3\sigma_v$			
A_{1g}	1	1	1	1	1	1	1	1	1	1	1	1			$x^2 + y^2, z^2$
A_{1u}	1	1	1	1	1	1	−1	−1	−1	−1	−1	−1			
A_{2g}	1	1	1	1	−1	−1	1	1	1	1	−1	−1		R_z	
A_{2u}	1	1	1	1	−1	−1	−1	−1	−1	−1	1	1	z		
B_{1g}	1	−1	1	−1	1	−1	1	−1	1	−1	1	−1			
B_{1u}	1	−1	1	−1	1	−1	−1	1	−1	1	−1	1			
B_{2g}	1	−1	1	−1	−1	1	1	−1	1	−1	−1	1			
B_{2u}	1	−1	1	−1	−1	1	−1	1	−1	1	1	−1			
E_{1g}	2	1	−1	−2	0	0	2	1	−1	−2	0	0		R_x, R_y	yz, zx
E_{1u}	2	1	−1	−2	0	0	−2	−1	1	2	0	0	x, y		
E_{2g}	2	−1	−1	2	0	0	2	−1	−1	2	0	0			$x^2 - y^2, xy$
E_{2u}	2	−1	−1	2	0	0	−2	1	1	−2	0	0			

8. Cubic Groups

T	E	$4C_3$	$4C_3^2$	$3C_2$	$\omega = \exp(2\pi i/3)$		
A	1	1	1	1			$x^2 + y^2 + z^2$
E	1	ω	ω^2	1			$x^2 + y^2 - 2z^2$
	1	ω^*	ω^{*2}	1			$x^2 - y^2$
F	3	0	0	-1	x, y, z	R_x, R_y, R_z	xy, yz, zx

T_d	E	$8C_3$	$3C_2$	$6S_4$	$6\sigma_d$			
A_1	1	1	1	1	1			$x^2 + y^2 + z^2$
A_2	1	1	1	-1	-1			
E	2	-1	2	0	0			$x^2 + y^2 - 2z^2, x^2 - y^2$
F_1	3	0	-1	1	-1		R_x, R_y, R_z	
F_2	3	0	-1	-1	1	x, y, z		xy, yz, zx

T_h	E	$4C_3$	$4C_3^2$	$3C_2$	i	$4S_6^5$	$4S_6$	$3\sigma_h$	$\omega = \exp(2\pi i/3)$		
A_g	1	1	1	1	1	1	1	1			$x^2 + y^2 + z^2$
A_u	1	1	1	1	-1	-1	-1	-1			
E_g	1	ω	ω^2	1	1	ω	ω^2	1			$x^2 + y^2 - 2z^2$
	1	ω^*	ω^{*2}	1	1	ω^*	ω^{*2}	1			$x^2 - y^2$
E_u	1	ω	ω^2	1	-1	$-\omega$	$-\omega^2$	-1			
	1	ω^*	ω^{*2}	1	-1	$-\omega^*$	$-\omega^{*2}$	-1			
F_g	3	0	0	-1	3	0	0	-1		R_x, R_y, R_z	xy, yz, zx
F_u	3	0	0	-1	-3	0	0	1	x, y, z		

O	E	$8C_3$	$3C_2$	$6C_4$	$6C_2'$			
A_1	1	1	1	1	1			$x^2 + y^2 + z^2$
A_2	1	1	1	-1	-1			
E	2	-1	2	0	0			$x^2 + y^2 - 2z^2, x^2 - y^2$
F_1	3	0	-1	1	-1	x, y, z	R_x, R_y, R_z	
F_2	3	0	-1	-1	1			xy, yz, zx

O_h	E	$8C_3$	$6C_2$	$6C_4$	$3C_2(=C_4^2)$	i	$6S_4$	$8S_6$	$3\sigma_h$	$6\sigma_d$		
A_{1g}	1	1	1	1	1	1	1	1	1	1		$x^2+y^2+z^2$
A_{2g}	1	1	-1	-1	1	1	-1	1	1	-1		
E_g	2	-1	0	0	2	2	0	-1	2	0		$x^2-y^2-2z^2, x^2-y^2$
T_{1g}	3	0	-1	1	-1	3	1	0	-1	-1	R_x, R_y, R_z	
T_{2g}	3	0	1	-1	-1	3	-1	0	-1	1		xz, yx, xy
A_{1u}	1	1	1	1	1	-1	-1	-1	-1	-1		
A_{2u}	1	1	-1	-1	1	-1	1	-1	-1	1		
E_u	2	-1	0	0	2	-2	0	1	-2	0		
T_{1u}	3	0	-1	1	-1	-3	-1	0	1	1	x, y, z	
T_{2u}	3	0	1	-1	-1	-3	1	0	1	-1		

9. The Icosahedral Groups

$\mathbf{I}$	E	$12C_5$	$12C_5^2$	$20C_3$	$15C_2$			
A	1	1	1	1	1			$x^2 + y^2 + z^2$
F_1	3	a	b	0	−1	x, y, z	R_x, R_y, R_z	
F_2	3	b	a	0	−1			
G	4	−1	−1	1	0			
H	5	0	0	−1	1			$\begin{cases} x^2 + y^2 - 2z^2 \\ x^2 - y^2, xy, yz, zx \end{cases}$

$\mathbf{I}_h$	E	$12C_5$	$12C_5^2$	$20C_3$	$15C_2$	i	$12S_{10}$	$12S_{10}^3$	$20S_6$	15σ		
A_g	1	1	1	1	1	1	1	1	1	1		$x^2 + y^2 + z^2$
T_{1g}	3	a	b	0	−1	3	b	a	0	−1	R_x, R_y, R_z	
T_{2g}	3	b	a	0	−1	3	a	b	0	−1		
G_g	4	−1	−1	1	0	4	−1	−1	1	0		
H_g	5	0	0	−1	1	5	0	0	−1	1		$x^2 + y^2 - 2z^2$, $x^2 - y^2$, xy, yz, zx
A_u	1	1	1	1	1	−1	−1	−1	−1	−1		
T_{1u}	3	a	b	0	−1	−3	$-b$	$-a$	0	1	x, y, z	
T_{2u}	3	b	a	0	−1	−3	$-a$	$-b$	0	1		
G_u	4	−1	−1	1	0	−4	1	1	−1	0		
H_u	5	0	0	−1	1	−5	0	0	1	−1		

$a = (1 + \sqrt{5})/2 = 1{,}618034,$
$b = (1 - \sqrt{5})/2 = -0{,}618034.$

10. Continuous Groups for Linear Molecules

$C_{\infty v}$	E	$2C_\infty^\varphi$	...	C_2	$\infty\sigma_v$	$0 < \varphi < \pi$		
$A_1 = \Sigma^+$	1	1	...	1	1	z		$x^2 + y^2, z^2$
$A_2 = \Sigma^-$	1	1	...	1	-1		R_z	
$E_1 = \Pi$	2	$2 \cos \varphi$	...	-2	0	x, y	R_x, R_y	yz, zx
$E_2 = \Delta$	2	$2 \cos 2\varphi$	...	2	0			$x^2 - y^2, xy$
$E_3 = \Phi$	2	$2 \cos 3\varphi$	...	-2	0			
...	...	...	...	...	...			

$D_{\infty h}$	E	$2C_\infty^\varphi$	...	C_2	$\infty\sigma_v$	i	$2S_\infty^\varphi$	...	σ_h	$\infty C_2'$	$0 < \varphi < \pi$		
Σ_g^+	1	1	...	1	1	1	1	...	1	1			$x^2 - y^2, z^2$
Σ_u^+	1	1	...	1	1	-1	-1	...	-1	-1	z		
Σ_g^-	1	1	...	1	-1	1	1	...	1	-1		R_z	
Σ_u^-	1	1	...	1	-1	-1	-1	...	-1	1			
Π_g	2	$2 \cos \varphi$	...	-2	0	2	$2 \cos \varphi$	...	-2	0		R_x, R_y	yz, zx
Π_u	2	$2 \cos \varphi$	...	-2	0	-2	$-2 \cos \varphi$	...	2	0	x, y		
Δ_g	2	$2 \cos 2\varphi$	...	2	0	2	$2 \cos 2\varphi$	...	2	0			$x^2 - y^2, xy$
Δ_u	2	$2 \cos 2\varphi$	...	2	0	-2	$-2 \cos 2\varphi$	...	-2	0			
.	.	.	...	.	.	.	.	...	.	.			

Appendix B

Kernels and Co-Kernels of Degenerate Irreducible Representations*

The sym-ops preserved in the kernels and co-kernels are given in parenthesis when necessary to specify the alinement of the subgroup. They are labeled as in Appendix A.

Table B.1: $\mathbf{D}_n$ Groups

Group	irrep	Kernel	Co – Kernels
$\mathbf{D}_3$	E	$\mathbf{C}_1$	$\mathbf{C}_2$
$\mathbf{D}_4$	E	$\mathbf{C}_1$	$\mathbf{C}_2(C_2')$, $\mathbf{C}_2(C_2'')$
$\mathbf{D}_5$	E_1	$\mathbf{C}_1$	$\mathbf{C}_2$
	E_2	$\mathbf{C}_1$	$\mathbf{C}_2$
$\mathbf{D}_6$	E_1	$\mathbf{C}_1$	$\mathbf{C}_2(C_2')$, $\mathbf{C}_2(C_2'')$
	E_2	$\mathbf{C}_2(C_2)$	$\mathbf{D}_2$

Table B.2: $\mathbf{C}_{nv}$ Groups

Group	irrep	Kernel	Co – Kernels
$\mathbf{C}_{3v}$	E	$\mathbf{C}_1$	$\mathbf{C}_s$
$\mathbf{C}_{4v}$	E	$\mathbf{C}_1$	$\mathbf{C}_s(\sigma_v)$, $\mathbf{C}_s(\sigma_d)$
$\mathbf{C}_{5v}$	E_1	$\mathbf{C}_1$	$\mathbf{C}_s$
	E_2	$\mathbf{C}_1$	$\mathbf{C}_s$
$\mathbf{C}_{6v}$	E_1	$\mathbf{C}_1$	$\mathbf{C}_s(\sigma_v)$, $\mathbf{C}_s(\sigma_d)$
	E_2	$\mathbf{C}_2$	$\mathbf{C}_{2v}$

Table B.3: $\mathbf{C}_{nh}$ Groups

Group	irrep	Kernel	Co – Kernels
$\mathbf{C}_{3h}$	E'	$\mathbf{C}_s$	–
	E''	$\mathbf{C}_1$	–
$\mathbf{C}_{4h}$	E_g	$\mathbf{C}_i$	–
	E_u	$\mathbf{C}_s$	–
$\mathbf{C}_{5h}$	E_1'	$\mathbf{C}_s$	–
	E_2'	$\mathbf{C}_s$	–
	E_1''	$\mathbf{C}_1$	–
	E_2''	$\mathbf{C}_1$	–
$\mathbf{C}_{6h}$	E_{1g}	$\mathbf{C}_i$	–
	E_{2g}	$\mathbf{C}_{2h}$	–
	E_{1u}	$\mathbf{C}_s$	–
	E_{2u}	$\mathbf{C}_2$	–

Table B.4: $\mathbf{D}_{nh}$Groups

Group	irrep	Kernel	Co – Kernels
$\mathbf{D}_{3h}$	E'	$\mathbf{C}_s(\sigma_h)$	$\mathbf{C}_{2v}$
	E''	$\mathbf{C}_1$	$\mathbf{C}_2$, $\mathbf{C}_s(\sigma_v)$
$\mathbf{D}_{4h}$	E_g	$\mathbf{C}_i$	$\mathbf{C}_{2h}(C_2')$, $\mathbf{C}_{2h}(C_2'')$
	E_u	$\mathbf{C}_s(\sigma_h)$	$\mathbf{C}_{2v}(C_2')$, $\mathbf{C}_{2v}(C_2'')$
$\mathbf{D}_{5h}$	E_1'	$\mathbf{C}_s(\sigma_h)$	$\mathbf{C}_{2v}$
	E_2'	$\mathbf{C}_s(\sigma_h)$	$\mathbf{C}_{2v}$
	E_1''	$\mathbf{C}_1$	$\mathbf{C}_s(\sigma_v)$, $\mathbf{C}_2$
	E_2''	$\mathbf{C}_1$	$\mathbf{C}_s(\sigma_v)$, $\mathbf{C}_2$
$\mathbf{D}_{6h}$	E_{1g}	$\mathbf{C}_i$	$\mathbf{C}_{2h}(C_2')$, $\mathbf{C}_{2h}(C_2'')$
	E_{2g}	$\mathbf{C}_{2h}(C_2)$	$\mathbf{D}_{2h}$
	E_{1u}	$\mathbf{C}_s(\sigma_h)$	$\mathbf{C}_{2v}(C_2')$, $\mathbf{C}_{2v}(C_2'')$
	E_{2u}	$\mathbf{C}_2(C_2)$	$\mathbf{C}_{2v}(\sigma_v)$, $\mathbf{D}_2$

* Adapted from Table 1 of P. Murray-Rust, H.-B. Bürgi and J.D Dunitz, Acta Crystallographica *A35*, 703 (1979).

Table B.5: $\mathbf{S}_{2n}$ Groups

Group	irrep	Kernel	Co − Kernels
$\mathbf{S}_4$	E	$\mathbf{C}_1$	−
$\mathbf{S}_6$	E_g	$\mathbf{C}_i$	−
	E_u	$\mathbf{C}_1$	−
$\mathbf{S}_8$	E_1	$\mathbf{C}_1$	−
	E_2	$\mathbf{C}_2$	−
	E_3	$\mathbf{C}_1$	−

Table B.6: $\mathbf{D}_{nd}$ Groups

Group	irrep	Kernel	Co − Kernels
$\mathbf{D}_{2d}$	E	$\mathbf{C}_1$	$\mathbf{C}_2(C_2')$, $\mathbf{C}_s(\sigma_d)$
$\mathbf{D}_{3d}$	E_g	$\mathbf{C}_i$	$\mathbf{C}_{2h}$
	E_u	$\mathbf{C}_1$	$\mathbf{C}_2$, $\mathbf{C}_s$
$\mathbf{D}_{4d}$	E_1	$\mathbf{C}_1$	$\mathbf{C}_2(C_2')$, $\mathbf{C}_s$
	E_2	$\mathbf{C}_2(C_2)$	$\mathbf{C}_{2v}$, $\mathbf{D}_2$
	E_3	$\mathbf{C}_1$	$\mathbf{C}_2(C_2')$, $\mathbf{C}_s$
$\mathbf{D}_{5d}$	E_1g	$\mathbf{C}_i$	$\mathbf{C}_{2h}$
	E_2g	$\mathbf{C}_i$	$\mathbf{C}_{2h}$
	E_1u	$\mathbf{C}_1$	$\mathbf{C}_2$, $\mathbf{C}_s$
	E_2u	$\mathbf{C}_1$	$\mathbf{C}_2$, $\mathbf{C}_s$
$\mathbf{D}_{6d}$	E_1	$\mathbf{C}_1$	$\mathbf{C}_2(C_2')$, $\mathbf{C}_s$
	E_2	$\mathbf{C}_2(C_2)$	$\mathbf{C}_{2v}$, $\mathbf{D}_2$
	E_3	$\mathbf{C}_3$	$\mathbf{C}_{3v}$, $\mathbf{D}_3$
	E_4	$\mathbf{S}_4$	$\mathbf{D}_{2d}$
	E_5	$\mathbf{C}_1$	$\mathbf{C}_2(C_2')$, $\mathbf{C}_s$

Table B.7: Axial and Cylindrical Groups

Group	irrep	Kernel	Co − Kernels
$\mathbf{C}_{\infty v}$	Π	$\mathbf{C}_1$	$\mathbf{C}_s$
	Δ	$\mathbf{C}_1$	$\mathbf{C}_s$
	etc	...	...
$\mathbf{D}_{\infty h}$	Π_g	$\mathbf{C}_i$	$\mathbf{C}_{2h}(C_2)$
	Δ_g	$\mathbf{C}_i$	$\mathbf{C}_{2h}(C_2)$
	etc	...	...
	Π_u	$\mathbf{C}_1$	$\mathbf{C}_{2v}(C_2)$
	Δ_u	$\mathbf{C}_1$	$\mathbf{C}_{2v}(C_2)$
	etc	...	...

Table B.8: Cubic Groups

Group	irrep	Kernel	Co − Kernels
$\mathbf{T}_d$	E	$\mathbf{D}_2$	$\mathbf{D}_{2d}$
	T_1	$\mathbf{C}_1$	$\mathbf{S}_4$, $\mathbf{C}_3$, $\mathbf{C}_s$
	T_2	$\mathbf{C}_1$	$\mathbf{C}_{2v}$, $\mathbf{C}_{3v}$, $\mathbf{C}_s$
$\mathbf{O}_h$	E_g	$\mathbf{D}_{2h}(C_4^2, C_4^2)$	$\mathbf{D}_{4h}$
	E_u	$\mathbf{D}_2(C_4^2, C_4^2)$	$\mathbf{D}_4$, $\mathbf{D}_{2d}$
	T_{1g}	$\mathbf{C}_i$	$\mathbf{C}_{4h}$, $\mathbf{S}_6$, $\mathbf{C}_{2h}(C_2)$
	T_{2g}	$\mathbf{C}_i$	$\mathbf{D}_{2h}(C_4^2, C_4^2)$, $\mathbf{D}_{3d}$, $\mathbf{C}_{2h}(C_2)$
	T_{1u}	$\mathbf{C}_1$	$\mathbf{C}_{4v}$, $\mathbf{C}_{3v}$, $\mathbf{C}_{2v}(C_2)$, $\mathbf{C}_s(\sigma_d)$, $\mathbf{C}_s(\sigma_h)$
	T_{2u}	$\mathbf{C}_1$	$\mathbf{D}_{2d}(C_4^2, C_2)$, $\mathbf{D}_3$, $\mathbf{C}_{2v}(C_2)$, $\mathbf{C}_2(C_2)$, $\mathbf{C}_s(\sigma_h)$

Appendix C

Group Correlation Tables*

$\mathbf{O}_h$	$\mathbf{O}$	$\mathbf{T}_d$	$\mathbf{D}_{4h}$	$\mathbf{C}_{4v}$	$\mathbf{C}_{2v}$	$\mathbf{D}_3$	$\mathbf{D}_{2d}$
A_{1g}	A_1	A_1	A_{1g}	A_1	A_1	A_1	A_1
A_{2g}	A_2	A_2	B_{1g}	B_1	A_2	A_2	B_1
E_g	E	E	$A_{1g}+B_{1g}$	A_1+B_1	A_1+A_2	E	A_1+B_1
T_{1g}	T_1	T_1	$A_{2g}+E_g$	A_2+E	$A_2+B_1+B_2$	A_2+E	A_2+E
T_{2g}	T_2	T_2	$B_{2g}+E_g$	B_2+E	$A_1+B_1+B_2$	A_1+E	B_2+E
A_{1u}	A_1	A_2	A_{1u}	A_2	A_2	A_1	B_1
A_{2u}	A_2	A_1	B_{1u}	B_2	A_1	A_2	A_1
E_u	E	E	$A_{1u}+B_{1u}$	A_2+B_2	A_1+A_2	E	A_1+B_1
T_{1u}	T_1	T_2	$A_{2u}+E_u$	A_1+E	$A_1+B_1+B_2$	A_2+E	B_2+E
T_{2u}	T_2	T_1	$B_{2u}+E_u$	B_1+E	$A_2+B_1+B_2$	A_1+E	A_2+E

$\mathbf{T}_d$	$\mathbf{T}$	$\mathbf{D}_{2d}$	$\mathbf{C}_{2v}$	$\mathbf{S}_4$	$\mathbf{D}_2$	$\mathbf{C}_{2v}$	$\mathbf{C}_3$	$\mathbf{C}_2$
A_1	A	A_1	A_1	A	A	A_1	A	A
A_2	A	B_1	A_2	B	A	A_2	A	A
E	E	A_1+B_1	E	$A+B$	$2A$	A_1+A_2	E	$2A$
T_1	T	A_2+E	A_2+E	$A+E$	$B_1+B_2+B_3$	$A_2+B_1+B_2$	$A+E$	$A+2B$
T_2	T	B_2+E	A_1+E	$B+E$	$B_1+B_2+B_3$	$A_1+B_1+B_2$	$A+E$	$A+2B$

	$C_2' \to C_2'$				C_2'	C_2''			C_2'
$\mathbf{D}_{4h}$	$\mathbf{D}_4$	$\mathbf{D}_{2d}$	$\mathbf{C}_{4v}$	$\mathbf{C}_{4h}$	$\mathbf{D}_{2h}$	$\mathbf{D}_{2h}$	$\mathbf{C}_4$	$\mathbf{S}_4$	$\mathbf{D}_2$
A_{1g}	A_1	A_1	A_1	A_g	A_g	A_g	A	A	A
A_{2g}	A_2	A_2	A_2	A_g	B_{1g}	B_{1g}	A	A	B_1
B_{1g}	B_1	B_1	B_1	B_g	A_g	B_{1g}	B	B	A
B_{2g}	B_2	B_2	B_2	B_g	B_{1g}	A_g	B	B	B_1
E_g	E	E	E	E_g	$B_{2g}+B_{3g}$	$B_{2g}+B_{3g}$	E	E	B_2+B_3
A_{1u}	A_1	B_1	A_2	A_u	A_u	A_u	A	B	A
A_{2u}	A_2	B_2	A_1	A_u	B_{1u}	B_{1u}	A	B	B_1
B_{1u}	B_1	A_1	B_2	B_u	A_u	B_{1u}	B	A	A
B_{2u}	B_2	A_2	B_1	B_u	B_{1u}	A_u	B	A	B_1
E_u	E	E	E	E_u	$B_{2u}+B_{3u}$	$B_{2u}+B_{3u}$	E	E	B_2+B_3

* Adapted from B.E. Douglas and C.A. Hollingsworth: Symmetry in Bonding and Spectra. An Introduction. Academic Press Inc., Orlando San Diego New York London Toronto Montreal Sydney Tokio 1985. Appendix 3, pp. 412–414

$\mathbf{D}_{4h}$ (*cont.*)	C_2, σ_v $\mathbf{C}_{2v}$	C_2, σ_d $\mathbf{C}_{2v}$	C_2 $\mathbf{C}_{2h}$	C_2' $\mathbf{C}_{2h}$	C_2'' $\mathbf{C}_{2h}$	C_2 $\mathbf{C}_2$	C_2' $\mathbf{C}_2$	σ_h $\mathbf{C}_s$	σ_v $\mathbf{C}_s$	$\mathbf{C}_i$
A_{1g}	A_1	A_1	A_g	A_g	A_g	A	A	A'	A'	A_g
A_{2g}	A_2	A_2	A_g	B_g	B_g	A	B	A'	A''	A_g
B_{1g}	A_1	A_2	A_g	A_g	B_g	A	A	A'	A'	A_g
B_{2g}	A_2	A_1	A_g	B_g	A_g	A	B	A'	A''	A_g
E_g	$B_1 + B_2$	$B_1 + B_2$	$2B_g$	$A_g + B_g$	$A_g + B_g$	$2B$	$A + B$	$2A''$	$A' + A''$	$2A_g$
A_{1u}	A_2	A_2	A_u	A_u	A_u	A	A	A''	A''	A_u
A_{2u}	A_1	A_1	A_u	B_u	B_u	A	B	A''	A'	A_u
B_{1u}	A_2	A_1	A_u	A_u	B_u	A	A	A''	A''	A_u
B_{2u}	A_1	A_2	A_u	B_u	A_u	A	B	A''	A'	A_u
E_u	$B_1 + B_2$	$B_1 + B_2$	$2B_u$	$A_u + B_u$	$A_u + B_u$	$2B$	$A + B$	$2A'$	$A' + A''$	$2A_u$

$\mathbf{D}_{2h}$	$\mathbf{D}_2$	$C_2(z)$ $\mathbf{C}_{2v}$	$C_2(y)$ $\mathbf{C}_{2v}$	$C_2(x)$ $\mathbf{C}_{2v}$	$C_2(z)$ $\mathbf{C}_{2h}$	$C_2(y)$ $\mathbf{C}_{2h}$	$C_2(x)$ $\mathbf{C}_{2h}$	$C_2(z)$ $\mathbf{C}_2$	$C_2(y)$ $\mathbf{C}_2$	$C_2(x)$ $\mathbf{C}_2$	$\sigma(xy)$ $\mathbf{C}_s$	$\sigma(yz)$ $\mathbf{C}_s$
A_g	A	A_1	A_1	A_1	A_g	A_g	A_g	A	A	A	A'	A'
B_{1g}	B_1	A_2	B_2	B_1	A_g	B_g	B_g	A	B	B	A'	A''
B_{2g}	B_2	B_1	A_2	B_2	B_g	A_g	B_g	B	A	B	A''	A''
B_{3g}	B_3	B_2	B_1	A_2	B_g	B_g	A_g	B	B	A	A''	A'
A_u	A	A_2	A_2	A_2	A_u	A_u	A_u	A	A	A	A'	A''
B_{1u}	B_1	A_1	B_1	B_2	A_u	B_u	B_u	A	B	B	A''	A'
B_{2u}	B_2	B_2	A_1	B_1	B_u	A_u	B_u	B	A	B	A'	A'
B_{3u}	B_3	B_1	B_2	A_1	B_u	B_u	A_u	B	B	A	A'	A''

$\mathbf{D}_{4d}$	$\mathbf{D}_4$	$\mathbf{C}_{4v}$	$\mathbf{S}_8$	$\mathbf{C}_4$	$\mathbf{C}_{2v}$	C_2 $\mathbf{C}_2$	C_2' $\mathbf{C}_2$	$\mathbf{C}_s$
A_1	A_1	A_1	A	A	A_1	A	A	A'
A_2	A_2	A_2	A	A	A_2	A	B	A''
B_1	A_1	A_2	B	A	A_2	A	A	A''
B_2	A_2	A_1	B	A	A_1	A	B	A'
E_1	E	E	E_1	E	$B_1 + B_2$	$2B$	$A + B$	$A' + A''$
E_2	$B_1 + B_2$	$B_1 + B_2$	E_2	$2B$	$A_1 + A_2$	$2A$	$A + B$	$A' + A''$
E_3	E	E	E_3	E	$B_1 + B_2$	$2B$	$A + B$	$A' + A''$

$\mathbf{D}_{2d}$	$\mathbf{S}_4$	$C_2 \rightarrow C_2(z)$ $\mathbf{D}_2$	$\mathbf{C}_{2v}$	$\mathbf{C}_2$	C_2 $\mathbf{C}_2$	C_2' $\mathbf{C}_s$
A_1	A	A	A_1	A	A	A'
A_2	A	B_1	A_2	A	B	A''
B_1	B	A	A_2	A	A	A''
B_2	B	B_1	A_1	A	B	A'
E	E	$B_2 + B_3$	$B_1 + B_2$	$2B$	$A + B$	$A' + A''$

$\mathbf{D}_{3d}$	$\mathbf{D}_3$	$\mathbf{C}_{3v}$	$\mathbf{S}_6$	$\mathbf{C}_3$	$\mathbf{C}_{2h}$	$\mathbf{C}_2$	$\mathbf{C}_s$	$\mathbf{C}_i$
A_{1g}	A_1	A_1	A_g	A	A_g	A	A'	A_g
A_{2g}	A_2	A_2	A_g	A	B_g	B	A''	A_g
E_g	E	E	E_g	E	$A_g + B_g$	$A + B$	$A' + A''$	$2A_g$
A_{1u}	A_1	A_2	A_u	A	A_u	A	A''	A_u
A_{2u}	A_2	A_1	A_u	A	B_u	B	A'	A_u
E_u	E	E	E_u	E	$A_u + B_u$	$A + B$	$A' + A''$	$2A_u$

				$\sigma_h \to \sigma_v(zy)$			σ_h	σ_v
$\mathbf{D}_{3h}$	$\mathbf{C}_{3h}$	$\mathbf{D}_3$	$\mathbf{C}_{3v}$	$\mathbf{C}_{2v}$	$\mathbf{C}_3$	$\mathbf{C}_2$	$\mathbf{C}_s$	$\mathbf{C}_s$
A_1'	A'	A_1	A_1	A_1	A	A	A'	A'
A_2'	A'	A_2	A_2	B_2	A	B	A'	A''
E'	E'	E	E	$A_1 + B_2$	E	$A + B$	$2A'$	$A' + A''$
A_1''	A''	A_1	A_2	A_2	A	A	A''	A''
A_2''	A''	A_2	A_1	B_1	A	B	A''	A'
E''	E''	E	E	$A_2 + B_1$	E	$A + B$	$2A''$	$A' + A''$

Subject Index

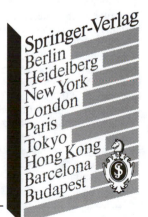

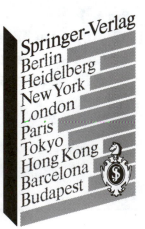